THE GOVERNMENT OF EMERGENCY

PRINCETON STUDIES IN

CULTURE AND TECHNOLOGY

Princeton Studies in Culture and Technology
Tom Boellstorff and Bill Maurer, series editors

This series presents innovative work that extends classic ethnographic methods and questions into areas of pressing interest in technology and economics. It explores the varied ways new technologies combine with older technologies and cultural understandings to shape novel forms of subjectivity, embodiment, knowledge, place, and community. By doing so, the series demonstrates the relevance of anthropological inquiry to emerging forms of digital culture in the broadest sense.

For a full list of titles in the series, go to https://press.princeton.edu/series/princeton-studies-in-culture-and-technology.

The Government of Emergency

Vital Systems, Expertise, and the Politics of Security

Stephen J. Collier
and Andrew Lakoff

PRINCETON UNIVERSITY PRESS

PRINCETON AND OXFORD

Published by Princeton University Press
41 William Street, Princeton, New Jersey 08540
6 Oxford Street, Woodstock, Oxfordshire OX20 1TR

press.princeton.edu

Library of Congress Cataloging-in-Publication Data

Names: Collier, Stephen J., author. | Lakoff, Andrew, 1970– author.
Title: The government of emergency : Vital systems, expertise, and the politics
    of security / Stephen J. Collier and Andrew Lakoff.
Description: Princeton, New Jersey : Princeton University Press, 2021. | Series: Princeton
    studies in culture and technology | Includes bibliographical references and index.
Identifiers: LCCN 2021011343 (print) | LCCN 2021011344 (ebook) | ISBN 9780691199283
    (paperback) | ISBN 9780691199276 (hardback) | ISBN 9780691228884 (ebook)
Subjects: LCSH: Emergency management—United States—History—20th century. |
    Disaster relief—United States—History—20th century.
Classification: LCC HV555.U6 C63 2021 (print) | LCC HV555.U6 (ebook) | DDC 353.9/5–dc23
LC record available at https://lccn.loc.gov/2021011343
LC ebook record available at https://lccn.loc.gov/2021011344

British Library Cataloging-in-Publication Data is available

Editorial: Fred Appel and James Collier
Production Editorial: Kathleen Cioffi
Cover Design: Layla Mac Rory
Production: Erin Suydam
Publicity: Kate Hensley and Kathryn Stevens
Copyeditor: Melanie Mallon

Cover image: National Security Resources Board, United States Civil Defense

This book has been composed in Adobe Text and Gotham

10 9 8 7 6 5 4 3 2 1

For our families

Masha, Sasha, and Nika (S. C.)

Daniela, Natalia, and Paloma (A. L.)

CONTENTS

Figures

Tables

A Vulnerable World

The evidence is clear: we live in an increasingly vulnerable world. Maps with isometric lines indicating flood zones tell us about the ever-growing likelihood that the places we live will be inundated in future hurricanes, torrential rains, or even high tides. Emergency exercises reveal that governmental response systems are not prepared to deal with future disease outbreaks. Stress tests isolate weak links in national and global financial systems that would be exposed in the event of a panic or an economic downturn. Network analyses point to alarming vulnerabilities to accidents or attacks (whether cyber or physical) on critical nodes of power systems. Models of climate change demonstrate the vulnerability of cities and critical infrastructures to heat waves, drought, floods, and landslides. This mountain of evidence points to a troubling contemporary reality: the vulnerability of the vital systems on which modern life depends to a startling range of potentially catastrophic events.[1]

Discussions of vulnerability and preparedness are often, understandably, caught up in the urgency of recent disasters and future threats: a looming hurricane; a critical system that is prone to failure; a virus that is a single mutation away from causing a deadly pandemic. And experts, policymakers, and scholars often search out the sources of vulnerability in relatively recent changes in the structure of our collective existence. Intensifying global flows of people, goods, and capital, they argue, make our world increasingly inter-dependent and therefore subject to sudden disruptions that spread through financial systems, electricity grids, information networks, or human bodies in rapid circulation and close proximity.[2] But there is another way to think about our vulnerable world. Rather than taking vulnerability for granted as a category of understanding—and investigating how our vulnerability became so acute and pervasive—we can ask how it became possible to think about our world in this way in the first place. More specifically, we can ask how we came to think of our world in terms of a particular *kind* of vulnerability. When did government first become concerned with the disruption or breakdown of life-sustaining vital systems? For what purposes were the techniques that, today, generate such an extraordinary profusion of evidence about our vulnerability originally invented? How did norms such as resilience and preparedness become political obligations, to which policymakers and officials are held accountable?

The Government of Emergency addresses these questions by turning to a period of American history in which this distinctive and now mostly taken-for-granted way of thinking about vulnerability was just taking shape. In the middle decades of the twentieth century, amid the Great Depression, World War II, and the Cold War, an array of technical experts and government officials developed a new understanding of the United States as a complex of vulnerable, vital systems. They also invented technical and administrative devices to mitigate the nation's vulnerability, as well as organizing a distinctive form of emergency government designed to prepare for uncertain future events that might catastrophically disrupt these systems. In doing so, these experts and officials did not, of course, solve the problem of vulnerability. Quite the contrary, they defined vulnerability as a particular kind of problem with which today's experts, officials, policymakers, and emergency managers are still grappling.

———

Our interest in these topics was initially sparked by the aftermath—at once troubling and disorienting—of the terrorist attacks of September 11, 2001. The most visible and controversial response by the federal government to these attacks was a series of aggressive security policies identified with the "war on terror." External security measures taken in the wake of the September 11 attacks included preemptive wars in Afghanistan and Iraq, drone strikes on suspected terrorist cells, and extrajudicial detentions, most notoriously in the prison complex at Guantanamo Bay. Domestically, new security measures included heightened border controls, domestic surveillance, and steps to protect large cities and transportation networks against attack. Many of these domestic measures were associated with a new federal agency, established soon after the attacks of 9/11, with an unfamiliar and Orwellian name: the Department of Homeland Security.

At one level, these new security measures challenged familiar conceptions of security. Externally, the focus of military and intelligence organizations on terrorist groups and other nonstate actors seemed distinct from the traditional framework of national security, related to struggles among sovereign states. Meanwhile, new domestic policies pointed to an ominous "securitization" of civilian life, with totalitarian overtones. In another sense, however, these widely discussed elements of the "war on terror" fit relatively comfortably with familiar understandings of security. These measures sought to identify and interdict enemies of the United States, employing traditional means of intelligence, surveillance, military force, border control, and policing.

But beneath the surface of these highly visible and contested measures, a different formation of contemporary security was consolidating. Its contours could be glimpsed by perusing the plans and strategic statements on problems such as "national preparedness" and "critical infrastructure protection" issued by the president, the Department of Homeland Security, the Department of Health and Human Services, and other parts of the US government in the years after 2001. The key norm articulated in these statements was not the deterrence, interdiction, capture, or defeat of an enemy. Rather, these statements laid out a strategy of preparedness for a range of uncertain future events—from natural disasters to disease outbreaks, blackouts, and terrorist attacks—that threatened to disrupt the vital systems that make contemporary life possible. They drew on forms of specialized knowledge and expert assessment that were quite different from those of domestic surveillance, foreign intelligence, and other approaches to understanding the plans and motivations of enemies. The evidence these documents adduced to assess vulnerability was produced by tools such as simulations of catastrophic events; scenario-based exercises to pinpoint gaps in preparedness plans; and evaluations of the "criticality" of particular facilities, such as ports, power plants, communication nodes, and transportation hubs. Finally, these statements of strategy proposed a distinctive set of preparedness measures: stockpiling critical supplies; securing vital facilities or creating redundant facilities; improving coordination among different parts of the federal government, and among federal, state, and local governments; and, perhaps above all, conducting more exercises to test readiness. The sudden consolidation of these norms, knowledge practices, and security measures was puzzling. Where had they developed and been cultivated before coming together so rapidly in new plans and practices?

In 2005, two years after the creation of the Department of Homeland Security, these questions were cast into starker relief, and inflected in new ways, by another domestic catastrophe: Hurricane Katrina, which inundated the city of New Orleans. Like the attacks of September 11, Katrina was followed by rancorous debate and finger pointing. Much of the blame fell on the Federal Emergency Management Agency (FEMA). FEMA had borne responsibility for disaster preparedness and response in the federal government for almost thirty years. But by 2005, it was incorporated into the new Department of Homeland Security, whose emphasis on counterterrorism, some observers charged, had left the agency unprepared for a massive natural disaster.[3] FEMA's failure to organize a competent response to Katrina raised doubts about the federal government's ability to prepare for a range of other disasters, such as the outbreak of a novel and dangerous infectious disease, a particularly acute concern for public health officials in 2005 given the reemergence of avian flu in Asia, one year earlier.[4] The failed response to Katrina also focused attention on the "distributed" structure of preparedness in the United States, which required complex coordination among local, state, and federal governments. This structure, too, had failed in spectacular fashion. Local governments proved poorly organized and ill equipped. State governments were unable to provide timely assistance.

But amid the arguments about where the failure lay, and who was to blame, a basic diagnosis was universally accepted: the government had been unprepared to deal with an event like Katrina and was obliged, in the future, to bolster preparedness, not only for natural disasters but also for a range of other future catastrophes. Thus, more questions: Where did this peculiar American structure of distributed preparedness come from? Why would an agency charged with anticipating terrorism also be responsible for natural disaster preparedness? And how had this unquestioned political responsibility—to prepare for events like Katrina—initially been established and entrusted to such a peculiar and apparently precarious governmental arrangement?

In a first stage of our research, we sought to address these questions by looking back to civil defense planning of the early Cold War.[5] Cold War civil defense was in one sense quite different from contemporary emergency management. Its primary concern was not preparedness for natural disasters, pandemic disease, or terrorist attacks. Rather, civil defense focused on strengthening the preparedness of local governments, communities, and households for a nuclear attack on the United States.[6] In another sense, however, the way civil defense planners identified problems and sought to address them was familiar: preparing to respond in the wake of a catastrophic event. Moreover, Cold War civil defense, particularly the Federal Civil Defense Administration (1950–1958), was a recognized part of the landscape of postwar history for scholars of American emergency management, who have identified it as the source of our current way of thinking about and organizing for emergencies.[7]

But as our research proceeded, our attention was increasingly drawn to another history—adjacent to but distinct from the history of civil defense—that turned out to be more germane to our concerns. Initially, we encountered a forgotten federal government office, the Office of Emergency Preparedness, that in the 1960s was charged with addressing many of the problems that have become so urgent and visible at the beginning of the twenty-first century.[8] The central concern of this office was the vulnerability of vital systems, such as oil pipeline networks, electricity and communication grids, and systems of economic circulation. And it sought to develop methods for anticipating the effects of various kinds of events— terrorist attacks, economic shocks, industrial strikes—that might disrupt these systems, as well as techniques for planning and testing a governmental structure capable of rapid, coordinated response. Digging into the history of this office and its predecessors, we found ourselves on a track that ran parallel to the story of civil defense (see figure 0.1). It led us to the National Security Resources Board and the Office of Defense Mobilization, which were established not to carry out the now-familiar functions of emergency management but to prepare for military-industrial mobilization. In contrast to the well-studied history of civil defense, the activities of these organizations have been largely neglected in the scholarship on the history of emergency management and, indeed, in the broader scholarship on American political development in the middle of the twentieth century. And yet, from 1947 (when the National Security Resources Board was created by the National Security Act) to 1958 (when the Office of Defense Mobilization was combined with the Federal Civil Defense Administration), these were the organizations working on the central problem of emergency government: preparedness for a nuclear attack on the United States. As we show in the chapters that follow, experts and officials working in these now obscure offices shaped current understandings and practices related to the vulnerability of vital systems, preparedness for future catastrophes, and the organization of emergency government.

Our research into the work of these mobilization planning offices opened up, in turn, a deeper history, which connected the history of emergency

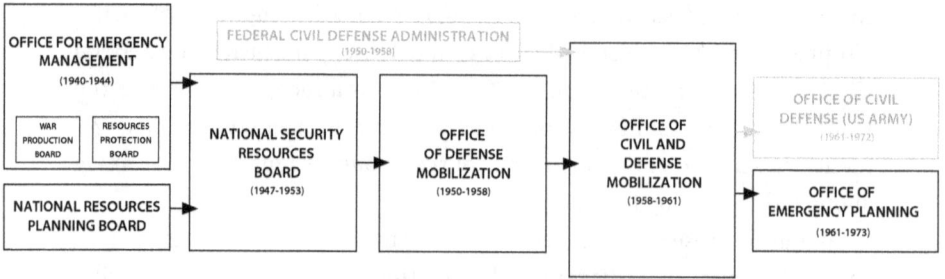

FIGURE 0.1. Organizational pathways of American emergency government. Many of the practices and institutions of contemporary American emergency government emerged from little-studied offices such as the National Security Resources Board and the Office of Defense Mobilization. The history of these organizations points to largely unexplored genealogical connections between emergency government as we know it today and major midcentury episodes in the development of American political institutions. *Credit*: Janice Yamanaka-Lew.

management in the United States to very different kinds of emergencies: the Great Depression and World War II. During these earlier episodes, we found, experts and officials working in domains such as mobilization planning, target selection for air war, and national economic planning developed new kinds of knowledge about flows of resources through the nation's vital systems and their vulnerability to catastrophic disruption. These were also the circumstances in which government reformers assembled the distinctive administrative and political mechanisms of American emergency government, with its small, centralized planning offices (the ancestors of FEMA), its complex arrangements for distributed preparedness across agencies and governmental units, and its often-fraught accommodations between democratic norms, expert control, and strong executive authority to address crisis situations. Thus, in the unexpected settings of depression and world war, we encountered the now-familiar norms and forms of US emergency government taking shape.

———

When we set out to write this book, we imagined that it would begin in the 1950s and move into the present, tracing how Cold War civil defense evolved into contemporary emergency management in its various guises of homeland security, pandemic preparedness, and natural disaster policy. But as this parallel history unfolded, the scope of our book shifted. What we had previously imagined would be the beginning of the story—nuclear preparedness in the 1950s—became its endpoint. Our question changed as well. The book's central concern was no longer the process through which nuclear preparedness expanded into preparedness for a range of other emergencies in the decades after the 1950s—a history that largely remains to be written. Instead, we traced how the knowledge practices, administrative devices, and governmental mechanisms originally invented to manage the emergencies of economic depression and world war were redirected to preparedness for uncertain future events that threaten vital systems.

This shift in empirical focus went hand in hand with a shift in, and significant expansion of, the conceptual and historical problems with which we were grappling. In the United States and elsewhere, the problem of the vulnerability of vital systems to catastrophic disruption is coeval with—and is indeed a crucial element in—the history of industrial and urban modernity itself. Thus, the process through which system vulnerability became such a prevalent governmental concern, and such a dominant feature of our politics, can only be described as one dimension of the broader emergence of a mass industrial and metropolitan society in the United States during the first half of the twentieth century. It is also linked to a significant mutation in political institutions. In contending with the "emergencies" of the Great Depression, World War II, and the early Cold War—all of which were understood as existential crises that demanded exceptional government measures—political reformers created new mechanisms of expert rule and expanded executive power. Thus, in investigating the genealogy of system vulnerability, we also address the process through which, as political scientist Clinton Rossiter put it in 1949, US government was "adjusted in all its ramifications to the mounting stresses of a protean, outward-looking, industrial society."[9]

———

As we completed this book, governments around the world were struggling to respond to a global crisis. In early 2020, the coronavirus outbreak that began in China was spreading rapidly. As the first wave of the pandemic arrived in the United States, officials faced a daunting prospect: the onset of a deadly disease with no effective biomedical countermeasures at hand, and an immunologically naïve population. Experts rushed to identify bottlenecks in health systems, such as shortages of masks, testing reagents, and medical personnel, that would limit the number of patients that could be treated. Policymakers argued about how to procure scarce materials and establish priorities for the allocation of limited resources. Local officials sought to identify essential functions—medical services, critical infrastructure, and the production and distribution of food, for example—whose operation would need to be secured as stay-at-home orders were imposed across the country. Epidemiologists updated pandemic models to anticipate surges in cases in particular areas and to estimate the demand that such surges would make on health resources. Debates flared up about the distribution of responsibility between federal agencies and the president, and between the federal government and the states.

As authorities sought to address multiplying breakdowns and bottlenecks in health systems, a distant episode of American history came to public attention. In spring 2020, Democratic lawmakers and a range of experts and interest groups urged then-president Donald Trump to draw on the emergency powers of the Defense Production Act to organize a forceful federal response to the pandemic. This Act, passed in 1950, gave the president authority to manage national economic resources in order to mobilize the industrial economy, initially for the Korean War.[10] By the late 1950s, mobilization planners had laid plans to use Defense Production Act powers to manage an array of other problems—including massive nationwide medical response—that would arise in the aftermath of a large-scale nuclear attack on the United States. These plans addressed many of the issues that health officials and policymakers would face, over half a century later, in spring 2020: ensuring adequate production capacity of essential medical supplies through government loans and production agreements; securing vital inputs to such production through priorities ratings and allocation controls; and managing the distribution of scarce medical resources, including personnel, to meet a medical emergency unfolding across the country.

The Trump administration made limited use of the Defense Production Act to procure items such as test kits and protective gear but was widely criticized for its unwillingness to employ it more expansively. "We're at war," proclaimed the former director of the Defense Production Act program division at the Federal Emergency Management Agency, "and the enemy is called Covid. The question is do we have the guts that our grandfathers had to mobilize the economy of the United States against the enemy."[11] Upon taking office in January 2021, President Joseph Biden issued an executive order that outlined a broad use of the Defense Production Act's emergency authorities. Priorities ratings would bolster vaccine manufacturers' access to equipment such as filling pumps and filtration units required to ramp up production. Loans and purchase agreements would spur investment in domestic plants to manufacture surgical gloves, whose production in other countries had been constrained by shortages of a vital input: nitrile butadiene rubber. Officials contemplated similar actions, such as issuing loans and purchase agreements, to expand the production of at-home coronavirus tests, N95 masks, and other critical supplies.[12]

As we show in this book, the powers of priorities ratings, allocation control, emergency loans, and purchase agreements are not the only elements of the government response to the Covid-19 pandemic that have roots in the emergencies of the mid-twentieth century. Indeed, many dimensions of the response can be traced back to attempts to manage national resources and to ensure the operation of vital systems during these prior emergencies. Perhaps like no other event in the last seventy years, the Covid-19 pandemic has thrust these problems to the center of attention. But a range of current issues—most notably the intensifying disasters that will result from climate change—ensure that this largely neglected dimension of emergency government will be increasingly central to contemporary politics.

ACKNOWLEDGMENTS

This book would not have been possible without the intellectual engagement and support of innumerable friends, colleagues, collaborators, and family members. We are particularly grateful to Ben Anderson, Carlo Caduff, Craig Calhoun, David Collier, Ruth Berins Collier, Deborah Cowen, Savannah Cox, Tyler Curley, Myriam Dunn, Lyle Fearnley, Andreas Folkers, Nils Gilman, Kevin Grove, Anke Gruendel, Frédéric Keck, Chris Kelty, Clay Kerchoff, Eric Klinenberg, George Lakoff, Robin Tolmach Lakoff, Sandy Lakoff, Turo-Kimmo Lehtonen, Brian Lindseth, Sven Opitz, Onur Özgöde, Paul Rabinow, Peter Redfield, Janet Roitman, Antina von Schnitzler, and Antti Silvast.

We are also grateful for support from the National Science Foundation under Grant no. 1058882, the Center for Advanced Study in the Behavioral Sciences, the Julien J. Studley Research Fund at The New School, and the Dean's Office of the USC Dornsife College of Letters, Arts, and Sciences.

THE GOVERNMENT OF EMERGENCY

The New Normalcy

During the past twenty years we have substituted for the normalcy
of the halcyon 1920s an almost unbroken series of emergencies:
depression, defense, war, inflation, cold war. Indeed, emergency
appears to have become the new kind of normalcy. National
emergencies tend to favor improvisation by government. Yet with all
our improvising, our "putting out of fires," our apparent activation
by events instead of deliberate activation *of* events, we have emerged
with a discernible pattern of domestic and foreign policy and, most
important, with an acceptance of the idea that government should
consciously plan a strategy for anticipating and meeting domestic
and foreign emergencies at the operational level.

—JAMES FESLER, SPEECH TO THE INDUSTRIAL COLLEGE OF THE
ARMED FORCES, SEPTEMBER 4, 1952

In 1954, the United States' Industrial College of the Armed Forces (ICAF)
published a massive multivolume tome, *Emergency Management of the
National Economy*.[1] The ICAF volumes collected a series of lectures that had
been delivered to military officers at the college, as well as a range of government documents that addressed ICAF's main concern: managing industrial
mobilization for war. The fourth volume, dedicated to *Principles of Administration*, reproduced a lecture by political scientist James Fesler, a veteran of
government reform during the New Deal and of mobilization planning during World War II.[2] Looking back on the previous two tumultuous decades,
Fesler observed that the United States had emerged from an "unbroken

series of emergencies"—"depression, defense, war, inflation, cold war"—with a "discernible pattern" of emergency government. Its hallmark was a new norm: "government should consciously plan a strategy for anticipating and meeting domestic and foreign emergencies at the operational level." In the "new kind of normalcy" Fesler described, emergency government was no longer confined to exceptional situations. Rather, ongoing emergency preparedness had become a part of governmental routine.

More than six decades later, it is taken for granted that government bears responsibility for continuously anticipating and preparing for emergencies. This assumption has been evident in efforts to assign blame and bolster readiness following disasters such as the terrorist attacks of September 11, 2001, Hurricanes Katrina and Sandy, and, most recently, the Covid-19 pandemic. It is noteworthy, then, that in 1952, when Fesler gave his lecture, this governmental norm was neither established nor taken for granted. Rather, it was new and required explicit statement and elaboration.

It is also noteworthy that Fesler's discussion addressed a set of problems and institutional contexts that seem distant from our contemporary understandings of emergency management. Today, government offices tasked with managing emergencies are concerned with preparedness for events such as natural disasters, disease outbreaks, and terrorist attacks, as well as with response and recovery in the aftermath of such events. But in 1952, the object of emergency management was the national economy, and its central aim was military-industrial mobilization—marshaling raw materials, industrial facilities, and manpower to build the tanks, planes, munitions, and other supplies necessary for total war. In this sense, Fesler's speech points us to the specificity of the historical conjuncture during which new norms for managing emergencies were first articulated in the United States and were connected to forms of expert knowledge, administrative practices, and legal mechanisms. The topics addressed in *Emergency Management of the National Economy* suggest some of the issues that, in this now unfamiliar landscape, were initially clustered around emergency government: resource planning, economic controls, internal security, economic intelligence, air targeting, government reorganization, domestic vulnerability, and non-military defense. And the government offices, commissions, and agencies whose work was either collected or discussed in the ICAF volumes—most long-since dissolved, and many virtually forgotten—provide a map of the institutional settings in which emergency government was addressed at this time. Among these were committees working on government reform and resource management during the New Deal; wartime and postwar mobilization planning offices; air-targeting and strategic intelligence units in the military; and offices of civil defense and domestic preparedness of the early Cold War.[3]

If *Emergency Management of the National Economy* situates the history of American emergency government in relation to economic management and military-industrial mobilization during the Great Depression and World War II, it also marks a point of inflection. In the early 1950s, emergency government was already in the process of becoming something different and, from our contemporary perspective, more familiar. In the foreword to the ICAF tome, another veteran of wartime mobilization planning, Arthur Flemming, described this new horizon of emergency government. At the time, Flemming was serving as director of the Office of Defense Mobilization (ODM). Created in 1950 to lead civilian mobilization planning for the Korean War, ODM had by 1953 become the most important domestic preparedness agency in the federal government. Surveying the landscape of the early Cold War, Flemming offered a grim assessment of the current world situation. The United States, he wrote, was in an "age of peril." The advent of long-range bombers and atomic weapons confronted national security strategists with the specter of a sudden "devastating attack on the continental United States." In the event of such a sudden attack, the United States would not have time to mobilize its "material and human resources" over the course of months or years, as it had in the prior two world wars. Rather, Flemming argued, the country would have to shift immediately to war footing and would be faced with managing the consequences of a crippling initial blow. Adequately preparing the nation for this eventuality could "save an untold number of human lives" and ensure that the United States could "continue a substantial portion of our war production and production essential for the holding together of our civilian economy."[4]

In light of these concerns about a devastating enemy attack, during the 1950s the civilian mobilization planning agencies turned their attention to a novel task. If earlier these agencies were concerned primarily with military-industrial production during a long war fought overseas, then increasingly their focus shifted to preparedness planning to ensure the survival of the national population and recovery of the economy in the aftermath of a domestic catastrophe. It is indicative of this shift that, by the early 1960s, the Office of Defense Mobilization had evolved into the Office of Emergency Planning, which was in turn renamed the Office of Emergency Preparedness. In 1962, the director of this office, Edward McDermott, outlined the aims and means of emergency government as they had come to be understood by this time. Citing a draft executive order issued by President John F. Kennedy, McDermott reported that he had been charged with coordinating the "national preparedness program," whose goal was to maintain a "state of readiness with respect to all conditions of national emergency." This meant, first and foremost, maintaining an "emergency management organization" that would be prepared to "handle the myriad of resource and economic problems necessary to save lives and sustain survival and expedite recovery." Reviewing these "resource and economic problems"—related to electric power, transportation, communications, food, and medical care—McDermott pointed to the vast scope of his office's concern. "We are really talking about the fundamentals of life on this earth," he intoned, "the elemental problems of safeguarding the food we eat, the fuel we consume, the transportation to maintain a steady flow of commerce, an intricate telecommunications system which will continue to function under all conditions, and perhaps most important, the foundation of constitutional government which underpins our way of life."[5] In sum, the Office of Emergency Planning was charged with sustaining the very biological and associational life of the American population during a future emergency.

In the decades since McDermott's speech, practices for anticipating and managing emergencies have continued to evolve, and the organization of emergency government has been frequently reshuffled. But McDermott's 1962 description of the task of governmental preparedness for emergency is strikingly similar to contemporary understandings. Emergency preparedness continues to focus on reducing the vulnerability of vital systems in anticipation of a range of potentially catastrophic future events, and on preparing for life-saving response and recovery in their aftermath. Thus, the Federal Emergency Management Agency's 2015 *National Preparedness Goal*—which currently guides governmental preparedness for events ranging from terrorist attacks to hurricanes and pandemics—refers to a "secure and resilient Nation with the capabilities required across the whole community to prevent, protect against, mitigate, respond to, and recover from the threats and hazards that pose the greatest risk."[6] The emphasis now, as in 1962, is on what the Department of Homeland Security's 2017 guidance on critical infrastructure protection refers to as "the essential services that underpin American society and serve as the backbone of our nation's economy, security, and health"; "the power we use in our homes, the water we drink, the transportation that moves us . . . and the communication systems we rely on."[7] Today, as in the early 1960s, emergency preparedness aims to ensure governmental functions relating to "health and safety," "infrastructure systems," "hydration, feeding, and sheltering," that, in the wake of a future disaster, will be essential to "rapidly meeting basic human needs," "restoring basic services," "establishing a safe and secure environment," and "supporting the transition to recovery."[8] And as has been true since the beginning of the postwar period, emergency government today is not an *exception* to the normal operation of the state. Rather, it encompasses the management of unfolding emergencies and ongoing preparedness for future emergency situations as permanent functions of *normal* government.

A Genealogy of Emergency Government

This book examines the formation of American emergency government in the middle decades of the twentieth century. It follows the process through which a governmental apparatus initially assembled to manage economic depression and industrial mobilization for war mutated into an apparatus of emergency preparedness for domestic catastrophe. The account presented in this book is a genealogy of emergency government that traces how now-familiar forms of knowledge, practices, and norms first came into being.[9] It is only relatively recently, we suggest, that we have come to understand and organize emergency government as a matter of reducing the vulnerability of vital systems, and it is only recently that preparedness for events that might disrupt these systems has become a basic obligation of government.

This genealogical approach to the study of emergency government can be usefully distinguished from histories of the field of disaster preparedness and emergency management, which follow the changing forms of knowledge and governance that have been applied to a certain class of phenomena—disasters. For example, in *Acts of God*, historian Ted Steinberg traces how the US government has understood and managed (or failed to manage) natural disasters such as floods, earthquakes, and storms, from the early days of the American republic to the present.[10] Scott Knowles, in *The Disaster Experts*, constructs what he calls a "disaster chronology" over roughly the same period, tracking how experts have made "the knowledge and control of disasters their special concern."[11] In contrast to such historical studies of disaster and disaster management, a genealogical approach asks how a range of seemingly disparate phenomena, from nuclear attacks and economic shocks to hurricanes and disease outbreaks, have been *constituted* as common types of events that present similar kinds of problems. Thus, the title of this book—*The Government of Emergency*—does not refer to the way that a pregiven class of events or situations has been governed. Rather, it refers to a form of political rationality, which we understand, following sociologist Nikolas Rose, as an "intellectual machinery or apparatus for rendering reality thinkable in such a way that it is amenable to political programming."[12]

As Rose suggests, political rationalities have both normative and epistemological dimensions. On the one hand, a given political rationality entails specific assumptions about the "proper distribution of tasks between different authorities" and the "ideals or principles to which government should be addressed." Thus, it implies certain presumptions (however contested and unstable) about what government is, what it should do, and what its limits should be. On the other hand, a political rationality involves a distinct "style of reasoning," that is, a body of "intellectual techniques for rendering reality thinkable and practicable, and constituting domains that are amenable—or not amenable—to reformatory intervention." Importantly, a style of reasoning entails specific "conceptions of the objects to be governed," whether the national economy, the population, or the vulnerable, vital systems on which the economy and the population depend.[13]

One strategy of genealogical research is to paint a "before and after" picture that aims, as Ian Hacking has put it, "to permanently fix in the mind of the reader the fact that some upheaval has occurred"—a momentous shift in ways of thinking and governing.[14] Our account is framed by such a conceptual and political "upheaval," in which new objects, aims, and practices of government came into being over a relatively brief period. But we also present a detailed account of *how* this momentous shift unfolded. We focus on specific organizations and on historically situated actors as they took up existing ways of knowing and intervening, or invented new ones, to address novel problems.[15] Through these often-mundane practices, a new political rationality—and indeed, we suggest, a new dimension of political modernity—took shape over the period spanning roughly from the Great Depression through the early Cold War.

The first part of the book examines the period from the 1930s to the early 1940s, in which the federal government faced two conditions of "national emergency": the Great Depression and World War II. During this period, emergency government largely involved *economic* interventions to ameliorate the Depression and to manage industrial production for total war. Chapter 1 follows the work of experts in a succession of domains—from city and regional planning to economic management, wartime mobilization, and air targeting—as they constituted vital systems as objects of systematic knowledge and as targets of intervention. Chapter 2 describes a parallel process through which government reformers invented administrative devices and organizational forms to address the economic emergencies of depression and war. It focuses in particular on how these reformers addressed the tensions between liberal constitutionalism and crisis government by assembling what they called an "administrative machinery" to organize and prepare for emergency situations.

The book's second part is situated in the years immediately after World War II, a period of heightening concern about the prospect of an enemy attack on the continental United States that would cripple military-industrial production systems. Chapter 3 shows how civilian experts and military officers developed systematic knowledge about American economic and infrastructural vulnerability and devised practices and understandings that would constitute a new kind of expertise—and a new kind of expert, the "vulnerability specialist."[16] Chapter 4 turns to the first efforts to develop techniques for reducing this vulnerability and preparing to manage the consequences of a massive attack. It examines postwar mobilization planning agencies, where experts and officials reoriented the existing institutions and practices of emergency government. If previously these institutions had focused on economic management of the unfolding emergencies of depression and war, their objective now shifted to preparing for a future war. Emergency government was thus becoming a matter of ongoing *peacetime* preparedness.

Part III traces a further shift in American emergency government that took place during the 1950s. As nuclear weapons and delivery systems grew increasingly powerful, mobilization planners deemphasized readiness to ramp up industrial production for a long war. Instead, they turned to the task of ensuring the continuous functioning of vital systems that would be required to sustain human life, economic activity, and governmental operations in the unprecedented conditions that would result from a thermonuclear attack. Chapter 5 examines the practices of "administrative readiness" developed by mobilization planners to prepare for government operations in a future emergency, culminating with a description of Mobilization Plan D-Minus (1957)—the first plan for national emergency preparedness in the United States. Chapter 6 focuses on one dimension of such national preparedness planning: the management of resources such as food, medical supplies, and services that would be essential to the population's postattack survival. The chapter traces how mobilization planners used the new tool of computer simulation to envision and prepare for an unprecedented future event—a catastrophic nuclear attack.

By the late 1950s, emergency government, which had previously focused on alleviating economic depression and mobilizing for war, had mutated into emergency preparedness for a future domestic catastrophe. A coherent set of understandings, practices, and organizational forms had consolidated into an apparatus that continues to structure emergency government—in the United States and beyond—to the present day. In the next two sections, we outline the broader conceptual and theoretical significance of this mutation in governmental rationality. First, we introduce the concept of vital systems security as a form of "reflexive biopolitics," oriented to the management of uncertain and potentially catastrophic future events. We argue that, beginning with the midcentury episodes we examine, securing the nation's vital systems has become a central norm of modern government. Second, we describe how American emergency government took shape as a response to the challenge that increasingly common use of emergency powers during war and economic crisis posed to democratic government. In these contexts, reformers assembled a political technology for governing emergencies that, they thought, would make it possible to avoid recourse to exceptional measures that would undermine constitutional democracy.

Vital Systems Security

In 1984, applied mathematician and security expert Robert Kupperman published *Technological Advances and Consequent Dangers*, a working paper for the Center for Strategic and International Studies, a think tank based in Washington, DC.[17] Kupperman's essay was a far-reaching reflection on the vulnerability of vital systems as a central problem of national security. For our purposes, Kupperman's paper indicates how system vulnerability was linked to a broader problematization of risk and security in modern societies.

For millennia, Kupperman argued, human beings had faced relatively localized and "self-extinguishing" threats that were "dissipated by the distribution of cultural assets, by the existence of physical and psychological 'hinterlands,' and by the cushioning function of institutional diversity and independence." Even the cataclysm of World War I was a contained event. "Diversities, distances, and differences, systematic inefficiencies of civilization in themselves," he argued, "provided the recuperative forces necessary to maintain continuity." But in the intervening years, the "extension of technology in the service of civilization" had enabled human beings to move "into every suitable niche, and even into some not so suitable." The increasingly "efficient, economical infrastructure" required to sustain this process carried with it an unacknowledged price. "Modern technological efficiency in the provision of food, water, energy, medicine, transport and

communication," he wrote, has been "oriented toward economic afford-
ability without much attention to complex network fragility." Pointing to
the "interlocking technologies" that underpin the "fragile dynamic cycle
of production, transportation, and consumption" in contemporary societies, Kupperman argued that the "greater a society's dependence for survival on its technological infrastructure, the greater its vulnerability to a
collapse triggered naturally or artificially at a key point." Like biological
organisms, contemporary human societies could not manage "fundamental
system failures multiplying at a biological rate." "A critical point is reached,"
Kupperman warned. "A cascade of organ-system failures ensues, and death
comes quickly." Modern civilization, in developing technologies oriented
to furthering the "ends of human life," had created a system whose "success
and importance to social survival make it, ironically, one of society's greatest weaknesses."[18]

In the 1970s and 1980s, the kinds of hazards that Kupperman identified—
what sociologist Ulrich Beck describes as "modernization risks"[19]—were
taking on a new kind of public and political life. Economic and energy
shocks, environmental crisis, and terrorism garnered increasing attention
alongside the paradigmatic specter of catastrophic risk, thermonuclear war,
which raised the prospect, for the first time, of self-inflicted human extinc-
tion.[20] Kupperman's reflections are especially significant for our story given
his career trajectory, which passed through some of the mostly forgotten
technical domains in which, we show in this book, the vulnerability of vital
systems was identified and addressed as a matter of governmental concern.
In 1980, Kupperman served the incoming Ronald Reagan administration as
the head of the transition for the Federal Emergency Management Agency
(FEMA), which President Jimmy Carter had created by executive order
in 1979. Prior to that, during the 1960s and early 1970s, Kupperman had
worked in one of FEMA's predecessors, the Office of Emergency Preparedness (OEP). As director of the Systems Evaluation Division within OEP,
Kupperman oversaw studies on "the impact on the Nation's security and
economy created by emergency contingencies of both military and nonmilitary nature," examining issues such as natural disaster assistance, the
continuity of government, damage assessment, resource management, and
the "survivability of networks related to national preparedness."[21]

The arc of Kupperman's career points us to a broader question: How did
it become possible to understand collective existence in the United States
as dependent on a complex of vital and vulnerable systems, and how did
the protection of such systems come to be a taken-for-granted obligation

of contemporary government? In the chapters that follow we show that, for nearly a century, a persistent discourse has examined collective life from a particular point of view: the vulnerability of modern society and economy to disruption of the vital systems on which they depend. And since at least the early Cold War, the federal government has been concerned with ensuring the continuous functioning of such systems in the face of catastrophic threats. Today, this problem of "vital systems security" is a central object and aim of government, defined in legislation, executive orders, and broad statements of security strategy.

REFLEXIVE BIOPOLITICS

We analyze the emergence of vital systems security as the product of a mutation in the government of modern life. Specifically, it marks a *reflexive* moment in the history of "biopolitics"—that is, the government of human beings in relation to their biological and social existence. Michel Foucault famously coined the term "biopolitics" to mark a shift, dating roughly to the late eighteenth century, in the aims and objects of government in European countries: from the "classical sovereignty" of the European territorial monarchies to a new governmental concern with ensuring the health and well-being of national populations.[22] Classical sovereignty, Foucault argued, ruled "from the standpoint of the juridical-political notion" of the legal subject. Diplomatic, military, and police apparatuses—elements of what might be called "sovereign state security"—aimed to ensure the security of the state itself in the face of foreign and domestic threats. By contrast, biopolitical government is exercised over the population—a collection of living beings understood as a "technical-political object of management." Foucault traced the "birth of biopolitics" to late eighteenth- and early nineteenth-century Europe, when government authorities sought to manage the health and welfare of populations in growing urban centers. The rapid growth of towns, the expansion of industry, the intensification of trade, and increasingly crowded living conditions posed "new and specific economic and political problems of governmental technique." In response, officials, planners, and experts in the nascent human sciences invented new forms of knowledge about—and devices for governing—the "fine materiality of human existence and coexistence, of exchange and circulation."[23] As Foucault emphasized, the point is not that the birth of biopolitics displaced prior mechanisms of sovereignty; indeed, particularly with the advent of total war, threats to sovereignty were a key catalyst for the development of biopolitics. Rather, the theme of biopolitics designates the interplay between the exercise of juridical power over legal subjects and the technical management of living beings.

Building on Foucault's analysis, scholars have traced the development of biopolitical government in a range of domains from the early nineteenth century. In efforts to reduce the toll of epidemics, organize conscription for war, or manage economic fluctuations, government bureaucracies generated vast amounts of data about phenomena such as birth, illness, and death; suicide and crime; and levels of production and employment.[24] This "avalanche of numbers," as Hacking puts it, made possible a new, statistical understanding of collective life.[25] The technical and political category of risk played a central role in this development, enabling experts and government officials to quantitatively analyze how phenomena such as crime, illness, accident, and poverty were distributed over a given population, and to assess the costs and benefits of measures to minimize these risks.[26] New governmental apparatuses in areas such as economic regulation, urban planning, and public health specified and managed these problems. As Foucault describes this complex process, a "constant interplay between techniques of power and their object" served to "carve out" the population and its specific phenomena (birth and death rates, disease processes, etc.) as a "field of reality."[27]

We take up this story of biopolitical modernity at a later conjuncture and in a different locale. Beginning in the early twentieth century, American planners and policymakers in various domains argued that with the development of mass industrial and metropolitan societies, the interdependencies that made modern collective life possible also rendered it vulnerable to catastrophic disruption from events such as economic shocks, industrial accidents, or wars. Over the following decades, experts and officials addressed this vulnerability by devising new ways to anticipate and mitigate the effects of such events, to reduce the vulnerability of vital systems, and to make society resilient to shocks.[28]

The first governmental apparatus for securing vital systems was assembled in the 1950s. In the early Cold War, planners and officials working on nuclear preparedness brought together a set of elements—knowledge forms, techniques of intervention, and organizational arrangements—that constituted system vulnerability as a target of governmental intervention. Like the demographers, public health experts, and urbanists of the nineteenth century, mobilization planners produced an "avalanche of numbers" about collective existence, not through statistical analysis of populations but by using scenarios, catastrophe models, and vulnerability assessments. Through this process, society became vulnerable in a novel way. Like the figure of population a century earlier, a new figure of collective life—the vulnerable, vital system—was "carved out" as an object of expert knowledge, technical intervention, and political concern.

By the late twentieth and early twenty-first centuries, this apparatus of vital systems security had been extended into new domains, including natural disaster response, pandemic preparedness, the management of economic crises, and homeland security.[29] This is not to say that vital systems security displaced prior forms of security or became the dominant form of collective security. As we will show, vital systems security emerged and consolidated in complex relation to sovereign state security and population security. Thus, the officials and planners in the 1950s-era Office of Defense Mobilization viewed the task of ensuring the functioning of vital systems in the wake of a nuclear attack as a matter of sovereign state security—prevailing in a future war.[30] Meanwhile, vital systems security has become central to many domains of biopolitical government, including the provision of population security in areas such as public health, urban planning, and economic governance. Indeed, we suggest that vital systems security should be understood as a form of "reflexive biopolitics." It shares the aim of population security: ensuring the health and welfare of populations. But these two forms of biopolitical security differ in their objects of concern, knowledge practices, and norms (see table 1). Whereas population security addresses regularly occurring events that can be managed through the distribution of risk, vital systems security deals with events whose probability cannot be precisely calculated, but whose consequences are potentially catastrophic. Vital systems security does not rely on statistical analysis of past events, but rather employs techniques of enactment such as catastrophe models and scenario-based exercises to simulate potential future events and thereby generate knowledge about present vulnerabilities.[31] Its interventions seek to increase the resilience of critical systems and to bolster preparedness for future emergencies.

A NEW POLITICAL RATIONALITY

Our claim is not that governmental concern with vital systems is itself novel. Governments have long been concerned with vital systems like roads, communication networks, and large systems of water management. The construction and control of transportation, energy, and communication systems—what has only recently come to be called "infrastructure"—is

TABLE 1. Three forms of security

	Sovereign state security	Biopolitical security	
		Population security	Vital systems security
Moment of emergence	Seventeenth century—absolutist states	Nineteenth century—social insurance, public health	Mid-twentieth century—nuclear preparedness
Aim	Secure sovereign power against internal and external threats	Manage regularly occurring threats such as endemic disease, poverty, and infirmity	Reduce vulnerability, prepare for future emergencies
Object of concern	Sovereign power: military strength, internal order, wealth	Social processes: economic production, circulation of goods and people	Vital systems: webs of industrial production, critical infrastructures, essential services
Forms of knowledge	*Raison d'état*: external balance of power, internal bases of sovereign power	Statistics, demography, epidemiology, social sciences	System-vulnerability thinking, catastrophe models, scenario-based exercises
Characteristic apparatuses	Diplomatic corps, militaries, mercantilism	Social insurance, infra-structure development	Governmental pre-paredness, emergency management

found in all large-scale complex societies.[32] Territorial empires have for centuries recognized what were referred to as "communications" as essential to prosperity and security. And military strategists have long been concerned with the importance of transportation and communication for military lines of supply; the military tactic of blockade goes back millennia.[33] But from the late nineteenth century to the mid-twentieth century, we observe a significant intensification and modulation of these concerns. In particular, three features distinguish vital systems security as a political rationality and delimit the conceptual and empirical scope of this book: first, its relationship to biopolitics; second, the emergence of specialized expertise about vital systems; and third, the consolidation of a new political norm—that governments must ensure the ongoing functioning of vital systems in the face of catastrophic threats.

Vital systems and modern biopolitics. First, we can refer to vital systems security in the sense we use the term here only with the emergence of modern biopolitics. Electricity networks, railroads, and complex chains of production became "vital systems" when they were linked to newly constituted problem domains such as the national economy or social welfare.[34] Although this development can be traced to the late nineteenth century, particularly in European contexts,[35] our narrative begins in the United States in the first decades of the twentieth century. We focus on two apparently disparate fields: regional planning and strategic bombing theory.[36] Experts in these fields initially used biological metaphors to illustrate the dependence of collective existence on what Muir Fairchild, an instructor at the US Army's Air Corps Tactical School in the 1930s, called "life-sustaining vital systems."[37] Fairchild's term suggested that, like the failure of vital organs or the breakdown of circulatory systems in a biological organism, the disruption of such systems would be catastrophic to the social body. As another Air Corps instructor put it in 1938, as the United States had "grown and prospered in proportion to the excellence of its industrial system," it had become "more vulnerable . . . to wartime collapse caused by the cutting of one or more of its essential arteries."[38] The use of such biological metaphors would fade over time (though never disappear, as Kupperman's 1984 report demonstrates). But from the case studies of the Air Corps Tactical School and the quantitative analyses of "criticality" and "essentiality" in wartime and postwar facilities ratings to contemporary assessments of critical infrastructure vulnerability or resilience, experts have defined the "vitality" of vital systems, and the threat posed by their disruption, in terms of these systems' role in the health and well-being of populations—the central concerns of biopolitical government.

System vulnerability expertise. Second, vital systems security is distinguished by the development of specialized knowledge that constitutes vital systems and their vulnerability as objects of expert analysis and rational-technical intervention. By the mid-twentieth century, technical specialists and officials working in mobilization and air-targeting agencies had devised new practices for assessing vulnerability and preparing for future events that might disrupt the nation's vital systems. This new form of expertise rested on the accumulation of a vast amount of information about American natural resources, productive facilities, and public works—what President Franklin Delano Roosevelt referred to in 1935 as an "inventory of our national assets."[39] Such expertise also drew on techniques for analyzing the interrelationships among the elements that this "inventory" comprised. Although specialists from many fields were involved in constituting vital systems—and the vulnerability of these systems—as objects of systematic knowledge, economists played a particularly prominent role. Economists first appear in our account during the New Deal, inventing a "science of flows" to analyze how shocks would propagate through the economic system, whether these shocks resulted from a plunge in demand during economic downturns or from a surge in demand caused by government stimulus policies or wartime mobilization. A number of these New Deal economists then migrated to air intelligence offices during World War II, where they developed an "economics of strategic target selection" to assess the vulnerability of enemy production systems and to recommend bombing targets.[40] A decade prior to the development of "systems analysis" at the RAND Corporation in the 1950s, these mobilization planners and air intelligence specialists established methods for the quantitative analysis of military-industrial complexes as ensembles of interlocking vital systems.[41]

In the closing years of World War II and the early Cold War, technical experts coupled the analysis of vital systems with new methods for modeling how a catastrophic event—such as an incendiary bombing attack on a city (during World War II) or an atomic detonation (after the war)—would unfold in space. As we show in chapter 3, these experts produced a new kind of knowledge about vital and vulnerable systems. Initially, military analysts in air intelligence units used graphical techniques such as maps and transparent overlays to generate assessments of urban and industrial vulnerability. By the mid-1950s, vulnerability experts had replaced maps and physical overlays with digital computers and geographically tagged data sets—a precursor of geographic information systems (GIS). The advent of computer simulation added another dimension to vulnerability analysis. By incorporating randomization procedures and multiple simulated runs in their models, vulnerability specialists could account for uncertainties about how a future attack would unfold. These simulation techniques—initially used as speculative "experiments" or "war games"[42] as part of nuclear preparedness planning (see chapter 6)—have come to be accepted in various domains as authoritative tools for generating knowledge about uncertain future events.[43]

Vital systems security as political obligation and norm. Third, and finally, vital systems security refers to an increasingly taken-for-granted norm of politics. After World War II, the task of ensuring the continuous operation of vital systems and managing the risk of catastrophic disruption came to be accepted as a basic obligation of sovereign government. This was not the first time that the US government was expected to deal with the consequences of domestic catastrophes. As Michele Landis Dauber has documented, there is a long American tradition of federal relief following disasters.[44] But prior to the middle of the twentieth century, these governmental responses were ad hoc, organized in the wake of what were understood to be unforeseeable "acts of god."[45] Only in the last several decades has government been held responsible for preparing *in advance* of future catastrophes that can be anticipated if not precisely predicted. And only in the last several decades has this obligation been addressed, at least in part, by technical measures that aim to ensure the functioning of vital systems.

The first statutory mention of this new governmental obligation (discussed in chapter 4) was in the 1947 National Security Act. The Act created a new peacetime mobilization agency—the National Security Resources Board (NSRB)—and charged it with undertaking measures to protect "industries, services, Government and economic activities" whose "continuous operation" Congress deemed "essential to the Nation's security."[46] The NSRB was a defense mobilization agency, in which the norm of "preparedness" still referred to military-industrial readiness for war. But planners working in government agencies charged with preparedness gradually adapted these techniques to address other kinds of potentially catastrophic events, such as hurricanes, floods, and infectious disease outbreaks. By the 1960s, the norm of preparedness could refer to any event that might catastrophically disrupt the nation's vital systems. The organization of responsibility for emergency preparedness has shifted almost constantly over the subsequent decades, and attention to this problem has ebbed and flowed. But the task of ensuring the continuous operation of vital systems is now a virtually unquestioned—if not always successfully met—obligation of contemporary government.

An "Administrative Machinery" for Governing Emergencies

The prior section described how experts and officials constituted system vulnerability as an object of specialized knowledge and a target of governmental intervention during the Depression, World War II, and the early Cold War. But on its own, this description of expert knowledge and technical interventions is too serene. It is too serene, in part, because these "interventions" into vital systems were closely linked to projects—whether war mobilization, strategic air targeting, or nuclear preparedness—that involved the mass slaughter of civilians, the annihilation of cities, and, after World War II, the prospect of nuclear holocaust.[47] It is also too serene because the developments we have described corresponded to an upheaval in American government. Technical experts and government officials often instituted the mechanisms of vital systems security through "emergency" measures that challenged American political traditions, such as deference to legislative prerogative and judicial precedent, as well as a diffuse and decentralized pattern of sovereignty. An account of the emergence and consolidation of vital systems security must, therefore, address the fraught relationship between emergency powers and constitutional democracy.

As a point of entry into these questions, we turn to the writings of a prominent midcentury American commentator on crisis government, political scientist Clinton Rossiter. Rossiter began his seminal study *Constitutional Dictatorship*, published in 1948, with a question that President Abraham Lincoln had posed at the outset of the American Civil War. "Is there in all republics," Lincoln asked, "this inherent and fatal weakness? Must a government be too *strong* for the liberties of its people, or too *weak* to maintain its own existence?" Had Lincoln been alive on the eve of World War II, Rossiter observed, he could have "framed his question in more modern terms." Was it possible for a democracy to "fight a successful total war and still be a democracy when the war is over?" For Rossiter, writing just after the end of World War II, the "incontestable facts of history" had provided an answer. "We have fought a successful total war," Rossiter declared, "and we are still a democracy." In this "severe national emergency," the US government had employed "devices and techniques" that made it "strong enough to maintain its own existence without at the same time being so strong as to subvert the liberties of the people it has been instituted to defend."[48]

In what follows, we show that the "devices and techniques" Rossiter referred to were the product of efforts by governmental reformers who, during the New Deal and World War II, sought to meet the challenge that, they thought, emergency situations posed to constitutional democracy. These reformers assembled what Rossiter called an "administrative machinery" that would enable the US federal government, especially its executive branch, to manage emergency situations through expert rule without recourse to an extra-constitutional state of exception. They believed, like Rossiter, that in an era of pervasive doubt about the prospects for democracy, they had successfully responded to the "taunt of the dictators" that "democracies cannot meet the demands of the modern world and still remain democratic," as the reformer Luther Gulick put it in 1941.[49] Our aim in describing these reformers' efforts is not to assess the validity of such claims. Rather, it is to reconstruct how they formulated and sought to address the problem that emergencies posed to democratic constitutionalism. Their responses shaped a distinctive political technology for governing emergency situations.

DEMOCRACY, EMERGENCY, AND THE MODERN AMERICAN STATE

Our account begins in the early twentieth century. At this time, Progressive reformers argued that, as Charles Merriam put it in 1933, governments had "to undertake new activities" to address intensifying processes of urbanization and industrialization. Among these new activities were the management of "public welfare, including education, recreation, health, social relief, and welfare planning"; the construction of public works, such as "highways and aid to communications"; and the "central control over social and economic forces."[50] The challenge, Merriam and other reformers held, was that American governmental institutions, which were set up when the United States was a largely rural and sparsely populated country, were ill suited to the functions required of what they referred to as a "positive state" that was involved in managing the health, well-being, and conditions of existence of a rapidly growing and an increasingly urban population. Merriam described this mismatch as "social lag" and argued for governmental "adjustment."[51] On the one hand, technical experts would have to play an expanded role in political administration. On the other hand, such an "adjustment" would require a significant shift in the locus of political authority: centralization to address issues that crossed local jurisdictional boundaries and decisive executive leadership to manage urgent social and economic problems.

In the early decades of the twentieth century, administrative reformers succeeded in instituting significant changes along the lines Merriam and other Progressives prescribed. Initially, their efforts focused on state and local governments, as they sought to deal with the growing pressures of urban growth and industrial expansion. By the 1930s, in the context of the New Deal, these reformers turned their attention to the national level and the federal government, where they confronted the "emergency" situations of the Great Depression and World War II. Between 1933 and 1945, federal agencies took on a vast range of new functions relating to the provision of social welfare, economic management, and industrial mobilization.[52] To better equip the federal government—particularly the executive branch—to meet these new demands, Progressive reformers working in and around the Roosevelt administration pushed through a series of laws and administrative changes. Partly as a result of their efforts, the American presidency, which began the 1930s as a solitary office with a small staff, emerged from the war as a powerful office that oversaw an array of agencies, wielding formidable discretionary powers.[53] New expert bodies were scattered throughout the executive branch, and new mechanisms of rational-technical administration were woven into laws and regulations.

Political commentators of the 1940s and 1950s were acutely aware that the economic and military emergencies of the period had wrought dramatic changes in the structure of US government. In 1950, Rossiter wrote that the "startling succession of major emergencies" had produced an "extraordinary expansion in the authority of the national executive, in both relative and absolute terms." The presidencies of 1933 and of 1945, he observed, were two "perceptibly different offices, in fact as well as constitutional theory."[54] In 1952, James Fesler also linked the "unbroken series of emergencies" of this period to a dramatic transformation of the American state.[55] "One of the most striking changes occurring in the form and functions of the American Government in the present century," Fesler argued, "has been the rapid growth . . . of governmental administrative activities." The federal government had "entered into a new world of administrative empires, alphabetic agencies, organizational charts, high and low levels or echelons, coordinators and expediters—all explained in strange terms of technical official rhetoric."[56]

In the last seventy years, scholars have continually returned to these episodes in which, to modify Charles Tilly's phrase, the emergencies of economic depression and total war (and later, Cold War) made the modern American state.[57] Our book addresses a more specific question that has received less attention: How did these events shape the American *emergency* state with which we are familiar today, whose major concern is reducing the vulnerability of vital systems and preparing for events that threaten to disrupt the operation of these systems? To answer this question, we turn from the broad problem of governmental "adjustment" to the specific challenge that, reformers thought, emergencies posed in the first half of the twentieth century.

CRISIS GOVERNMENT: "RATIONALISM, TECHNICALITY, AND THE EXECUTIVE"

Historians have documented a significant shift in the range of situations in which governments invoked emergency powers in the early twentieth century, both in Europe and in the United States. Previously, governments most frequently drew on emergency powers to address wars, rebellions, and other threats to state sovereignty. By contrast, in the first decades of the twentieth century, governments increasingly invoked emergency powers in response to events such as labor strikes, financial crises, and economic downturns, in which a direct threat to sovereignty was absent.[58] Notably for our purposes, governments often drew on emergency powers to address threats to the functioning of vital systems in urban and industrial societies. The British Emergency Powers Act of 1920, for example, authorized actions to limit strike activity that interfered "with the supply and distribution of food, water, fuel, or light, or with the means of locomotion."[59] Emergency measures in the United States addressed similar problems. As historian Harold L. Platt has documented, the surge in demand produced by industrial mobilization during the World War I resulted in "terrifying famines of food and fuel" in cities, which were "exacerbated by a virtual gridlock of the nation's transportation."[60] Wartime emergency measures such as price and production controls addressed such breakdowns in vital systems.

If threats to the functioning of vital systems presented governments with novel technical problems, they also presented a political challenge. Was constitutional liberalism compatible with the decisive executive action and rational-technical administration required to manage crisis situations? This question was most famously posed by German jurist Carl Schmitt in his writings of the 1920s and 1930s. American reformers were aware of and on occasion referred to Schmitt's arguments in their own reflections on emergency powers and democratic government. For our purposes, Schmitt's analysis of political authority in crisis situations allows us to pinpoint the fundamental problem that these reformers identified with the exercise of emergency powers in a democracy.

In his 1921 study *Dictatorship*, Schmitt outlined the challenge that emergency situations posed to liberal constitutional government.[61] With their emphasis on deliberation, legislative prerogative, democratic rule by the governed, and deference to precedent and legal norms, Schmitt argued, liberal constitutional governments were rigidly oriented to the past. This orientation was adequate to normal politics, when governments were dealing with familiar situations whose contours could be anticipated based on prior experience. But it was inadequate when governments faced economic shocks, political insurrections, and wars, which demanded a future-oriented form of executive power that could decisively respond to the ever-changing and unforeseeable demands of an emergency situation.[62] "If the concrete means of achieving a goal can, under normal circumstances, be predicted with regularity," Schmitt wrote, then in "cases of emergency" government had to "do everything that is appropriate in the actual circumstances."[63] Emergency government, for Schmitt, could be conducted only through discretionary executive authority based on the rational-technical—rather than charismatic and political—"needs" of a situation. It required rule by "dictate" and, crucially, *according to* the "dictates" of a given crisis as it unfolded. It is this model of emergency government that Schmitt referred to as "dictatorship," based on the model of the Roman "commissarial" dictatorship, which was appointed for the duration of an emergency. For Schmitt, "dictatorship" did not imply an absence of constitutional or legal constraints. Rather, it referred specifically to the rational-technical character of discretionary executive authority: the actions of a dictator could be judged only by asking "whether the means, in a very technical sense, are appropriate or not—that is, whether they have achieved their goal." In this sense, dictatorship was for Schmitt a "political technology" of crisis government, a particular way of arranging "rationalism, technicality, and the executive."[64]

The question that American reformers raised about the "adjustment" of governmental institutions to an urban and industrial society resonated deeply with Schmitt's analysis: How, in liberal democracies, could technical rule and executive power be mobilized to address the distinctive challenges that confronted modern states? And some of these reformers, in seeking out models for emergency government under these circumstances, followed Schmitt in looking to the Roman model of the commissarial dictatorship. This was true not only of academic observers like Rossiter—who wrote extensively on the Roman institution in the 1940s—but also of the administrative reformers who were directly involved in assembling the institutions of American emergency government in the 1930s. Thus, Charles Merriam analyzed the Roman conception of dictatorship as early as 1900 in his study *The History of the Theory of Sovereignty Since Rousseau*. He returned to the concept again in *The New Democracy and the New Despotism*, written in 1939, when, significantly, legislation to adjust the American executive to meet the demands of emergency situations was under debate. Defending the concept "in its historic sense" (and citing Schmitt's 1921 study), Merriam wrote that dictatorship was a "temporary device to meet an emergency," one that was fully compatible with democratic government. "Pestilence, war, famine, flood, panic, depression," he explained, "are crisis moments when decisionism is concentrated in the hands of one or a few who may act before it is too late."[65]

If Schmitt's analysis in *Dictatorship* helps us to pinpoint how American reformers framed the problem of crisis government—as a matter of finding an accommodation between executive power, rational-technical rule, and constitutional democracy—it also casts these reformers' distinctive response to this challenge in relief. Schmitt had drawn a distinction between a "commissarial" dictatorship—based on the Roman model—and what he called a "sovereign" dictatorship. A commissarial dictator was created by legislative decision, hemmed in by the constitution, and limited to the duration of a given crisis. Meanwhile, sovereign dictatorship—which Schmitt soon came to favor—stood entirely outside of law and the constitution. For American governmental reformers, in contrast, the problem was not one of choosing *between* these two models of dictatorship, since they considered "sovereign" dictatorship to be unacceptable in the American governmental system. Rather, they sought to design a form of commissarial dictatorship that was compatible with US political institutions. Rossiter, who wrote extensively on this problem after World War II, argued that in drawing a broad distinction between commissarial and sovereign dictatorship, Schmitt had lumped together a vast range of "heterogeneous offices under the former category."[66] Referring to debates about emergency powers in Weimar Germany (a key point of reference for Schmitt[67]), Rossiter found it strange that so much "energy should have been expended on this question of how much of the Constitution could be disregarded by a President in the use of emergency powers and so little in working out a law that would have settled many of the uncertainties and ambiguities" about the nature of emergency powers and how they would be marshaled.[68]

This precise problem was the focus of American reformers' attention beginning in the late 1930s. In contrast to Schmitt's "latitudinarian" view of the emergency powers implied by a commissarial dictatorship,[69] American reformers labored to define the particular techniques and organizational forms of emergency government, and to specify how these would be constrained by statutory provisions, governmental checks, and constitutional restraints. In this sense, as Kim Lane Scheppele has noted, these mechanisms of emergency government were "crucially non-Schmittian" because they were never "outside the law." Instead, they were based on "alternative forms of legality" that were lodged within the "processes of normal governance."[70]

The American approach to emergency government was not planned all at once as an abstract blueprint. Rather, it gradually evolved through a series of political and administrative struggles that we examine in the chapters that follow: over executive branch reform in the late 1930s, over the control of

TABLE 2. The American political technology of emergency government

	Characteristic features	Key development(s)
Administrative machinery of emergency government	*Executive control* exercised by small planning and management offices working under the president that provide a center for preparedness, coordination, and command	Reorganization Act; Office for Emergency Management (1939–1941)
	Delegatory statutes that temporarily transfer legislative authorities to the executive for the duration of an emergency	Lend-Lease, Stockpiling Act; War Powers Act (1939–1942)
	Distributed structure of emergency government among various executive branch agencies and among state and local governments	Delegations to emergency offices and executive departments (1942–1943; 1950s)
Techniques of administrative readiness	*Emergency government planning* for emergency organization, essential functions, and action steps	NSRB and ODM work on preparedness (1949–1955)
	Planning for uncertain future catastrophes using scenario-based exercises and catastrophe models to formulate and test preparedness plans	ODM work on D-Minus Process (1955–1957)

mobilization planning during World War II, over planning for urban and industrial dispersal in the late 1940s, and over nuclear preparedness planning in the 1950s. By the mid-1950s, a political technology of emergency government had consolidated that is more or less recognizable today. On the one hand, this political technology involved an "administrative machinery" of emergency government, through which the executive was organized, and the power to rule was distributed among its parts. On the other hand, it involved techniques of administrative readiness to prepare the government to assume the form and the functions that would be required to manage an emergency (see table 2).

The structure of emergency government. One distinctive feature of the American political technology of emergency government is an "administrative machinery" designed to establish strong executive authority to manage emergency situations without undermining civilian rule. The recommendations of Progressive reformers working on a 1937 Committee on Administrative Management (described in chapter 2) laid the groundwork for this structure of emergency government. In combination with a number of "delegatory statutes" that transferred certain legislative powers to the president, a 1939 Reorganization Act allowed Roosevelt to wield discretionary authority during emergency situations.[71] Roosevelt drew on this authority to address the ongoing Depression and to manage the large-scale mobilization needed to prepare for an anticipated war.[72] Following the German invasion of France, Roosevelt created an Office for Emergency Management within the Executive Office of the President. The first director of the Office for Emergency Management, William McReynolds, described it as a "device through which [the president] can exercise immediate supervision and control over emergency situations."[73] During World War II, the Office for Emergency Management served, following Rossiter, as an "administrative sky-hook" on which Roosevelt could suspend a succession of emergency agencies—such as the Supply Priorities and Allocations Board and the War Production Board—in which technical experts managed the war production effort. The Office for Emergency Management was the prototype for subsequent executive branch emergency planning and management offices, from the 1950s-era Office of Defense Mobilization to today's Federal Emergency Management Agency.

Another feature of American emergency government, as initially assembled under the Office of Emergency Management during World War II, was its "distributed" character. It established a central locus for coordination and control while preserving the diffused sovereignty of the US constitutional system. Within the federal government, American emergency planning and management since the 1950s has been distributed across federal agencies. Since these agencies are empowered by legislation, they are subject to congressional oversight, thus preserving, at least in principle, the balance of power among the branches of government. Emergency planning and management has also been distributed between the federal government and states, based on a coordinative structure that maintains state sovereignty while enabling states to request assistance from federal authorities when overwhelmed.

Techniques of administrative readiness. A second distinguishing feature of the American political technology for governing emergencies is a practice that Cold War preparedness planners called "administrative readiness." In their work on administrative readiness, planners sought to address what Schmitt identified as a particular limitation of liberal constitutional regimes in dealing with emergency situations: their reliance on legislation that "codifies a series of expectations drawn from the experiences of legislators" based on the past, rather than a future-oriented anticipatory planning for unexpected contingencies.[74] By using techniques of anticipatory knowledge, the practice of administrative readiness created new kinds of "expectations" and "experiences" about uncertain future events so that officials could prepare for emergencies within the framework of constitutional government.

These techniques of administrative readiness were developed in the period immediately after World War II, when mobilization planning offices shifted from operational tasks of wartime resource management to preparedness for a future emergency. One set of techniques involved advanced planning for the temporary government organization that would come into being in a future emergency. Among these were "blueprint" planning for emergency offices and standby legislation; lists of essential functions that emergency government offices would assume; and plans that detailed action steps to be taken in a future emergency. When mobilization planners began to work on these techniques of administrative readiness in the late 1940s, they assumed that a future war would look more or less like World War II, in which the government would manage a long period of military-industrial production. But by the mid-1950s, planners became convinced that, with the advent of ever-more powerful weapons and delivery systems, a future emergency would demand "entirely new and grotesquely different functions" that had "no human experience behind them," as mobilization planner Edwin George put it in 1956.[75] In response, they devised new techniques, such as scenario-based exercises and computer-based procedures for simulating nuclear attacks, to anticipate the governmental functions that would be required in a future emergency and to identify gaps in preparedness in the present.

The Politics of Contemporary Security

Today we are regularly confronted with evidence of our vulnerability to catastrophic events, and, certainly, with the toll exacted by natural disasters, technological accidents, disease outbreaks, and other events that disrupt vital systems or that challenge our collective capacity to organize emergency response. Expert bodies, government commissions, and media reports tirelessly document what sociologist Craig Calhoun has described as a "world of emergencies."[76] Meanwhile, emergency declarations are routine features of governmental practice, both in the United States and globally.[77] Beyond the specific political debates such events engender—What went wrong? Who is to blame? Are we prepared for the next emergency?—a number of social theorists and political commentators have argued that the apparent ubiquity of catastrophes and of governmental states of emergency is diagnostic of the political condition of the present, raising fundamental questions about security, rationality, and democracy. Our analysis in this book does not directly engage in such theoretical debates. But by investigating the forgotten contexts in which taken-for-granted ways of thinking and governing initially took shape, it may cast these debates in a new light.

Catastrophe and the limits of calculative rationality. One strain of critical analysis has examined how the specter of impending catastrophe challenges expert understandings of risk and the forms of social and economic security that became authoritative in industrial modernity. For instance, in his work on reflexive modernization, Ulrich Beck distinguishes between two phases of modernity. He argues that "first modernity," which arose in the late nineteenth to mid-twentieth century, was characterized by the establishment of institutions for managing risks such as unemployment, endemic disease, and accidental death. These risks were relatively predictable and bounded, Beck argues, and so security mechanisms such as social insurance or infrastructure provision could be used to distribute their effects over larger collectives.[78] "Second modernity," by contrast, is characterized by the proliferation of unpredictable and uncontrollable hazards that threaten to destroy "the very foundations of life."[79] Beck analyzes a range of such threats, from ecological catastrophe, global financial crisis, and the spread of chemical toxins to mass-casualty terrorism, nuclear war, and climate change. A key feature of these new hazards—and a defining element of the "reflexive" quality of second modernity—is that they have been generated by the very processes that modernization projects sought to foster: industrialization, urbanization, and technological innovation.[80] Because these reflexive risks are unpredictable and have potentially unbounded effects, according to Beck, they "escape the institutions for monitoring and protection" that have been embedded in governmental institutions over the past century.[81] There "is no expert" in managing such risks, Beck writes; even when technical specialists can estimate the probability or consequences of a given hazard, these assessments often cannot provide authoritative guides for political action to mitigate risk.[82]

For Beck, the ubiquity of catastrophic risk is a key to understanding the political condition of the present. Reflexive modernization, he argues, presents "totally new types of challenges to democracy," as established accommodations between security, expertise, and governmental action are thrown into disarray.[83] Here, Beck's argument converges with a broader literature that has examined what François Ewald refers to as the "deeply disturbed relationship" that exists today between democratic publics and "a science that is consulted less for the knowledge it offers than for the doubt it insinuates."[84] For some observers, this circumstance demands a new politics oriented to the precautionary avoidance of catastrophic risk, or the replacement of discredited technocratic institutions by reinvigorated democracy.[85] As Sheila Jasanoff has put it, "The problem we urgently face, is how to live democratically and at peace with the knowledge that our societies are inevitably 'at risk'. Critically important questions of risk management cannot be addressed by technical experts with conventional tools of prediction."[86] A more ominous prospect is what Beck refers to as a "totalitarianism of hazard prevention," in which democratic processes are suspended in the name of the "right to prevent the worst," and the "exceptional condition" produced by uncontrolled catastrophes "threatens to become the norm."[87]

State of emergency—the exception as norm. This prospect—that in contemporary democracies the "exceptional condition threatens to become the norm"—is approached from a very different perspective by a number of critical thinkers who have analyzed the relationship between emergency powers and liberal constitutional government.[88] Much of this work was written in response to the expanding emergency powers marshaled by the US government following the attacks of September 11, 2001, and focuses on surveillance policies, the treatment of terrorism suspects, and other aspects of the "war on terror." This work links the proliferation of emergency measures in the aftermath of 9/11 to a broader tendency in modern democracies to govern through emergency powers. For example, philosopher Giorgio Agamben argues that persistent and widespread recourse to emergency measures suggests that mechanisms of dictatorial rule, unbounded by juridical or legislative restraints, are transforming constitutional order "to varying degrees in all the Western democracies."[89] According to Agamben, "states of exception" to normal constitutional order exemplify the general condition of modern democracies.

In *Critique of Security*, political theorist Mark Neocleous passes through similar historical territory, examining how, in Europe and the United States in the nineteenth and twentieth centuries, liberal constitutional governments refashioned the institutions of the state of siege and martial law—originally invoked when states faced direct threats to sovereign power—as more general political instruments. Initially, these governments deployed such tools to wage class war against organized labor through disciplinary measures to break strikes and ensure economic flows. Today, Neocleous argues, liberal governments invoke emergency powers to address other phenomena, from the catastrophic to the apparently trivial: famines, drug abuse epidemics, football hooliganism, and natural disasters or "even just a bit of unusual weather."[90] For Neocleous, the elision of the distinction among different kinds of emergency undermines the very idea of normalcy. The result, he claims, is an insidious securitization and militarization of civil government, as the "state of emergency" has become the most "common prescription in the pharmacopoeia of statecraft" in liberal democracies.[91] A broader literature on "securitization" has analyzed similar dynamics, investigating how government authorities invoke the specter of existential threats to justify exceptional measures that undermine democratic norms. As Rita Taureck has described this dynamic, "by stating that a particular referent object is threatened in its existence," a strategic actor asserts a right to take "extraordinary measures" to ensure its survival. Securitization thus moves an issue "out of the sphere of normal politics" and into the realm of "emergency politics," where it can be dealt with "swiftly and without the normal (democratic) rules and regulations of policy-making."[92]

———

We share with such critical analyses an interest in (and concern with) the challenges that catastrophic threats pose to modern government: on the one hand, to experts' ability to assess and manage such threats; on the other hand, to mechanisms of democratic rule and distributed sovereignty. But our genealogical approach provides a different perspective on these questions. We begin from the observation that these critiques can be situated within broader problematizations of risk, security, and democracy.[93] Over the last century, the issues that are now raised as problems for political or social theory have been addressed by an array of reformers, experts, and government officials as urgent practical matters. Thus, since the 1930s, technical experts and government officials have been increasingly concerned with problems of "reflexive modernization": the appearance of threats to the very "foundations of life" that are systematically generated by modernization processes, and the difficulty of assessing these novel threats using established forms of assessment and mitigation.[94] Meanwhile, a range of administrative reformers, government officials, and legal experts worried about how the increasing "normalcy" of emergency government in the United States in the middle of the twentieth century might undermine democratic norms. In this light, our strategy is not to offer another theoretical analysis of catastrophic risk or the "state of emergency." Rather, we examine how historically situated actors initially formulated these questions, and how their responses have shaped contemporary emergency government.

This analytical strategy points to a more differentiated understanding of our current "world of emergencies" than the sweeping diagnoses presented in much recent social and political theory. Increasingly prevalent catastrophic risks may indeed challenge existing forms of expertise and existing security mechanisms. But this does not mean that they exceed all means of technical assessment and mitigation. In the episodes we examine, technical specialists and government officials assembled new forms of expert knowledge about vulnerability, and they invented mechanisms to ensure the continuous operation of life-sustaining vital systems in the event of future disasters. Regardless of whether these mechanisms have achieved the aims for which they were designed and deployed, they have become increasingly authoritative and pervasive across many domains of contemporary life. Moreover, the provision of vital systems security by means of such mechanisms has come to be widely accepted as a central obligation of government.[95] Our analysis also complicates the claim that increasingly pervasive states of emergency break down the distinction between emergency government and normal government, or that emergency decrees necessarily contain the seeds of authoritarianism. Indeed, American reformers during the Great Depression, World War II, and the early Cold War sought to invent devices and techniques of emergency government that would obviate the need for exceptional measures.

We do not mean to argue that there is no need for concern about "exceptionalism" or "securitization" in American politics. Constitutional norms are threatened by security measures in many domains; the "war on terror" and recent immigration policies provide obvious examples. Rather, the point is that it is not possible to deduce a general logic of emergency government from such examples.[96] There are forms of emergency government that are compatible with liberal constitutional government, and these forms predominate in many core domains of contemporary emergency management. There are ways of anticipating and mitigating uncertain and unprecedented catastrophes that are grounded in authoritative knowledge, even if that knowledge is itself uncertain and is the subject of controversy and contestation. And there are, finally, different ways that a threat can be "securitized." It makes a significant difference whether a particular threat is addressed by reducing the vulnerability of vital systems to disruption or by imposing disciplinary controls and extrajudicial measures that undermine civil liberties.[97] The point of analyzing these alternatives is to sharpen our discernment, to bolster our ability to assess whether particular emergency measures truly threaten our norms of government, and, perhaps, to better equip ourselves to craft a politics of emergency that better accords with our collective aspirations for the future.

The Objects of Genealogical Analysis

The last section of this introduction describes the methods of inquiry we used to construct this genealogy. In piecing together this account, we have mainly drawn on primary documents, including bureaucratic reports, memoranda, technical studies, and plans.[98] Some of the texts we examine, such as the forty-seven-volume *United States Strategic Bombing Survey*, are relatively well known among scholars of US political and military history. Most, however, are obscure documents—in some cases, recently declassified—that were intended for narrow audiences of officials, experts, and, in some cases, policymakers. They are significant for us not because they were necessarily influential. Rather, they provide us with insight into the formation of a schema or diagram of emergency government, through which a particular range of situations was constituted as a problem that called out for certain kinds of analysis and remedial intervention. In working with these primary documents, we have examined first, the *styles of reasoning* through which specific domains of expert practice are defined; second, the *knowledge infrastructures* that make it possible to constitute targets of governmental intervention; and third, *sites of technical practice*, in which experts and officials confront and formulate solutions to immediate practical problems.

In describing the emergence of novel styles of reasoning, we do not mean to suggest that a domain of practice that had previously been irrational became more rational.[99] Nor do we mean to treat the history of expert thought as the progression of ever-more accurate approximations of an objective reality. Rather, our goal is to examine the conceptual and pragmatic structure of particular forms of knowledge, and the coming-into-being of things—such as "vital systems" or "national resources"—that, as Ian Hacking puts it, "do not exist in any recognizable form" until they have become objects of expert analysis.[100] More concretely, an analysis of styles of reasoning focuses on experts and officials whose authority is grounded in technical knowledge and formalized (and often, but not always, quantified) demonstration.[101] Although this approach shares something with traditional intellectual history, it is distinct in its focus on the material and institutional circumstances of knowledge production. Its emphasis, thus, is not on the original insight of individual thinkers. Rather, it examines how, in response to specific intellectual and practical challenges, historically situated actors have responded to problematic situations by taking up disparate elements and organizing them into a coherent ensemble.[102] What makes a particular ensemble of such heterogeneous elements interesting and worth studying is that it assumes a stable form that later comes to be taken for granted as a common-sense and self-evident way of understanding and acting in a range of other domains. As Paul Rabinow writes, an "initial response to a pressing situation can gradually be turned into a general technology of power applicable to other situations."[103]

Knowledge infrastructures, following Paul Edwards, are "robust networks of people, artifacts, and institutions that generate, share, and maintain specific knowledge about the human and natural worlds."[104] We focus in particular on protocols and practices for data collection, aggregation, and analysis through which experts constituted new objects of knowledge and intervention, such as "system vulnerability" and "resilience." As Edwards has argued, a focus on knowledge infrastructures may be particularly illuminating when studying objects—in Edwards's case, "global climate," in our case, "national resources" or "vital systems"—that experts can take up only by accumulating enormous amounts of data and analyzing these data using complex modeling procedures. Thus, for example, chapter 1 shows how New Deal economists constituted novel epistemic objects such as national resources and the national economy by assembling "inventories" of data (about raw materials, industrial firms, and utilities) and by inventing a "science of flows" to anticipate how disturbances (whether from economic depression, fiscal stimulus, or enemy attack) would propagate through the vital systems of the American industrial economy. Once assembled in a particular context, these knowledge infrastructures—and the practices, analytical techniques, and accumulations of data of which they are composed—are often taken up in other settings, sometimes in modified or augmented form. Thus, for example, system vulnerability expertise was patched together from elements of knowledge infrastructures that had been developed for other purposes: economic planning, mobilization policy, air targeting, and damage assessment.

Finally, our analysis focuses on sites of technical practice where officials, experts, and policymakers were confronted with urgent governmental problems and sought to link styles of reasoning to policies, regulatory interventions, and administrative mechanisms. This focus on forms of expert knowledge and technical practices does not indicate a narrow interest in the mundane concerns of bureaucrats and technical specialists. The sites of technical practice we study were based in government offices and agencies whose activities had immediate political stakes.[105] In the episodes we examine, the relationships between politics and technology were fraught; controversy and struggle led to the invention of novel governmental forms and practices. For example, many attempts to expand expert rule or discretionary executive power from the 1930s to the early 1950s, whether to deal with economic depression and wartime mobilization or nascent Cold War, provoked broad public disputes about the scope of executive power in the US constitutional system. Characteristic features of American emergency government—such as its organization through coordinating bodies rather than hierarchical organizations—should be understood, at least in part, as the result of attempts to accommodate the need for authoritative action in emergency situations with the traditions of constitutional democracy. Similarly, domestic preparedness measures after World War II provoked fractious debate about the distribution of authority in the federal system, most prominently in struggles over the scope of federal civil defense organization and the place of martial law in managing a domestic emergency. The pattern of "distributed" preparedness and the tools for anticipatory planning developed in the 1950s were the product of an attempted accommodation between political norms and expert rule. Even the focus of technical experts and government officials on the protection of vital systems is, in part, the product of disputes over the extent to which government can, in the name of security, interfere with existing economic relations and forms of collective life.

Our argument is not that these attempts to resolve the contentious problems of democracy, expert rule, and executive authority were necessarily successful. Thus, we do not claim that the completion of Plan D-Minus—which culminates our account of the shift in the early Cold War from mobilization planning to emergency preparedness—provided an effective schema for managing a nuclear catastrophe. The evidence, indeed, points in the opposite direction. Only five years after D-Minus was circulated, the director of the Office of Emergency Planning, Edward McDermott, referred to the "tortuous evolution" of American emergency preparedness, and the "painful" efforts of emergency preparedness agencies as "at best an example of our inability to discard outdated concepts and to keep pace with the threat of increasingly devastating weapons."[106] Equally gloomy assessments have recurred throughout the history of American emergency government, particularly following some of its more spectacular failures. Nor did new forms of emergency government settle the fraught relationship between security and democratic norms. It is true that emergency government proved to be compatible with liberal democracy to an extent that many observers in the 1930s and 1940s thought impossible. But controversies about emergency preparedness and response—whether related to the US government's reaction to events such as the 9/11 attacks and the Covid-19 pandemic or to immigration restrictions justified by reference to unspecified threats to security—demonstrate that the tensions between national security and democratic norms are still with us. Far from being a history of successful solutions, then, our account is a genealogy of the formulation of problems, whose significance lies in the fact that they remain urgent, vexing, and contentious.[107]

A final word should be said about the relationship of our genealogical account to existing historical literature on the midcentury United States. In tracing techniques and knowledge practices as they migrate across different domains, our research took us to a number of sites that have attracted little attention from historians. For example, the activities of the Office of Defense Mobilization between 1953 and 1958, the setting for chapters 5 and 6, have received only passing treatment in secondary accounts. Its records from this period in the National Archives are a disorganized mess—unceremoniously dumped in file boxes decades ago, still awaiting their historian. Similarly, the efforts of offices such as the Enemy Objectives Unit and the Resources Protection Board during World War II, or the postwar National Security Resources Board, have been addressed only in specialized literature, if at all. Part of the pleasure and intrigue of conducting research for this book lay in discovering antecedents to contemporary emergency government in the practices of these obscure agencies.

In other cases, however, genealogy led us to some of the best-studied topics of midcentury American history, such as economic policymaking during the New Deal, mobilization planning during World War II, and Cold War civil defense. Although our treatment of such well-known episodes may seem cursory to experts who have studied them in detail, our hope is that, by approaching them with novel questions in hand, we can provide fresh insight into these established topics of scholarly inquiry. Take, for example, the history of Cold War civil defense. Perhaps due to its public visibility, a large literature has traced contemporary emergency management to the Federal Civil Defense Administration, which was created in 1950 and then abolished in a complex reorganization in 1958.[108] Meanwhile, little attention has been paid to two other domestic preparedness planning offices of the early Cold War that we focus on here—the National Security Resources Board (NSRB) and the Office of Defense Mobilization (ODM)—despite their significance in the formation of contemporary emergency government. The NSRB has been examined only for its brief and somewhat accidental dalliance with civil defense planning in 1949–1950 (described in chapter 4). Its work on nonmilitary defense has been almost entirely neglected. Meanwhile, the activities of the ODM between 1953 and 1958 have also been mostly ignored, even though this office shaped important aspects of American emergency government, such as the complex structure for distributing responsibility for preparedness across multiple federal agencies, or the practice of iterative planning through scenario-based exercises and computerized damage simulation.[109]

Our analysis also suggests a new perspective on the now-extensive literature on post–World War II science and technology. In the last thirty years, this literature has cast light on the military origins of the digital computer, systems analysis, and a new theory of rational choice. Historians have typically traced these developments to wartime operations research, and then to a powerful postwar nexus of prestigious think tanks, elite engineering universities, and military research groups.[110] The conjuncture of mobilization planning, vulnerability assessment, and emergency preparedness we describe in this book is marginal at best to this historical research, but we show that many of these high-profile technical developments grew out of wartime mobilization planning and air targeting, as well as postwar nuclear preparedness. The techniques of resource management used in wartime mobilization planning, for example, were important predecessors of systems analysis (see chapter 2). Vulnerability specialists of the 1940s and early 1950s invented a range of techniques that would later become key elements of GIS and catastrophe modeling. And postwar mobilization planning and vulnerability assessment were among the settings in which digital computers were first used (see chapter 3).

At the same time, our account suggests that familiar landmarks in Cold War history—typically associated with the work of prestigious civilian experts—may be better understood by bringing into view the humbler and largely forgotten work of mobilization planners, vulnerability specialists, and midlevel government officials. Such familiar points of reference as the US Strategic Bombing Survey, Project East River, and the Gaither Committee were initiated by invisible bureaucrats who sought to put a scientific imprimatur on their proposals and plans. In this respect, the significance of a legendary figure like Herman Kahn of the RAND Corporation is not that he was the first to propose methods for "thinking about the unthinkable" prospect of nuclear catastrophe. Mobilization planners had for years been working on techniques of scenario planning and catastrophe modeling to address precisely this problem. Rather, Kahn's fame is the result of his promotion of such techniques to a broad audience outside of classified government work.[111]

Stepping back to survey a broader historical canvas, our account illuminates a generative web of connections among mobilization planning, air intelligence, domestic preparedness, and emergency management that has been largely unexamined. In describing this web of connections, we hope that the *Government of Emergency* casts light on one of the less-explored legacies of the New Deal, World War II, and the Cold War: a new way of understanding and governing the United States—from the perspective of a future catastrophe.

PART I

Crisis Government in the Great Depression and World War II

1

Vital Systems

In 1926, American airpower theorist William C. Sherman published *Air Warfare*, a compendium of lectures delivered to future air officers at the US Army's Air Corps Tactical School (ACTS). The book was one of several texts written during the period by self-appointed apostles of airpower who argued that the invention of the airplane—which had made its military debut toward the end of World War I—entailed a fundamental transformation in the nature of warfare. Traditional military thinkers envisioned air forces as merely an adjunct to the activities of the Army and Navy; they would provide "services" such as aerial reconnaissance. The airpower theorists, in contrast, were convinced that airplanes introduced a new military capability that was especially important given the contemporary phenomenon of total war. Strategic thought had previously focused on the front line, where enemy forces were arrayed against each other. But in the era of total war, they argued, the boundary that separated military forces from civilian life was blurred. A nation's population and industrial capacity were now the source of military power, making their destruction or disruption a primary military objective. As Sherman put it, "Under the modern conception of the 'nation in arms,' with every member of it a 'war worker' of some kind or other there is no sound reason for granting immunity from attack to any class of enemy subjects."[1] The distinctive characteristic of the airplane, in this light, was its ability to fly over front lines into an enemy nation's industrial and urban centers to attack the very bases of its military power.

Sherman's better-known contemporaries, the Italian Giulio Douhet and the American Billy Mitchell, envisioned strategic bombing campaigns as indiscriminate and terrifying attacks on urban population centers designed to break civilian morale. Mitchell, for example, imagined airplanes shooting air torpedoes at cities from a hundred miles away. Sherman's work, by contrast, stood out for its sober and, in his view, scientific approach to air strategy. He emphasized the importance of applying a "principle of economy of force," and the section of his book devoted to aerial bombardment focused on the selection of targets whose destruction would have the greatest impact on the military capacity of the enemy. Sherman predicted—incorrectly, as it turned out—that "international norms" would preclude indiscriminate attacks on civilians. Thus, rather than breaking the will of the population through terror bombing, the "military objective of bombardment aviation, *par excellence*, is the hostile system of supply."[2]

This concern with disrupting the "system of supply" has a long history in military reflection on logistics, and in the ancient tactics of blockade and siege. But Sherman held that, given the advent of industrialized war, "the modern system of supply is a thing more complex than in former days, and perhaps even more vital." It now encompassed the entire chain of military-industrial production: the extraction of raw materials from natural deposits; their combination with labor and energy inputs to make intermediate industrial products and military end items; and the transportation of these end items, finally, to the front lines. The question for air war planners was how this modern system of supply could be made a target of military operations. It was impossible "to attack every factory of the enemy" or every point along a transportation system. But it was not necessary to do so, since "the system is not vital in all its members." "Industry," Sherman explained, consists of "a complex system of interlocking factories, each of which makes only its allotted part of the whole." It is "this very quality of modern industry"—its complexity and the interdependence of its parts—that "renders it vulnerable." To exploit this vulnerability, the enemy's supply system should be studied "in precisely the same way that the intelligence service attempts to ascertain the strength, dispositions, and intentions of the hostile combatant elements." Based on such analysis, "the sensitive points of the system may be ascertained, and concentrated upon."[3] The implication was that air war planners would need to generate detailed knowledge about the collective life of an enemy nation, understood as a complex of vital systems that were vulnerable to catastrophic disruption through carefully targeted aerial bombardment.

Sherman's book provided only the most general outline of a theory of target selection for strategic bombing of the vital, vulnerable systems of an enemy nation. His inquiries were brought to an end by his death from sudden illness just a year after the publication of *Air Warfare*. It was not until World War II that a formalized economic approach to target selection was introduced into military planning. But just when Sherman was formulating his theory in the 1920s, the vital elements of modern life that constituted his "system of supply" were being analyzed from another perspective. In the wake of breakdowns in urban provisioning and industrial production during World War I, and in response to overcrowding and congestion that accompanied metropolitan growth, American city and regional planners began to investigate the interlocking transportation, energy, and production systems that made modern urban life possible.[4] Like Sherman and later airpower theorists, these planners often used organic metaphors, conceptualizing metropolitan economies, in their complex totality, as living beings, and understanding the communication, energy, and transportation networks that linked their disparate parts as akin to the circulatory systems of the body. To generate practical knowledge about such vital systems, regional planners recruited technical experts who developed sophisticated procedures for mapping the interdependencies and resource flows that made up an urban and industrial form of life.

This chapter describes the constitution of a new object of knowledge—the vital system—and a new conception of collective life as a complex of vital systems susceptible to catastrophic disruption. It focuses on the period from the 1920s through World War II, following two parallel forms of reflection on vital systems. One, which we trace through urban and regional planning, New Deal economics, and wartime mobilization, concerned the construction and rationalization of vital systems as central to urban growth and prosperity and to national social welfare and economic production. The other, which we trace from the Air Corps Tactical School, where Sherman taught, to wartime air-targeting and air intelligence organizations, was concerned with the vulnerabilities of such systems, and with discovering how best to disrupt them. These parallel lines of development mirrored each other. The very elements that made particular systems vital from the perspective of regional planning or national economic policy made them inviting targets for air war planners seeking out bombing objectives. This chapter thus traces an early moment in the emergence of reflexive biopolitics, in which the mechanisms that were essential to the functioning of modern collective life—that is, to the biopolitical concerns of social welfare, public health, and economic prosperity—were identified as sources of vulnerability to catastrophic interruption.

Governing Vital Flows: From the Region to the Nation

Our story begins in the first decades of the twentieth century, in the emerging field of urban and regional planning. During this period, American city planners, engineers, economists, and civic leaders began to conceptualize cities and their surroundings as organic wholes comprising functionally interdependent parts. They argued that the relevant unit of urban planning could not be delimited by the political boundaries of a city. Rather, planning would have to encompass the hinterlands, both suburban communities— whose growth was seen as one solution for overcrowding and congestion in central districts—and a city's natural surroundings, understood as a place of recreation and as a source of power and of raw materials for urban industry. Thus, the 1929 *Plan of New York and Its Environs*, a landmark of early regional planning, referred to its subject as a "great whole" that was "vast in scale and complex in detail"; "a living thing, with a certain spirit of its own, a sort of anatomy, and something like a functional physiology."[5] Urban planners, engineers, and economists invented new ways to generate, assemble, and analyze data about the interdependencies that bound this "great whole" together. They focused on transportation and power networks as essential "arteries" that enabled the efficient flow of resources, goods, and people within a region. And they saw the construction and rationalization of such systems as privileged instruments for managing the problems of metropolitan areas and fostering future growth. In the early years of the New Deal, many leading advocates of regional planning moved into federal planning agencies, applying their understanding of regions as complexes of interdependent systems and flows to the nation as a whole.

FROM CITY PLANNING TO REGIONAL PLANNING

Historians of urbanism generally trace the beginnings of regional planning to Chicago in the early twentieth century.[6] As Harold L. Platt has documented, at this time urban planners and civic leaders began to look to "the expert advice of engineers to build the necessary infrastructure of a modern 'networked city.'"[7] During a period of rapid growth of the city and its environs, a collection of business leaders, led by Charles Norton, enlisted leading architects and urban planners to create a city plan. Along with Norton, the Chicago Plan Commission included two figures who would contribute in significant ways to the broad influence of regional planning, both on other metropolitan areas and on national planning during the New Deal. The first was Frederic Delano, a Harvard-trained engineer (and uncle of the future President Roosevelt) whose ideas about railroad rationalization were incorporated into the Chicago Plan.[8] The second was Charles Merriam, a University of Chicago political scientist who was among the first advocates of metropolitan government, which he saw as a "practical response to the fiscal and jurisdictional problems brought on by urban-industrialization."[9]

The Chicago Plan (1909) differed in crucial respects from prior American city plans. In contrast to the previous emphasis on the aesthetic unity of central districts, the Chicago Plan focused on "improving the transit and transportation of the metropolis" as a means to incorporate future growth, to provide an accommodating environment for private industry, and to extend municipal services to expanding suburban populations.[10] These planning efforts converged with initiatives to integrate regional power systems, most notably in business magnate Samuel Insull's project to develop a regional power grid for Chicago Edison. Insull sought to link the expanding consumption of electrical power—first in Chicago itself, then in the affluent North Shore, and ultimately in rural areas—to sources of fuel such as coal reserves and grist mills converted to hydropower in distant river towns.[11]

The World War I mobilization drive, which created a massive surge in demand for military equipment, led to widespread disruptions of industrial production and urban provisioning. These disruptions focused the attention of planners and officials on metropolitan power and transportation systems. As Platt has documented, during the war, city dwellers were confronted with "terrifying famines of food and fuel, exacerbated by a virtual gridlock of the nation's transportation. . . . [C]ritical shortages . . . threatened the very survival of urban America." Even in Chicago, where Insull had stockpiled coal to prepare for power shortages, a blizzard in the winter of 1918 made life in the city a "grim struggle for survival" as "city and suburban residents suffered without any renewal of fuel and food supplies." Meanwhile, the federal government's efforts to organize industrial mobilization for the war were nearly paralyzed by breakdowns, bottlenecks, and shortages. Army Corps engineers, who were responsible for supplying power to war industries, "faced defeats throughout the industrial heartland" as "electrical shortages reduced production levels."[12] The system of transportation through New York harbor, which handled a significant portion of foreign trade in the United States and was a key point of embarkation for American troops, "virtually collapsed, crippling the whole American war effort."[13]

The wartime experience provided a vivid lesson about the dependence of modern society on the reliable operation of industrial and infrastructural systems, and about these systems' susceptibility to disruption. According to historian Mel Scott, the experience taught regional planners to see urban problems that had previously been addressed in isolation, such as congested highways, inadequate water supplies, and overloaded sewers, as "indications of future emergencies and warnings that some of the interlocking systems of the urban complex were in danger of breakdown."[14] To mitigate such vulnerabilities, planners advocated the development and rationalization of vital transportation networks, energy systems, and urban services.[15] For example, Frederick Law Olmsted Jr. took from the wartime experience the importance of "the few big things" such as transportation and power systems that are essential to urban life.[16] The regional planner Morris Knowles numbered among these railroads, which he referred to as "the arteries through which the lifeblood of our country's trade flows," and highways—"the veins through which the individuals engaged in [trade] quicken their lives"—as well as "water transportation," "water and sewerage systems," "stream regulation," and light and power.[17]

In this context, the vision of a regional energy system that Insull had pursued in prewar Chicago—in which electricity grids connected city centers to new suburban areas and to sources of power—spread to other parts of the country.[18] For example, utility industry lobbyists persuaded the federal government to support a "superpower" system for the industrial corridor stretching from Boston to Washington, DC, that would supply industry and transportation with inexpensive power.[19] Even as Progressive reformers attacked Insull as a rapacious monopolist, they promoted similar ideas for integrated regional power planning, beginning with the Muscle Shoals power plant in Alabama, built during World War I to power nitrate plants that were essential for making explosives. Proposals for the integrated planning of urban and industrial development around the power plant circulated through the 1920s (one of notable ambition was advanced by Henry Ford) and would later provide a model for integrated planning in New Deal agencies such as the Tennessee Valley Authority.

THE REGION AS ECONOMIC MACHINE: THE PLAN FOR NEW YORK

In the interwar period, proposals for building and rationalizing vital transportation and power systems were tied to systematic analyses of the elements that made up regional economies, and the interdependencies among these elements. The pre–World War I Chicago Plan, as one contemporary observer noted, was "not based upon any series of special statistical investigations made for the purpose."[20] By contrast, regional planning of the 1920s drew on detailed investigations of the flows of goods, resources, and people that made up a metropolitan economy. To conduct these investigations, regional planners recruited new kinds of experts—most notably engineers, statisticians, and economists—who introduced techniques of analysis that made it possible to constitute transportation and power systems as objects of knowledge and intervention.

The *Plan for New York and Its Environs* (1929) was the most prominent and influential example of these efforts to generate systematic knowledge about the circulation of resources through a region. The New York Plan, which was shaped by Chicago Plan veterans Norton and Delano, involved an unprecedented effort to collect information about the social and economic life of the New York metropolitan region. Thomas Adams, a leading exponent of regional planning who served as the Director of Plans and Surveys for the New York Plan, emphasized the importance of studying "each element in the community structure in relation to other elements in order to get proper insight into the composite and complex problem of urban growth."[21] Adams commissioned Columbia University economist Robert Murray Haig to assemble a team of social scientists to conduct an "elaborate social and economic survey of the New York region."[22] The resulting 1927 *Regional Survey of New York and Its Environs* was itself a landmark in the production of metropolitan knowledge. Adams noted its "unique character and scope," drawing particular attention to its "experimentation in methods of collecting and analyzing facts at all stages."[23] The economic section, for example, drew on data from factory inspection agencies to compile a quarter million handwritten cards, which detailed the character, location, and number of employees of each industrial establishment. By representing information from these cards on maps imprinted on transparent overlays, the distribution of activity in a particular economic sector could be visually juxtaposed with other information such as land values and transportation networks.

According to Scott, the *Survey* that Adams organized for the New York Plan was the most detailed analysis ever conducted of a metropolitan economy.[24] It enabled planners to identify the source of urban problems such as congestion and overcrowding, and to focus interventions on the systems that were most essential to the functioning of the regional economy. For example, one volume of the *Survey* analyzed transportation as an underlying structure upon which other functions of urban life depended. "Transit and transportation services," it observed, "have become more and more necessary in promoting the growth of modern city aggregations. . . . While not in themselves basic industries, these services are essential to the success of industries of every sort, and the further development of the city, whether vertically or horizontally, depends on their expansion."[25] Upon transportation, it continued, "hinges the efficiency of industry, to which the city dweller must look for livelihood, and the reliable and economical supply of his daily needs."[26]

Given the importance its authors assigned to transportation in sustaining urban prosperity and well-being, the situation depicted by the *Survey* was worrisome. Robert Murray Haig and Roswell C. McCrea noted that New York had historically possessed transportation advantages "in a unique degree," and that these had made the city "a center of industry—related to manufacture, commerce, marketing, finance, as well as to the development of the transportation facilities themselves."[27] But congestion and bottlenecks were crippling economic activity and limiting the city's potential growth. William J. Wilgus, a military engineer who contributed a chapter to the transportation volume of the *Survey*, warned that "failure to solve the vastly complex problems" that attend a "massing of humanity" like New York would "simply bring the growth of the community to a standstill," adding that "this is not a matter for the future . . . the crisis is upon us."[28] Everything "practicable should be done," the *Survey* concluded, "to promote the means of [transportation]."[29] To identify targets of intervention, the survey team traced regional flows of resources and goods, such as grain and mill products, other foodstuffs, fuel and ore, and building materials. Then, employing data on the production and consumption of these items in various parts of the New York region, the team compared the actual flow of goods to that which would be theoretically desirable. Which materials would have to be routed through the most congested central nodes of the system? Which could instead be sent through bypasses, thus relieving congestion? The economic survey also projected future trends of production and consumption in different parts of the city for critical commodities such as coal. Based on such analyses, planners formulated a comprehensive plan to expand and rationalize the regional system of transportation.

In sum, the *Survey* offered a minutely detailed picture of the region as complex totality, with interlocked, highly specialized industries, in which patterns of daily life were increasingly dependent on industrial processes and precisely coordinated flows of goods, and in which power and transportation networks—what would later be called infrastructures—were seen as strategic targets of planned intervention, both because they were the source of current problems and because they were crucial determinants of New York's future status as a center of industry, transportation, and commerce. "The metropolis, in one of its aspects," Haig and McCrea wrote, "is essentially a piece of productive economic machinery competing with other metropolitan machines. It will prosper or decline as compared with other metropolises roughly in proportion to the relative efficiency with which it can do economic work—that is, produce goods and services." The integration of vital transportation and power systems, they concluded, was the lynchpin of future growth: "The greatest cities in [the] future are likely to continue to be where coal, oil, and water power can be conveniently brought together in conjunction with the best facilities for transportation."[30]

NEW DEAL PLANNING: FROM THE REGION TO THE NATION

In the early months of the New Deal, many leading exponents of regional planning moved into federal planning agencies, where they advocated for investment in public works such as electrical power and transportation systems to ameliorate the Great Depression. Following the appropriation of 3.3 billion dollars for public works in 1933, President Roosevelt created a Public Works Administration (PWA) to undertake "as much public construction . . . as possible in the shortest possible time."[31] The PWA was led by Secretary of the Interior Harold Ickes, the former head of the National Capital Parks Commission. Recognizing the urgent need to establish criteria for choosing targets of public investment, Ickes formed the National Planning Board, which was made up of "advisory planning committees . . . established for continuous study of public works projects, in such fields as water, roads, or buildings, and for the formulation of long-range plans and programs."[32] The National Planning Board was repeatedly reconstituted and renamed throughout the New Deal, as President Roosevelt sought to find the resources and the authority required to support its activities.[33] Following the convention in the literature, we will refer to this succession of boards using the name of its last incarnation: the National Resources Planning Board (NRPB).[34] When it was first constituted, Ickes selected Frederic Delano as chair of NRPB's powerful advisory committee. Delano was joined on the committee by Charles Merriam, his colleague from the Chicago Plan, and the eminent economist Wesley Mitchell.

These leading planners of the early New Deal viewed regional planning experiments such as the *Plan for New York* as models that could be applied to other parts of the country, and to the nation as a whole.[35] They understood the nation as a single "community," linked not only by shared political institutions but also by a collection of interdependent material flows, natural resources, and production processes, woven together through networks of power, communication, and transport. In a 1935 book that described the early years of the PWA, Ickes described "the country's vast system of public works"[36]—sewage, drainage, water, road and highway, power, electric distribution, and filtration systems—as the vital elements that constituted the "national economy."[37]

As we explore in greater detail in the next chapter, these New Deal planners were engaged in a difficult political balancing act in their advocacy of regional planning as a "microcosmic" model for "a larger concept of a planned society."[38] They feared—correctly, as it turned out—that centralized planning would lead to charges that the New Deal was undermining American traditions of limited government and private enterprise. They therefore emphasized the carefully targeted and strategic nature of their interventions—restricted to the most essential or strategic points—while arguing for the efficacy of such measures in combating the Depression and improving general welfare. "Wise planning," as a 1934 NRPB report put it, would be based on "control of certain strategic points in a working system," adding that the number of these points was not as important as "their strategic relation to the operation of the society in which they work." Public works, the report suggested, presented precisely such "a strategic point offering large possibilities of planning accomplishment."[39] Ickes illustrated this point through a discussion of highway transportation. "Upon roads," he wrote in 1935, "depends the prosperity of millions." Roadside gas stations, tourist camps "with their proprietors and helpers," and a "thousand varieties of other stands selling everything from dolls to town lots" were all "assets of the roadside industry." Meanwhile, the provision of cheap electricity through hydropower projects would raise "almost to a revolutionary extent, the standard of living in underprivileged homes," bringing "electric lights to families who have lived with lamps and candles," and providing refrigeration and heat, as well as "power for washing machines, sewing machines, lathes and pumps for the farmer and his wife."[40] President Roosevelt promoted hydropower development in similar terms, alluding to the World War I–era Muscle Shoals water control project, which powered an ammonium nitrate plant. "This power development of war days leads logically to national planning for a complete river watershed involving many states and the future lives and welfare of millions," he proclaimed in his 1933 message to Congress proposing the creation of the Tennessee Valley Authority. "It touches and gives life to all forms of human concerns."[41]

The Science of Flows

In their advocacy for the integrated planning of large regions, and in their emphasis on public works as a key instrument of government intervention, early New Deal planners drew directly on the lessons of regional planning. Moreover, many techniques developed by regional planners—such as massive surveys of economic systems that tracked interdependencies among their parts—were directly translated into New Deal national planning institutions like NRPB. But national economic policy during the New Deal also presented distinct and novel problems. In the 1920s, regional planners had focused on the congestion and overcrowding of burgeoning industrial cities, which they sought to relieve through local interventions. Amid the Great Depression, by contrast, New Deal planners were concerned with the national economic downturn. As Ickes put it, it was necessary to address the "odd conjunction of idle labor and unused materials . . . the anomalous situation which showed, on one hand, relief rolls crowded with the names of building trade workers, and yards and warehouses filled with lumber, steel and cement; and, on the other, communities in insanitary housing, inadequate sewer systems, and unsatisfactory roads." Public works programs, therefore, were instruments not only for approaching local or regional problems but also for mitigating these national effects of the Great Depression by triggering a "flow of building materials and stimulating industrial activities."[42] New Deal planners anticipated that the economic effect of such expenditures would be "much greater than [their] relative volume suggests," as an NRPB report on public works put it.[43] In order to model these effects, they investigated how federal expenditures on public works would translate into processes of construction and production, or flows of natural resources, intermediate products, and finished commodities; and how these material processes would affect such aggregate quantities as general demand, unemployment, inflation, or the supply of money.

This section describes these efforts, detailing how New Deal economists developed what we refer to as a "science of flows." Like the analyses conducted by regional planners, the science of flows drew on detailed study of the elements that an economic system comprised, and of the flows of resources that connected them. It added to this picture a dynamic understanding of how changes in one area would be "transmitted through the economic system," as the economist John Maurice Clark put it in a study commissioned by NRPB.[44] New Deal economists, statisticians, and officials built a knowledge infrastructure that incorporated both vast amounts of data about the resources that constituted the American economic system and new techniques for analyzing how government interventions could be targeted to have the maximum effects.

PATHWAYS OF PROPAGATION: VITAL FLOWS AND THE BUSINESS CYCLE

The New Deal science of flows was based on new forms of economic analysis that had been developed in the 1920s to understand the business cycle and control economic fluctuations. As Evan Metcalf has argued, the business cycle was "the single most important issue of domestic economic policy during the first half of the 20th century." And during the 1920s, elected officials and economists came to regard downturns not only as a problem for business, but also as the cause of "mass unemployment for an increasingly interdependent population."[45] They therefore sought to go beyond traditional policies of emergency poverty relief to identify government interventions that would make possible the "better control of economic forces," as the secretary of commerce and future president Herbert Hoover put it in 1921.[46] Economists such as John Maurice Clark and Wesley Mitchell studied the business cycle as both a monetary phenomenon and a feature of the production and circulation of commodities, which could be analyzed by examining levels of production and plant utilization, the accumulation of stocks, and patterns of capital investment in productive equipment.

Following a steep deflationary downturn in 1920–1921, a succession of government economic commissions—many enlisting leading academic economists as well as business leaders—studied the business cycle and proposed means to manage it. To understand the sharp downturn, these commissions analyzed the dynamics of stocks and flows of resources in the economy. A post–World War I boom had led to the accumulation of excess stocks in industry. This glut of inventory led to a particularly violent contraction, as businesses cut back on production and investment and laid off workers when demand fell.[47] At this time, direct government intervention to address the business cycle remained politically unacceptable. Therefore, the government commissions emphasized the collection and dissemination of information about stocks and flows in the economy, which, they thought, would encourage private businesses—and in some cases, state and local governments—to act in ways that would counteract the business cycle. As the prominent study *Waste in Industry* argued in 1921, the way to control "dangerous overproduction was to inaugurate a system of frequent and accurate statistics" about commodity production and inventories, which would give businesses a clear picture of the material balance between supply and demand.[48] A manager in possession of "the fundamental facts of the relationship of his particular business to his industry and . . . general business conditions," argued the President's Committee on Unemployment and Business Cycles two years later, would be able "to avoid the difficulties of 1921." For example, managers could reduce capital costs by deferring investment until downturns, when interest rates were low. Alternatively, given the high ratio of fixed capital to variable costs in many businesses at the time, managers might find it economically rational to accumulate stocks during downturns as a way to maintain levels of production and avoid layoffs. By smoothing out production and employment across fluctuations in the business cycle, profit-maximizing firms would act countercyclically in a manner that was consistent with the interest of the economy as a whole.[49]

The construction industry—important to our story given its role in public works policy—was central to this vision of what was referred to as "regularization."[50] Economists assumed that, because of the industry's highly cyclical character, construction activities played an important role in "causing fluctuations in the economy as a whole," as Metcalf puts it—amplifying both booms and busts.[51] But provided with "knowledge of short-run supply constraints" and "guidance as to the periods in which construction should be deferred or should be initiated," both private sector builders and state and local public works agencies might be induced "to spread construction into slacker periods."[52]

In the early 1930s, some of the economists who had worked over the prior decade to develop this way of understanding economic cycles—in terms of the analysis of stocks and flows—joined New Deal planning offices.[53] Wesley Mitchell sat on the National Resources Planning Board (NRPB) advisory committee, while John Maurice Clark was among a group of economists that NRPB commissioned to study public works. In their work in the Roosevelt administration, these economists continued to be concerned with managing fluctuations in the business cycle, arguing that, as a 1934 NRPB report put it, public works might function as a "'balance wheel' to counteract the cyclical oscillations of business activity."[54] But in contrast to the government position in the 1920s—which had been to strictly resist a direct role in such countercyclical actions—New Deal planners argued that government should take direct "control of certain strategic points in a working system." They pointed to public works such as dams, roads, or power systems as among the most important "strategic points" that government should control, "both to relieve [economic] distress and to promote business revival by providing employment and stimulating the demand for products of the heavy industries."[55]

To design such interventions, New Deal planners combined the knowledge practices of economic analysis with those of regional planning. On the one hand, they drew on the tools developed by regional planners for tracking flows and interdependencies. For example, in the early days of the New Deal, the PWA examined the material flows that would be triggered by federal expenditures on public works projects. As part of this project, the Bureau of Public Roads conducted a detailed study of federal expenditure on paved roads that traced the flow of materials "back through the manufacturers of equipment, repair parts, gasoline, explosives and lubricating oil to the cement and steel mills, the stone quarries, and the gravel pits."[56] But New Deal economists were not interested only in the material flows that directly resulted from a particular project. The NRPB's vision of federally financed public works as an instrument of countercyclical policy rested on the proposition that, as a 1934 NRPB report put it, "construction work exercises an influence upon business cycles much greater than its relative volume suggests."[57] Thus, on the other hand, they addressed questions of 1920s economics: How should planners direct limited resources to projects that would maximize such "multiplier" effects? And how rapidly could such effects be expected to move through the economic system?

In 1934, NRPB commissioned a study by John Maurice Clark to address these questions.[58] In the 1920s, Clark had provided an account of the multiplier in terms of the velocity of money and what he referred to as the "propagation period"—the speed with which money cycled through the economy from consumption to production, wages, and savings, and then back to consumption. In his study for NRPB, Clark investigated whether multiplier effects might result from the chain of material flows and labor processes that would be initiated by government expenditure on public works. He began by tracking production processes and material flows, as earlier studies for the PWA had done. "For every ten workers engaged, for example, in actually building a dam," Clark wrote in illustration, "there are others engaged in making cement and excavating machinery and steel, and in hauling materials to the site of the dam and in numberless incidental services." The additional question Clark posed was whether this expenditure on public works would "produce an effect considerably larger than itself" as it moved through chains of material flows, investment decisions, labor processes, business transactions, and consumption and production choices. Clark drew on prior economic thinking about stocks and flows of commodities, about slack or tightness in the production apparatus, and about the behavior of consumers and producers. He distinguished between "primary effects," which could be traced through the chains of production processes, labor, and material flows that were directly triggered by federal construction projects, and "secondary effects," which "arose as these substantive flows were translated into the money economy and affected economic aggregates such as general demand and the price level." Thus, Clark sought to link the dynamics of material flows and production processes to the aggregate phenomena of what was beginning to be understood as the "national economy."[59]

In some cases—when, for example, public works programs triggered new production or investment, or workers immediately spent their wages on durable goods that they had previously forgone—government expenditure might have multiplicative effects. But in other cases—if, for example, there was significant unused production capacity, if new orders were filled out of inventory, or if workers had a low propensity to spend new income—the effects of federal spending might be damped down. In place of the regional planners' view of the metropolis as an "economic machine," the New Deal economists were developing an understanding of the economy as a complex adaptive system.

"AN INVENTORY OF NATIONAL ASSETS": NEW DEAL STATISTICS

Clark undertook his NRPB study in 1934, early in the New Deal. At the time, few public works projects had been completed, and quantitative information about the nation's economic activity—indeed, about any aspect of national economic life—was sparse. As Clark acknowledged at the time, his work remained highly speculative. But over the next decade, economists, statisticians, and other technical experts produced a massive volume of data about national economic life: statistical information about natural resources, industrial facilities, and supporting services (what would later be called infrastructures), and about the flows of materials that linked them. Employing techniques similar to the ones that Clark had laid out, economists working both inside and outside government administration drew on this statistical information to model the likely effects of government intervention on specific sectors and on the economy as a whole. This accumulation of data and modeling of economic processes made it possible to constitute the national economy—understood as a complex of vital systems and resource flows—as an object of expert knowledge and as a field of governmental intervention.[60]

From its inception, New Deal reformers had conceived NRPB's role as akin to a "general staff" that would pursue the "systematic, continuous, forward-looking application of the best intelligence available to programs of common affairs in the public field."[61] In particular, they argued that the Board's central function of public works planning depended "upon the adequacy and integration of . . . factual data and surveys," and they lamented "yawning gaps in our information."[62] In response to this deficiency, NRPB conducted extensive inquiries into the spatial, demographic, material, and productive structure of the national economy. President Roosevelt referred to a 1934 report by the Board as the first attempt to create an "inventory of our national assets."[63] A series of NRPB studies of the late 1930s and early 1940s—*Patterns of Resource Use*; *Energy Resources and the National Economy*; *Public Works Planning*; and others—are considered landmarks of New Deal economics.

On one level, NRPB's studies examined the national economy as a macroeconomic aggregate, defined by abstract variables such as the money supply, inflation, or the overall rate of employment. But they also linked such abstract concepts to an analysis of substantive activities and flows of resources through the economy.[64] NRPB devised techniques for studying production systems as linked activities that connected inputs to end products. For example, a study of steel and iron production (see figure 1.1) traced the transformation of bituminous coal and iron ore into rolling mill products (such as sheets, bars, plates, rails, and other forms) that were eventually used in a range of industrial sectors, from automobile production to mining and construction. The result of these studies, as a 1939 NRPB study put it, was a picture of "the organized activity through which the 130 million people in this country obtain their daily living." From this perspective, the national economy was the "huge and highly complex producing organization" through which "the Nation's resources of manpower and materials are used to satisfy human wants."[65]

NRPB's work was one part of what Joseph Duncan and William Shelton describe as a "revolution" in government statistics during the New Deal.

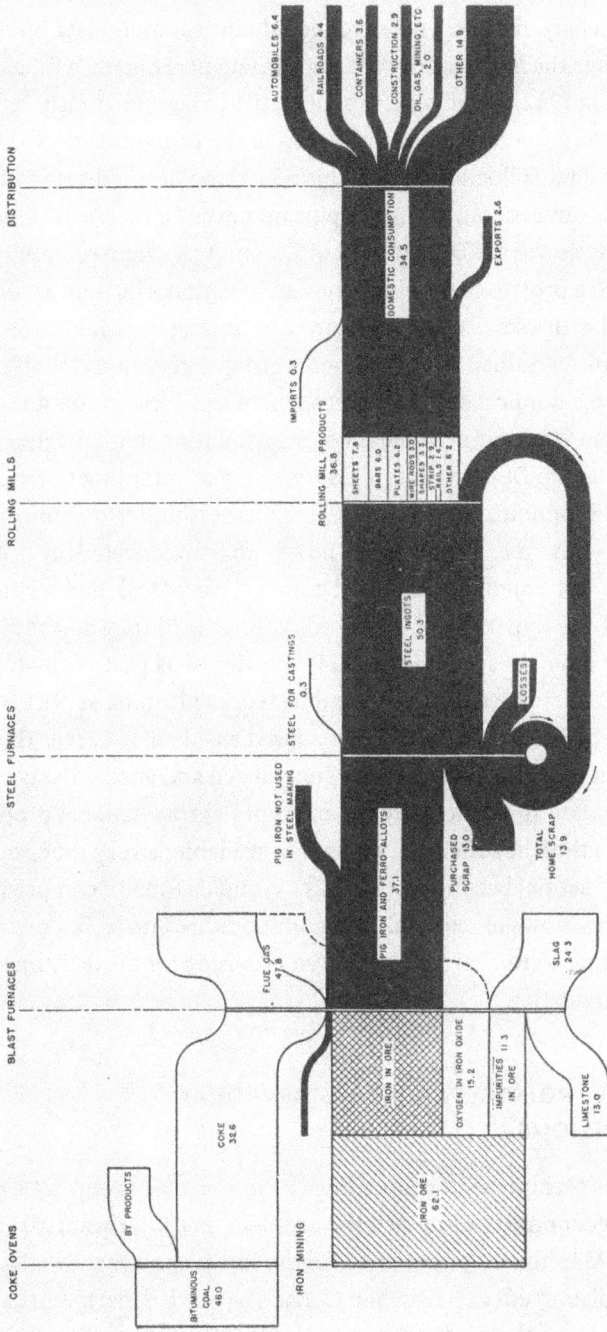

FIGURE 1.1. Physical flow from raw materials to finished products. New Deal expert bodies such as the National Resources Planning Board conducted studies that traced the flow of materials through different sectors of the economy. This chart illustrates the flow of materials through the iron and steel industry, from inputs such as coal, ore, and limestone to outputs used in the automobile, railroad, construction, and oil and gas industries. *Source:* National Resources Committee, *The Structure of the American Economy, Part 1: Basic Characteristics* (Washington, DC: Government Printing Office, 1939).

Agencies across the federal government dramatically expanded their data-gathering activity, recruiting scores of statisticians into government service.[66] Crucially, these agencies collected not just more data but different kinds of data. The first head of the Central Statistical Board, Winfield Riefler, recounted in 1942 that if government statistics had previously focused on the "the interpretation of general trends in the economy," then during the New Deal, data collection efforts shifted to "the detailed study of specific segments of our economy."[67] An important nexus for these new kinds of statistical analysis was the Department of Commerce, where economist Simon Kuznets led a project to develop a system of national income accounting.[68] He worked with a small team of researchers, among them his former student at Penn, Robert Nathan. The Commerce group gathered and analyzed a vast amount of economic data from heterogeneous and scattered sources to create the framework for national income accounting in the United States.[69] After leaving the Department of Commerce and returning to the National Bureau of Economic Research in 1934, Kuznets undertook another massive study—this time of commodity flows—that investigated the production of primary, intermediate, and final goods.[70] New Deal planners used this account of the dynamics of commodity flows (did they add to final consumption? did they accumulate as inventories or as part of the total stock of capital?) in areas ranging from public works planning at NRPB to macroeconomics research at the Treasury and the Federal Reserve. This work is best known for contributing to the generalized and abstracted concepts of macroeconomics—particularly the concept of gross national product. But at the same time, it rested on, and made available, a very different understanding of national economic life—as a complex totality comprising natural resources, flows of materials and commodities, and labor and industrial processes that were vital to the material provision for the daily needs of the national population.

STRATEGIC POINTS: TARGETING NEW DEAL INTERVENTIONS

Using these techniques for modeling the national economy as a whole or by sector, economists addressed the problem that John Maurice Clark had posed in 1934: how to target government interventions to have the maximum stimulatory effect.[71] Lauchlin Currie's research department in the Federal Reserve was the most important site for efforts to model the effects of government interventions in the economy.[72] At the Federal Reserve, and, subsequently, as a special advisor to President Roosevelt, Currie directed a group of Harvard-trained economists that included Emile Despres, Charles Kindleberger, Chandler Morse, and Walter Salant. All these economists later worked to apply the science of flows to strategic intelligence during World War II to address a different problem of targeting: the selection of enemy "resources" for air bombing. In the context of the New Deal, when the question was how to manage the ongoing economic downturn, their work filled a vacuum, providing, as Currie's biographer put it, both the "theoretical terms" and the "quantitative magnitudes" needed for a program of compensatory fiscal and monetary policy.[73]

Currie and his colleagues in the Federal Reserve argued that the programs of the early New Deal had been haphazard, and that the effects of such programs were contradictory, simultaneously providing stimulus and imposing austerity.[74] They sought to formulate a more consistent and coherent policy by analyzing the "net contribution" of specific interventions to desired outcomes, such as higher national income or lower unemployment. Currie's group was particularly concerned with the effect of fiscal interventions (such as borrowing, taxation, or expenditure) that shifted money from one part of the economy to another. Would additions to personal income (due to tax cuts, for example) be spent, thus adding to national income? Or would they be saved, thus blunting the effects of government policy to stimulate the economy? Would government expenditures on relief programs or public works have a multiplicative effect as they moved through successive cycles of production, employment, consumption, and further production? Or would such expenditures crowd out private initiative?

In part, the New Deal economists addressed these questions by analyzing relationships among macroeconomic aggregates (the money supply and general demand, for instance) or microeconomic behaviors (through concepts such as the consumption function). But analysis of material flows and productive processes was a key element in these assessments. Public works remained a primary instrument of countercyclical policy. Currie and other New Deal economists sought, therefore, to understand the relationship between material flows and broader economic aggregates. In this sense, the work of these New Deal "macroeconomists" converged with the research being conducted at NRPB on the "substantive economy."

A 1940 study conducted by Harvard economist John Kenneth Galbraith for NRPB, which examined the effects of federal public works in the period 1933–1938, provides an example of this convergence. Galbraith was recruited to government service in the late New Deal. He worked on mobilization planning during World War II and was one of the directors of the United States Strategic Bombing Survey, which assessed the effects of the American air war on Germany's war-making capacity. In his NRPB study, Galbraith argued that to understand the multiplier effects of public works expenditures, it was necessary that any particular project, whether a dam, a road, or a power plant, "be broken down by individual materials" whose flow could be traced through producers, consumers, and distributors.[75] If in 1934 John Maurice Clark suffered from a paucity of concrete information about these interrelationships, six years later Galbraith was able to draw on the accumulated experience of New Deal public works programs and on an increasing (though still limited) amount of statistical information produced by government agencies.

In the first step of his analysis, Galbraith traced the use of materials and machinery in federal public works projects: cast iron pipe and fittings, cement, concrete products, crushed stone, brick tile and clay products, lumber and timber products, various kinds of construction equipment, and so on. The result was a model of the material interconnections through which the effects of federal expenditure on public works would propagate. To this point, Galbraith's mode of analysis was no different from studies conducted by regional planners of the 1920s, or by the Public Works Administration in the early 1930s. But Galbraith cautioned that material interrelationships did not tell one anything, directly, about the effects of public works projects on employment or output. Rather, one had to consider a series of economic factors—among them "certain problems with respect to inventories, capacities, and level of output"—to understand both the size of these effects and the speed at which they would propagate through the economic system. If material orders were filled "out of stocks rather than by increased production," Galbraith noted, the "stimulating effect of the public works orders would be modified or delayed."[76] By investigating available data on changes in inventories, he concluded that, in general, government orders for materials were filled out of new production, not out of inventory changes—thus they were stimulative. Galbraith also considered the effects of what the New Deal economists called "slack" in the production apparatus. "Output in construction supply industries," Galbraith observed, "has not approached the levels of capacity," making it "quite unlikely that public works expenditures have induced any significant volume of new capital investment in construction supply industries."[77] Through this detailed analysis of flows and their dynamics, and drawing on a growing body of specific data, Galbraith provided a much more precise understanding of what kinds of public works to target for investment, and what specific effects might arise from government expenditures.

THE ECONOMICS OF TOTAL WAR

As war broke out in Europe, New Deal economists became leading advocates for a massive military-industrial mobilization, thus opening a new front in the fractious politics of the New Deal (see chapter 2). In part, they saw the prospect of US entry into the war as an opportunity to garner support for the kind of fiscal stimulus that had been so difficult to push through under other circumstances.[78] They were also convinced that the war's outcome would be determined by the production of military equipment and materiel, which depended on mobilizing the totality of a nation's natural resources, industrial facilities, and labor power for total war. Winfield Riefler, the first head of the New Deal's Central Statistical Board, who later worked in the Board of Economic Warfare in London during World War II, described this totality of elements as a "military-industrial complex"—years before President Dwight D. Eisenhower gave this term a different meaning. "For a military-industrial complex to function," Riefler explained, "it must possess not only an appropriate assortment of factories, industrial man power and technical skills; in addition, it must have access to the foodstuffs necessary to sustain its civilian population and its armed forces, and to the various raw materials required for a minimum level of civilian consumption and a maximum level of munitions of war."[79] The central economic questions of total war were, thus, What level of military-industrial production could be sustained given available resources and productive facilities? How should resources be allocated among different parts of the civilian and military economy to maximize military strength? And what use of limited military means would maximize destruction of an adversary's military-industrial production? Beginning in 1939, New Deal economists were at the forefront in addressing these questions, drawing on the science of flows to address the demands of total war.

Initially, the New Deal economists focused on industrial mobilization planning, adapting the style of analysis they had used to understand and intervene in the 1930s economic downturn to address the problem of coordinating the American economy for massive military production. They conducted the first studies of industrial mobilization while still working in peacetime civilian agencies—such as the Department of Commerce, the Federal Reserve, and planning offices such as NRPB. These initial studies employed the same tools that had been developed for planning government interventions to ameliorate the Great Depression. For example, Lauchlin Currie, who left the Federal Reserve in 1939 to become a special economic advisor to the president, saw war mobilization as "but one special case of the overall problem of securing and maintaining full employment," adding that this conclusion followed from the line of investigation he had initiated at the Federal Reserve.[80] V. Lewis Bassie, using Kuznets's national income data, examined defense expenditures as one of the "possible initiating factors" that would bring about a change in gross national product, focusing on the effects of increased purchase of selected war items.[81] Walter Salant and a group of economists at the Department of Commerce estimated the production potential of the economy at full employment to determine the maximum level of US military production, as well as the levels of deficit spending that would be required to achieve full employment.[82]

The institutional center of gravity for work on the economics of war mobilization shifted in 1940 amid a broader "exodus from the permanent agencies to the defense agencies."[83] Many New Deal economists relocated to executive branch offices that were created during the early years of the war, initially to various parts of the National Defense Advisory Committee. For example, Robert Nathan, who had worked with Kuznets on national income at the Department of Commerce, was named chief of military requirements. Harvard economist Edward S. Mason was appointed to direct the Committee's Economics Research Section. John Kenneth Galbraith was appointed head of the Price Division in the Office of Price Administration and Civilian Supply. And V. Lewis Bassie was appointed chief of the Civilian Requirements Section.[84]

One group of New Deal economists continued to work in the emergency mobilization agencies, both before and after US entry into World War II. As we describe in the next chapter, they reached the peak of their influence in the early years of the war, establishing the science of flows as an authoritative means to determine the maximum feasible level of military-industrial production, and to manage the flow of materials through the American industrial economy. Meanwhile, another group migrated into the world of air intelligence and air targeting. In wartime organizations like the Office of Strategic Services and the Enemy Objectives Unit, these economists applied the science of flows to a question that emerged from the world of military strategy: not how to mobilize the domestic economy but how to disrupt the vital systems on which an enemy's military production depended.

The Vulnerability of Vital Systems: Airpower Theory

In a 1935 lecture at the US Army's Air Corps Tactical School (ACTS) at Maxwell Air Base in Alabama, instructor Hal George asked: "Has modern civilization reduced or increased the vulnerability of nations?"[85] He pointed to intensifying processes of urbanization and industrialization as a new source of vulnerability. "The trend in modern nations," George observed, "has been toward specialization in industry and agriculture, which makes for large territories which are not self-supporting." Urban life would be impossible without technological systems that linked the city to its hinterlands. "The city dweller is dependent upon other communities for nearly everything he consumes," he continued, "and the consumer and producer are brought together only through the medium of intricate systems of modern transportation." Indeed, the construction of large-scale public works had made the concentration of labor and industry in cities possible. "Nearly everyone is dependent upon systems of public works which provide elements essential to daily life: electric power, sources of fuel, and of water." "All industry," moreover, was "dependent upon electric power."[86] "It appears," George observed, "that nations are susceptible to defeat by the interruption of this economic web."[87]

George's reflections on the systems that made urban-industrial civilization possible were strikingly similar to those of regional planners of the 1920s and New Deal planners of the 1930s. But they were addressing distinct problems. Regional planners and New Deal planners focused on building infrastructural systems and sought to identify "strategic points" in industrial production systems that could be targeted to foster economic prosperity. George and other air war planners had a different concern: identifying points of vulnerability in an enemy's industrial economy that would be optimal targets of strategic bombing. During the 1930s, these air war planners served in ACTS, a center for thinking about strategic bombing and the vulnerability of industrial production systems, where many of the major airpower strategists of World War II were trained or served as instructors. ACTS theorists focused on what they referred to as "vital systems"—chains of industrial production, energy networks, and transportation grids. These systems were vital in that they were essential to the functioning of an urban and industrial society. They were also vital in a military sense: since modern war was conducted by fully mobilized urban and industrial societies, disrupting these systems would be strategically decisive in crippling an adversary's war-making capacity.

AIRPOWER AND INDUSTRIAL WAR

Interwar airpower theorists' conceptualization of the vital, vulnerable system as a strategic target was the result of two major transformations in the practice of warfare: the rise of total war and the use of the airplane as a weapon. In total war, each side mobilized the entirety of its population and economy in its military effort. Infrastructural and industrial systems were central to such mobilization, and were often described using the metaphor of a social organism. For example, British military theorist Ernest Swinton argued in 1906 that "railways are the arteries of modern armies," adding that "vitality decreases when they are blocked, and terminates when they are severed."[88] Airpower thinkers built on this conception. In 1917, British air war strategist Lord Tiverton noted that "a modern army cannot fight without an adequate supply of munitions and therefore the strategy will be directed along the lines which best destroy such supply," suggesting the profitability of an attack on "the bottleneck of production" in German industry.[89] Meanwhile, since an urban labor force was crucial to producing munitions and other supplies for the industrialized military, the civilian population's "will to fight" was also seen as a vulnerable point that might be targeted for attack.[90] In this conception, air forces would play a privileged role in modern warfare. Military strategists could now target enemy supply lines, reaching from industrial and agricultural production centers to the front, as well as the civilian population whose labor produced the materiel necessary for fighting a war, and whose morale was essential for the enemy's continued war effort.[91]

In the United States, the most prominent exponent of this theory was General William "Billy" Mitchell. Like his contemporary William Sherman, Mitchell focused on the ability of airpower to cripple the enemy's capacity to wage war by disrupting the systems necessary for economic and social life. "If the vital centers of the opposing nation are destroyed, or made incapable of carrying out their part in the upkeep of people," Mitchell wrote in a 1927 article entitled "Airplanes in National Defense," then "resistance is no longer possible, and capitulation or surrender is the outcome." Optimal targets of air attack might include "manufactories, means of communication, the food products, even the farms, the fuel and oil and the place where people live and carry on their daily lives." Mitchell noted that the concentration of industrial activity in cities made them especially likely targets of enemy attack. Pointing to the industrial core of the United States, he asked his readers to imagine that "New York, Chicago, and the railways south of Sandusky were paralyzed." Such a situation "would disrupt our whole existence as a nation," he argued. "Within a very short time the nation would have to capitulate or starve to death."[92]

Mitchell's arguments were part of a debate during the interwar period about the future military role of the airplane. Mitchell advocated the establishment of an autonomous air force, which would have a strategic mission of crippling an enemy's military-industrial complex. His position met strong resistance from the Army and Navy, which saw the future role of airplanes as limited to tactical support for infantry or ships and viewed Mitchell's advocacy for a freestanding air force as a direct challenge to their authority. Mitchell famously lost this battle and was court martialed for insubordination in 1925. But his strategic thinking about airpower was carried forward by a group of training officers at the Army's ACTS, initially based in Langley, Virginia, and relocated (in 1931) to Maxwell Air Base in Alabama. Mitchell's followers shaped the curriculum at ACTS, where they developed a doctrine of air warfare based on high-altitude, precision daylight bombing of enemy vital systems that would shape American airpower thinking in World War II and the early Cold War.

ACTS was staffed by a group of young officers, mostly West Point graduates, who had entered the service after World War I. While sharing Mitchell's emphasis on strategic (as opposed to merely tactical) bombing, the school's instructors honed a distinct approach to the selection of enemy targets. Mitchell and other early airpower theorists, ACTS instructor Haywood Hansell later wrote, "advocated destruction of factories and industrial centers and population centers" as a means of breaking the civilian population's will to resist. By contrast, ACTS thinkers maintained that "the idea of killing thousands of men, women and children was basically repugnant to American mores." But this aversion to terror bombing did not prevent them from developing an equally deadly, if less direct, approach to air war. Instead of inciting mass terror through direct attacks on population centers, Hansell recalled, "the school favored destruction or paralysis of national organic systems on which many factories and numerous people depended."[93]

This concept of the "organic" or "vital" system rested on an understanding of the enemy nation as akin to a living organism whose functioning could be crippled with precision attacks. Hansell and his colleagues argued that just a few bombs, dropped on carefully chosen targets, might "destroy vital civic systems, render the cities untenable, and force their evacuation." For example, a study of New York City revealed that "a very small number of hits on a few sensitive spots could cause collapse of the life-sustaining vital systems" that supported the urban population.[94] ACTS theorists understood these vital systems to be interdependent and interconnected parts of what they referred to as the "industrial fabric." Whereas Sherman's notion of the supply system was clearly an extension of traditional military logistical thought, the concept of the industrial fabric envisioned a nation's industrial production complex as composed of interwoven strands: if one were broken, the entire web would disintegrate. ACTS instructor Donald Wilson later described the doctrine in terms of total war strategy. "Future wars for survival," he argued, "would be between industrial nations; continuation of the war would depend on maintaining intact a closely-knit and interdependent industrial fabric."[95] The new technique of precision bombing, he posited, "gave us an instrument which could cause collapse of this industrial fabric by depriving the web of certain essential elements—as few as three main systems such as transportation, electric power and steel manufacture would suffice."[96]

For the strategy of high-altitude, precision bombing to succeed, air war planners had to identify the most critical nodes to destroy, so that, as a 1930s ACTS textbook put it, "the whole of the economic machine ceases to function."[97] To identify these critical nodes, Wilson later recalled, ACTS theorists "would look not only to the important systems, but for keys within these systems." ACTS officers were particularly interested in identifying essential nodes that—because of their physical properties—could be destroyed by aerial bombing, could not be readily repaired, and whose function could not be substituted by other facilities. For example, Wilson pointed out that US "railroads did not stock spare steel girders or trusses for bridges, and that some ten main lines carried all the through east-west traffic in the industrial area" of the country. These bridges presented an ideal precision target: they were essential, easily interrupted, difficult to fix, and no alternative routes were available. As a consequence, an attack that knocked out "ten important bridges on each side of the main lines in the area would be a very serious blow."[98] A 1936 flood in Pittsburgh provided ACTS instructors with another example of such a vulnerable node in a vital system. The flood damaged a factory that produced a spring necessary for making pitch propellers, thereby delaying aircraft production throughout the United States. Hansell noted the economy of force that could be achieved if, in a future war, enemy airpower focused on destroying such a factory: by cutting off the supply of these particular springs, "to all extent and purposes a very large portion of the entire aircraft industry in the United States" would be "nullified just as if a great many airplanes had been individually shot up, or a considerable number of factories hit."[99]

THE UNITED STATES AS MODEL OF VULNERABILITY

ACTS theorists understood that successful target selection in future air wars would require detailed knowledge about the industrial production systems of enemy nations. Yet, they had little or no access to the kind of information that could help them to specify the critical nodes in other national economies that should be the targets of future bombing missions. As an alternative, they scrutinized data on metropolitan areas, public works, and industrial production in the United States—information that was being produced in ever-growing volumes by government planning offices. "At the Tactical School," Hansell explained, "we had laid out the methodology and, since we had no foreign intelligence, we used the industrial structure of the United States as a working model."[100] Looking at cities and regions of the United States, they focused on what Hansell called the "critical organic systems whose destruction would paralyze a modern state," such as transportation, fuel, food distribution, and steel manufacturing.[101] Two lectures delivered by ACTS instructor Muir Fairchild in the late 1930s—one called "New York Industrial Area" and the other "National Economic Structure"— illustrate how airpower theorists used studies of the United States to advance their thinking about industrial vulnerability. Fairchild would become a major figure in air war planning during and after World War II and would prove influential in bringing ACTS concepts to bear as part of broader reflection about future war. In his ACTS lectures, Fairchild approached what regional planners had referred to as the "economic machinery" of American cities from a different perspective: What were the key vulnerabilities of urban areas to aerial bombing?

In his 1938 lecture on New York City, Fairchild described the vital nodes that made urban-industrial civilization in New York both possible and highly vulnerable to attack from the air. "What we are really concerned with," he began, "is an analysis of a great city. What is necessary for its continued existence? If we should find this out, we should be able to apply our air offensive pressure most effectively, by directing our air attacks at the most vulnerable part of the city structure in a planned campaign whose results should be quick and decisive because each blow would augment and reinforce all the other blows to produce an accumulative effect." After reviewing New York City's economic importance for the United States as a whole—the central role of its ports, its banks, and its industry—Fairchild turned to the vulnerability of the city's vital systems. He first considered its water system to "see what it is like and how vulnerable it is to interruption by air attack," asking: Where are the dams, the reservoirs, the aqueducts? How vulnerable are they? How long could the city survive if its supplies were cut off? Fairchild then considered the transportation of food—specifically, the problem of shipping the vast quantities of food necessary for New York's inhabitants into the city from its hinterlands. "It is apparent," he noted, "that this whole system of feeding this great metropolitan area depends upon the continuance of uninterrupted rail communications into the area."[102] Analyzing a map of the rail system, Fairchild showed that the best way to cut off food transport into the city would be to deny the use of rail terminals along the west side of the Hudson. The bottlenecks that regional planners had identified in the *Plan for New York and Its Environs* as impediments to economic development were, from the vantage of airpower strategy, prime targets of enemy attack.

ACTS studies devoted particular attention to systems that made possible a range of other activities.[103] Fairchild focused on one such underlying system in his analysis of the vulnerability of New York City: electricity. In New York, eight steam-generating plants supplied power for subways and intercity trains, also serving as transformer stations, where the voltage of power generated outside the city was stepped down for local distribution. "A direct hit from one bomb of the proper size" would "destroy one of these steam generating plants," noted Fairchild; replacement would take nine to eighteen months. In a figure titled "The Aerial Bomb versus Public Service Power in the New York City Area" (figure 1.2), Fairchild offered a "graphic picture" of what "a limited number of accurately placed bombs" would "mean to the industrial and public service power of the metropolitan area." The figure indicated that a mere three bombs would reduce the city's power by 50 percent. Four more direct hits would reduce it to 25 percent; and another four would bring electric power to only 10 percent of its preattack level.

"It would take no very large force," he argued, "to practically assure depriving this whole great metropolitan area of all sources of power for a period of many months." Such a disruption would mean no rapid transit, no vehicular traffic through tunnels, and no "vertical traffic" in great buildings; the port would be crippled; the water supply would be cut off, since pumping was critical to water distribution; and "we might expect wholesale spoilage of food and inability to maintain an adequate supply on hand." Following an air attack, vast fires would rage without the means to put them out; manufacturing would be crippled; streetlights would be out; household appliances would not work; there would be no newspapers or radio, so

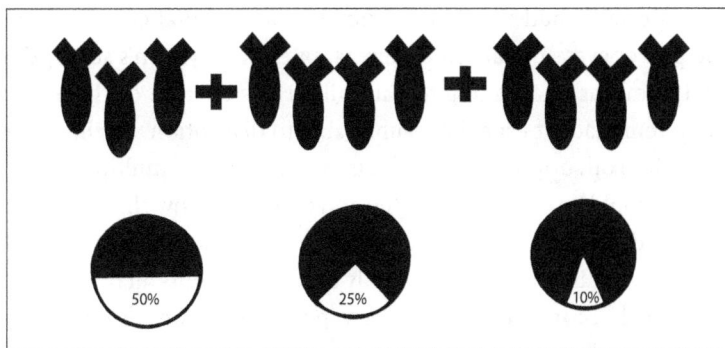

FIGURE 1.2. The aerial bomb versus public service electric power in the New York City area. Muir Fairchild, an instructor at the Air Corps Tactical School (ACTS) in the 1930s, presented a "graphic picture" of the effect of "a limited number of accurately placed bombs" on critical nodes of the electricity system in New York City. Lacking detailed information about adversary nations, ACTS instructors analyzed the vulnerabilities of vital systems in the United States. *Source*: Phil Haun, *Lectures of the Air Corps Tactical School and American Strategic Bombing in World War II* (Lexington: University Press of Kentucky, 2019), 175.

communication would be cut off; traffic would be blocked. "At one stroke," Fairchild concluded, "the industry, the home, the entire machine that we know as New York City, could not function."[104] As Hansell later recalled of Fairchild's analysis, "it soon became apparent that the very heart of our industrial system was the electric power system," on which all other industrial and economic functions depended.[105]

Fairchild further pursued this analysis of power system vulnerability—now on the scale of the entire nation—in a 1939 lecture, "National Economic Structure." As in his lecture on New York City, he began by emphasizing the dependence of collective life on electrical power.[106] "Whether for good or ill," Fairchild observed, "we are definitely committed to an electrical economy. Without its electrical life-blood our country can no longer continue to function as a great nation." Drawing on data compiled by the Federal Power Commission and the Army Corps of Engineers, Fairchild noted that the disruption of electric power in a future war could be worse than the World War I power shortages, which had crippled military production (and led to intensive efforts to integrate isolated power systems).[107] Given the essential role of electrical power in a modern economy, he argued, the vulnerability of the electricity system implied the vulnerability of the industrial economy as a whole: an enemy that could disrupt the electrical system would "be in good position to accomplish all of his probable war aims."[108]

Having established the dependence of the national economy on the electric power system, Fairchild then assessed this system's susceptibility to aerial bombing. According to Fairchild, the structure of the electrical power system made it especially vulnerable to disruption. To illustrate, he compared it to oil supply. The petroleum system drew on multiple sources of supply that could be easily substituted and that could flow through multiple pathways of distribution. Electrical energy, by contrast, "must move through a definite circuit, including many highly specialized units all built to handle a specific load." Moreover, since it "is impossible to store electricity as we store petroleum," activities dependent on electricity were also vulnerable to disruption: "Immediately upon the breakdown of any link in the chain the industry served by that system ceases to operate." Fairchild's analysis of the electrical power system was typical of the broader ACTS approach. To assess the susceptibility of the economic structure of a nation to precisely targeted attacks, ACTS thinkers engaged in a meticulous analysis of the technical properties of specific systems. These technical characteristics were crucial to understanding not only the physical vulnerability of a plant itself (could it be destroyed through aerial bombardment?) but also the indirect effects of disrupting a given plant on a broader system (transportation, electrical power, etc.) and on the "industry served by that system."

Based on his analysis, Fairchild concluded that the United States' electrical power system presented an ideal target to a potential enemy. In a "very short period of concentrated operations," Fairchild argued, a small force "should be able to knock out" the handful of facilities that were vital to the power system. The results of such an operation "would be immediate, cumulative and comparatively permanent." It was possible to "form an idea of the pressure that would be exerted" by an enemy attack on the power grid by imagining "section after section of our great industrial system ceasing to produce all those numberless articles which are essential to life as we know it." Such an attack would strain the national economic structure "to the breaking point, if it did not, indeed, prove to be conclusive."[109]

STRATEGIC AIR INTELLIGENCE: AWPD-1, 1939–1941

As Haywood Hansell later noted, ACTS studies of industrial vulnerability that used the United States as a "working model" demonstrated "the importance of certain systems and factories: electric power, rail transportation, fuel, basic materials such as steel; food supplies and processing; water supplies; and armaments and aircraft factories." But in the absence of detailed knowledge of foreign economies, this work on US vulnerabilities remained "an abstract exercise lacking in practical results." Without "foreign industrial analysis"—which Hansell regarded as "the sine qua non of strategic air warfare"—"there could be no rational planning for the application of air power." This lack of concrete information became more pressing as war with Germany approached. ACTS officers were eager to formulate proposals for a mobilized air force that had the types and quantities of planes and pilots required to carry out strategic campaigns that targeted the vital systems of the Axis powers. The US Army's intelligence service—still resistant to ACTS officers' claims about the strategic significance of airpower—refused to allow the Army Air Corps access to such intelligence.[110] In response, working under Henry "Hap" Arnold, the chief of the Army Air Corps, in 1940 Hansell set up a separate air intelligence branch to engage in "economic-industrial analysis and description of vital and vulnerable systems."[111]

The task of analyzing the structure of enemy economies presented precision bombing advocates with a distinctive challenge: they had to piece together widely scattered information about enemy economies. To gather such intelligence, Hansell recruited experts in European affairs and business, such as diplomatic historian James T. Lowe and financier Malcolm Moss, both of whom would play important roles in air intelligence during—and, in Lowe's case, after—World War II. For example, Moss drew on his Wall Street contacts to elicit detailed information about German utilities that was held by New York City banks that had financed their construction. "Using these sources," Hansell wrote, "together with scientific journals and trade magazines, we put together a comprehensive target study on the German electric power system and electric distribution system."[112] Another source of economic intelligence was an expert on the European oil production system who provided information that demonstrated the "extreme importance and vulnerability" of German synthetic oil plants—knowledge that would be put to use in 1944 Allied attacks. Based on these analyses, the intelligence unit prepared a set of "target folders, aiming points, and bomb sizes" for a range of potential target systems, including electrical power; fuel supply; steel production; the aircraft industry; railway, canal, and highway networks; and food processing and distribution.

During the lead-up to the American entry into World War II, airpower advocates made use of the work done by Hansell's air intelligence unit in arguing for a large strategic air force. In July 1941, President Roosevelt ordered the military branches to estimate their requirements for victory in Europe. According to Hansell, airpower advocates saw Roosevelt's order as "an opportunity for which 'Billy' Mitchell and the believers in his philosophy had been struggling since the conclusion of World War I—the privilege of drafting the specifications around which to create American air power."[113] At Arnold's instigation, the War Plans Department handed over to a small group of ACTS instructors the task of planning the air war and developing estimates of the amount of planes, munitions, personnel, and fuel that would be required for it. This group—which included Hansell and Harold George— composed a list of air war requirements over a short period in August 1941, with the assistance of Col. Richard D. Hughes, who had been part of Hansell's air intelligence team and would soon be put in charge of target selection for the Eighth Air Force in London. Hughes later reflected that the work of this group—summarized in a thick volume entitled *Munitions Requirements of the Army Air Force for the Defeat of Our Potential Enemies*—was the basis "upon which the American Air Forces were subsequently to plan not only their operations for the whole of World War II, but also their manufacturing, man power, and training programs."[114]

This air war plan, known as AWPD-1 (Air War Plans Division-1), was a template for industrial mobilization for air war. The plan's estimate of requirements was based on an analysis of the number of planes, bombs, and manpower that would be needed to successfully execute precision attacks that would "break down the capacity of the German nation to wage war."[115] AWPD-1 listed 154 "key node" targets in the German industrial production system. In selecting these targets, Hansell later explained, "we attempted to identify 'service systems,' i.e., systems which motivated or connected industries, rather than industries themselves"[116]—such as electrical power, the favored target for ACTS; the transportation network; and the petroleum and synthetic oil industries. Once the air war planners had selected their targets, they estimated the military forces required to destroy each target.[117] "Using this procedure," Hughes later recalled,

> they were able to plot almost all of our future requirements—such as what size training program should be embarked on to train the necessary air and ground crews, the scale of the manufacturing program for building the necessary planes, armaments, and equipment of all kinds, the bases it would be necessary for us to occupy before our planes would be able to reach those targets, and even the need to design and manufacture the B-29 if we were to be able to defeat the Japanese by air power at all.[118]

The hurriedly assembled plan called for a massive and entirely unprecedented expansion of US airpower to execute a full-scale bombing attack on the industrial capacity of Germany: more than 60,000 aircraft, almost 180,000 officers, and nearly two million enlisted men. It was approved as the Air Annex to the Victory program in September 1941 (see chapter 2).[119]

AWPD-1 was a minutely detailed mobilization plan for building an air force—"as thick as a New York city phone book," according to Hughes. But although AWPD-1 specified the German oil, aluminum, and aircraft industries as target systems, it did not estimate the military-industrial effects of bombing specific targets. Hughes later underscored the "immaturity of the art of target selection" at the time. "The very few weeks during which our little staff section had been in existence," he recalled, "had not permitted us even to begin thinking logically about such problems."[120] The strategy of precision bombing, historian Barry Katz notes, "rested upon two factors that lay outside the competence of military planners: a general theory of the complex interdependencies of the enemy economy, and detailed intelligence as to the location, organization, and layout of the specific targets of maximum vulnerability as identified by the theory."[121] Fairchild had made a similar point in his lectures on urban and national economic structure, acknowledging that his own analysis was somewhat "amateurish," and that the proper selection of targets would require systematic study by "the economist—the statistician—the technical expert—rather than a strictly military study or war plan."[122] Over the next few years, such experts would be enrolled in just this project, as a group of New Deal economists who had been working on problems of domestic economic management were recruited into air intelligence and targeting organizations.

Total War and Strategic Vulnerability

Despite fractious debates about priorities and strategic doctrine at the outset of World War II, a wide range of American military and civilian leaders and technical experts agreed that a total war would be decided by a country's ability to marshal national resources in order to maximize destruction of an enemy's military-industrial power. The task was one of simultaneously mobilizing military production and "demobilizing" an enemy's military production through the strategic bombing of industrial targets. New Deal economists played a central role in addressing one aspect of this problem in wartime mobilization planning offices, such as those set up in the National Defense Advisory Committee, where they applied the science of flows to the novel problems of military-industrial production. This section traces how another group of New Deal economists moved into economic intelligence

and air-targeting offices, where they drew on the science of flows to identify the vulnerabilities of enemy industrial production systems that could
be targeted in strategic bombing campaigns. In doing so, they invented a
new kind of expertise—what economist Joseph Coker would later call an
"economics of strategic target selection."[123]

OSS RESEARCH AND ANALYSIS

In July 1941 President Roosevelt charged William J. Donovan to reorganize
American intelligence-gathering efforts in advance of the war. Donovan
formed the Office for the Coordinator of Information, later the Office of Strategic Services (OSS), as a central intelligence clearinghouse. Convinced that
the most important wartime intelligence would be gleaned not from clandes-
tine operations but from the application of social scientific expertise to readily available information, Donovan gathered a collection of academic experts
in the OSS Research and Analysis Branch.[124] Edward Mason, who had taught
economics at Harvard along with Lauchlin Currie, was appointed head of the
OSS Economic Analysis Division. Mason recruited Emile Despres, Chandler
Morse, William Salant, and Charles Kindleberger from Currie's Division of
Research and Statistics at the Federal Reserve, alongside several other young
economists, including Wassily Leontief, Carl Kaysen, and Walt Rostow.[125]

The OSS Economic Analysis Division's initial research projects sought
out vulnerable nodes in the German industrial production system. As
Chandler Morse later recalled, "By intuition rather than plan, we focused
on what we perceived to be actual or potential areas of weakness that might
endogenously become militarily significant or be given an exogenous military push"—via an aerial strike—"in that direction."[126] OSS economists
analyzed supply chains—including material flows and the transportation
and power systems on which they depended. The goal was to identify constraints on Germany's military capacities through what Rostow described
as the "detailed analysis of the scale and character of supply requirements,
under combat conditions of various intensity, as well as close attention to
the carrying capacity of roads, railway lines, and ships."[127] One of the division's early economic intelligence projects investigated how the logistical
difficulties of supplying the basic needs of a vast standing army would constrain Germany's ongoing invasion of the USSR.[128] In its March 1942 report,
the group argued that given its depletion of reserves and its poor logistical
situation (the occupying army depended on a single rail line for supplies),
Germany would be in a weak position at the start of the 1942 campaign. The

OSS economists understood the various elements of an advancing army as a system of inputs and outputs and described the flows of men and materiel required for different kinds of activities. "Using all kinds of data taken from Army field manuals, [and] from World War I experience," Kaysen later recalled, "this group of about four or five people constructed a great big matrix which showed various kinds of fighting, rapid advance, static defense, heavy fighting, and so on, and estimates of the tonnage of food, ammunition, POL [petroleum, oil, and lubricants], spare parts, and so on, that would be consumed by the different sorts of German divisions there in a day under these various conditions."[129]

In their initial research projects, OSS economists developed an understanding of total war that would underpin their later work in air-targeting offices. In Riefler's terms, they conceptualized the war as a competition among military-industrial complexes, each trying to maximize its own production and to reduce that of the enemy. For example, in late 1941 Despres investigated whether an Allied army could be prepared to invade Continental Europe within two years. His report on the investigation, according to Katz, described "a deadly race to bring American resources up to capacity while simultaneously reducing those of Germany to a level that favored an Allied invasion force."[130] To reduce German capacity, Despres advocated harassing operations, a political program to instigate unrest in Germany and in German-occupied territories, and, in his own area of expertise, a program for precision bombing of economically significant targets. The OSS economists' investigations thus complemented and mirrored those of New Deal economists who continued to work on mobilization planning. The economists working on mobilization were concerned with organizing the country's total resources for a coming war; for the economists in OSS, in contrast, the problem was how to use available resources to maximize the damage inflicted on the enemy's military production capacity.[131]

The OSS economists approached the task of planning for total war with Germany by constructing a picture of the German economy similar to the picture of the American economy that had been pieced together through surveys of national resources and production processes during the New Deal. Since no such data collection efforts could be organized to investigate the German economy, they developed techniques for reconstructing material flows and industrial production processes based on obscure and often fragmentary information. For instance, by analyzing published rail freight schedules, they were able to determine the location of German oil refineries—findings that were later confirmed by aerial reconnaissance.[132]

Another example was the "serial number analysis" technique developed by Sidney Alexander of the Industrial Resources Section to estimate rates of German weapons production.[133] Alexander traveled to Tunisia to gather abandoned and damaged German hardware, since military equipment such as tanks had markings indicating the place and date of their manufacture. Using this information, the OSS economists were able to establish the rate of German tank production and to identify industrial plants where critical parts were made.[134]

THE ECONOMICS OF TARGET SELECTION

With the United States' entry into the war in December 1941, military strategists faced the practical challenge of planning the bombing campaign in Europe. Combat units of the Eighth Air Force arrived at a base just north of London in June 1942 as the first contingent of American bombers to enter the war. Their mission was to use airpower to prepare the way for an invasion of Continental Europe from England. Many officers of the Eighth Air Force, such as Haywood Hansell, Larry Kuter, and Muir Fairchild, had been ACTS instructors and brought with them the school's doctrine of high-altitude precision bombing of industrial targets. Upon their arrival in England, however, they found themselves ill-equipped to identify priority bombing targets for specific operations since they lacked the economic expertise and knowledge of the structure of German industrial production necessary to develop a systematic approach to target selection.

In the summer of 1942, Colonel Richard D. Hughes (who had worked with Hansell on AWPD-1) was transferred from the Office of the Chief of the Air Corps in Washington, DC, to the Plans Section of the Eighth Air Force in London. There, his task was to assimilate the vast amount of economic intelligence data being produced by the British Ministry of Economic Warfare into American air war plans and, based on this analysis, to select targets for the Eighth Air Force. Given the Air Force's lack of relevant experts, he requested that the US embassy's civilian economic section provide him with "suitable personnel" from Washington to help him make sense of this material.[135] The first economists to arrive from the United States—Charles Kindleberger and Walt Rostow of the OSS Economic Analysis Division—were the founding members of the Enemy Objectives Unit (EOU), which was charged with helping the Air Corps select air targets and assess bomb damage.[136] EOU was administratively housed in the Economic Warfare Division within the US embassy in London, headed by the former director of the Central Statistical Board, Winfield Riefler. The unit's initial leader, Chandler Morse, recruited several OSS economists, including Harold Barnett, Carl Kaysen, and William Salant, to develop a "rigorous science of air warfare."[137]

Though the unit was based in the US embassy and staffed by civilian economists, EOU staff reported to Colonel Hughes. The unit was a "military-civilian, bastard, set-up," as Hughes later put it, but one that eventually, "for the first time in any [Air Force], began producing, in writing, a clear cut, logical, philosophy of target selection for the economical employment of strategic air power in an enemy country."[138] Initially, Hughes set the EOU group to work "amplifying the British target maps and files" on potential targets that he thought might someday be attacked by US air forces.[139] The resulting "aiming point reports" contained detailed information about particular facilities. This information included a given facility's exact location and layout as well as the key points in the facility that were most physically vulnerable to aerial bombing, and that would be the most costly to repair.[140] EOU staff then turned to the selection of bombing targets, a task they conceptualized in economic terms: What was the optimal way to direct limited military resources to maximize disruption of enemy military production? As Walt Rostow summarized this problem, to assess the value of a given target of attack, one had to ask "how large its effect would be within its own sector of the economy or military system; how quickly would the effect be felt; how long it would last; and what its direct military (as opposed to economic) consequences would be." These questions were addressed in a series of "target potentiality reports" that, as Rostow put it, identified "target systems where the destruction of the minimum number of targets would have the greatest, most prompt, and most long-lasting direct effect."[141]

The EOU approach to target selection was laid out in a 1942 memo, "The Selection of Industrial Bombing Targets," written by William Salant, which described a quantitative method for ranking bombing targets according to their military value.[142] Salant was another Harvard-trained economist who had previously worked with Lauchlin Currie as a member of the White House staff. In his work with Currie, he had developed an account of the Keynesian multiplier that examined how government interventions (that, for example, shifted money from savings to government consumption) could be targeted to maximize stimulus of the national economy. Salant drew on this approach in his work at EOU, asking which targets of aerial bombing would have the most strategically significant effects on the total German war effort. Although EOU was a relatively obscure civilian planning office whose recommendations were often ignored by military commanders, Salant's memo, according to Katz, laid the "intellectual foundations" for a "philosophy of air power that would survive well into the nuclear age."[143]

Salant argued that the prioritization of industrial targets required consideration of factors that linked a particular target to the broader military-industrial economy. His analysis drew on economic concepts that had been developed as part of the New Deal science of flows. "It is better," Salant wrote, invoking the concept of the multiplier, "to attack a factory the loss of whose output will have widespread effect in causing stoppages elsewhere than one which is a relatively isolated unit in the industrial system." "It is better to attack at bottleneck points in the industrial structure," where there was tightness in the system of supply, than at points where there is slack due to "surplus capacity or abundant supplies." "It is better to attack production of a material that is essential than one for which substitutes can be found."[144] "It is better to attack a factory producing something the entire output of which goes into essential uses than one which has a wide variety of uses, varying in importance (e.g., armor plates rather than steel ingots)." Taking all these factors into account, the memo advised, targets could be prioritized based on "how much harm can be inflicted on the enemy per unit cost to us" in bombers, munitions, fuel, and pilots.[145]

The application of Salant's approach required a detailed understanding of the structure of enemy industrial production. By examining input-output relationships so that the flow of materials and the operation of production facilities could be linked to military end items, it would be possible to model how the shock of a bombing attack would propagate through an economic system: What parts of military production would be affected? How—and how soon—would disruptions affect fighting capacity? And how might an enemy adaptively adjust its production in response? To capture these effects, EOU economists drew on a number of novel concepts proposed by Salant in his memorandum. For example, the concept of "depth" referred to the length of time required for damage from a particular bombing raid to have an impact on the enemy's war-making capacity—the interval between the production of a given item and its eventual contribution to a military engagement. The concept of "cushion" indicated the enemy's capacity to absorb losses in a particular industry by lowering inventories, shifting stocks from civilian to military use, or finding substitutes for specific manufactured goods or materials. The EOU economists also used capital theory to assess cushion, estimating dynamics such as the rate of substitution of labor for capital that had been destroyed by bombing.

In spring 1943, EOU analysts began to produce what they called "target potentiality reports" to determine how the air force might maximize damage to German military resistance at the moment of an Allied invasion.[146] These reports contained information on resource use and production patterns in a particular industrial sector, the vulnerability and recovery time of facilities in this sector, and its "depth"—that is, how long it would take for a disruption in the production system to affect the fighting front. Thus, in the case of rubber (described in figure 1.3), use in military-industrial production was largely devoted to vehicles and aircraft; vulnerability to fire was relatively high; recovery times were around nine months; and disruptions could be expected to affect the operation of aircraft, tanks, and other vehicles within two to three months. As EOU analyst Charles Kindleberger explained, these reports helped air war planners select which industrial production system to target for attack. "Such industries as steel and electric power lay too deep to be worthy of attack," he noted. "In the winter of 1943–1944 all the steel that would be used in defending against the invasion had already been produced." Similarly, "electric power stations were both difficult to destroy and had the 'cushion' of standby local power at most industrial plants producing army equipment."[147]

Kindleberger explained that EOU's approach to target assessment made it possible to understand "bombing a particular industry in an enemy territory as a means of bringing the economy to a halt by depriving the user industries, including the military, of the selected industry's output." Specifically, it provided "the rationale for bombing such basic industries as ball bearings and oil, and for recommending bombardment for many other industries, such as carborundum grinding wheels, needed for polishing metal."[148] Rostow later described how this approach led EOU to push for targeting the German oil industry, precipitating a fierce debate with British air war planners late in the war. The disruption of the oil system promised "not only to affect the whole German war production structure but also to limit the fighting capacity of the ground and air forces." The oil industry, meanwhile, "was a sufficiently limited target system to offer a chance of cutting deep within a reasonably short period of time."[149]

In sum, drawing on the New Deal science of flows, EOU economists analyzed material flows and interdependencies in order to understand how shocks would be transmitted through the enemy war production system. While their approach echoed Sherman's concerns with the "system of supply," it also incorporated new methods for analyzing the national economy as a whole, conceived as a complex of interdependent activities. Rostow

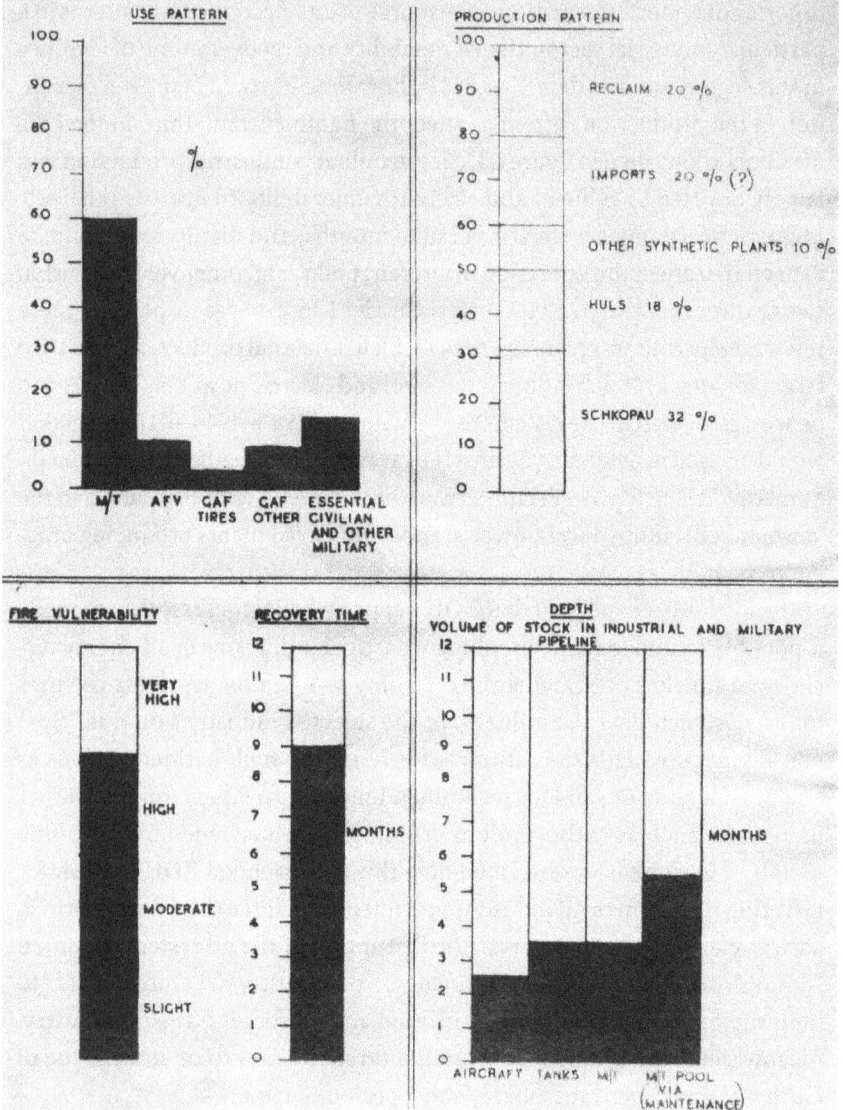

FIGURE 1.3. Enemy Objectives Unit: Target summary diagram—rubber. Economists working in the Enemy Objectives Unit during World War II prepared reports on potential target systems in the German military-industrial economy. These reports examined production and use patterns in specific industries, the vulnerability of facilities and the time required for recovery, as well as the time it would take for disruptions in the supply chain to affect fighting on the front. *Source*: Target Potentiality Reports, US Embassy, United Kingdom, 1943, Iris number: 00215632, Air Force Historical Research Agency, Maxwell Air Base.

later reflected on the heterogeneous sources of the EOU approach: "Our knowledge as economists of the structure of production, buttressed by what we had learned from the logistical studies in Washington and aiming-point reports, converged with the military principles of Hughes, Anderson, and the British Air Staff." In addition, "there was something of the Austrian theory of capital and Leontief's input-output matrix in our way of looking at things."[150] This fusion of interwar airpower theory and New Deal economics proved to be a durable form of expert understanding. As Hughes recounted after the war, "The basic principles hammered out by the 8th Air Force and this group in London were subsequently used by every American Strategic Air Force in every theater, and the same principles must hold good today in the selection of targets within the U.S.S.R. and its satellites by our present Strategic Air Command."[151]

The United States Strategic Bombing Survey: American Vulnerability

EOU's economics of target selection was just one among many competing frameworks for defining the proper aims and targets of the European air war, which were hotly contested among different parts of the American military establishment and between British and American commanders. Partly with the aim of weighing in on these debates, as the European war was reaching its conclusion Fairchild and other leading advocates of strategic airpower proposed a detailed assessment of the efficacy of the Allied Combined Bomber Offensive in Europe. This assessment project would eventually become the first stage of the United States Strategic Bombing Survey.[152] Officially, the Survey's task was to gauge the efficacy of strategic bombing in destroying German war-making capacity. Airpower advocates hoped to use the results of the survey to make the case for precision targeting in the Pacific war, as it entered its last stages. As we discuss in chapter 3, this effort was ultimately not successful. After Curtis LeMay replaced ACTS officer Hansell as air commander for the Pacific theater, US air forces waged a brutal campaign of incendiary bombing in Japanese cities.[153] But over the longer term, the Strategic Bombing Survey—which was extended to study the Pacific war and included the atomic bomb attacks on Hiroshima and Nagasaki—strongly influenced strategic thinking after World War II. Among other things, it provided evidence to support the establishment of an autonomous Air Force and served as a key channel through which the EOU's economics of target selection was transmitted to air intelligence and target planning in postwar strategic thought. For our purposes, the Survey is also significant for another reason. Namely, it applied EOU's style of analysis to a problem that would take center stage in discussions of national security after World War II: US vulnerability to enemy attack.

Air force leaders planned the Strategic Bombing Survey's organization and objectives over the period from April to October 1944. The Survey enlisted hundreds of civilian and military investigators, eventually producing more than three hundred published reports. Civilian leadership and a vast amount of technical analysis lent the survey an air of objectivity and impartiality. Its official chair, Franklin D'Olier, the CEO of Prudential Insurance, was seen as an "amiable figurehead."[154] Its more influential members included financier Paul Nitze, who would become a leading national security advisor of the early Cold War; lawyer George Ball, who had served in the Lend-Lease Administration; and economist John Kenneth Galbraith, who had studied public works planning for the NRPB and mobilization planning for the National Defense Advisory Committee.

Advocates of precision bombing within the Air Force leadership shaped the questions addressed by the Survey. Paul Nitze, before beginning his work on the project, met with Muir Fairchild (whom Nitze later referred to as "the clearest theoretician with respect to air strategy that the U.S. Air Force had") to learn what questions the Survey should ask. Nitze described the Survey's aims in terms that clearly reflected the ACTS approach. "One of the essential questions we had to answer," he recounted, "was whether or not the concentrated attack on one essential component of an industry would, in fact, bring that industry to a halt."[155] To understand the relation between physical damage to specific targets and overall economic effects, the Strategic Bombing Survey drew on conceptual tools that EOU economists had developed in planning for the European air war: the analysis of the depth, cushion, substitutability, and resilience of various target systems. If the economics of target selection remained relatively marginal in World War II, then the Survey established it as an authoritative form of expertise for approaching airpower strategy in the postwar period.

Survey teams inspected specific target locations and wrote up standardized damage assessment reports. Division directors then used these reports to evaluate the effects of targeted bombing on a given industrial sector. These sector reports were the basis for an overall evaluation of the effectiveness of strategic bombing. The Survey's European war summary—written by Nitze, Ball, and Galbraith in June 1945—broadly affirmed the strategic preeminence of airpower in modern warfare. "Allied airpower was decisive," they concluded, "in the war in Western Europe." Though it was deployed too late in the war to realize its maximum effects, the "power and superiority" of Allied airpower made possible "the success of the invasion." Air bombing had brought the economy that "sustained the enemy's armed forces to virtual collapse," although "the full effects of this collapse had not reached the enemy's front lines when they were overrun by Allied forces."[156] Echoing EOU doctrine, the Survey argued that the Air Force had been most effective in bombing basic infrastructures like transportation and oil production—vital, vulnerable systems whose disruption could bring a modern economy rapidly to a halt.[157] As Nitze later noted, these systems, "once severely damaged, could not be quickly restored to full production nor could stocks be readily replaced."[158] "Throughout the war," reported the Survey's Oil Division, Germany's "oil stocks, particularly critical items like aviation and motor gasolines[,] were so tight that her whole military effort in the air and on the ground would have collapsed like a pricked balloon in three or four months had her oil supply been dried up."[159]

If the Strategic Bombing Survey championed the offensive possibilities of strategic airpower as a means to exploit the vulnerabilities of modern military-industrial production systems, it was also concerned with what its authors called the "resilience" of modern economic systems. One puzzle the Survey's analysts tried to solve was why Germany had been able to keep increasing its war production even during the most devastating phase of the air onslaught in early 1944. The Economic Division, led by John Kenneth Galbraith, found that the German war economy had been much more "resilient under air attack" than target planners had anticipated.[160] One source of this resilience was significant slack in the war economy. Due to a combination of overconfidence and mismanagement, German war production had never achieved its full potential. "Plant and machinery" were "plentiful and incompletely used," and the German economy had a "substantial cushion of production potential." As a result, it was "comparatively easy to substitute unused or partly used machinery for that which was destroyed." German resilience was also bolstered by what the Survey described as "passive defense" measures—such as industrial dispersal and hardening of targets—that did not involve the use of military forces. "The recuperative and defensive powers of Germany were immense," the Survey reported. "The speed and ingenuity with which they rebuilt and maintained essential war industries in operation clearly surpassed Allied expectations. Germany resorted to almost every means an ingenious people could devise to avoid the attacks upon her economy and to minimize their effects."[161] The summary report for the Pacific war, which arrived at similar findings about the effectiveness of nonmilitary defense measures, concluded, "The experience of both the Pacific and European wars emphasizes the extent to which civilian and other forms of passive defense can reduce a country's vulnerability to air attack." Even faced with attack using atomic bombs, the report insisted, civilian casualties could be dramatically reduced through measures such as "progressive evacuation, dispersal, warning, air-raid shelter, and postraid emergency assistance program." Similarly, economic vulnerability could be "enormously decreased by a well worked out program of stockpiles, dispersal and special construction of particularly significant segments of industry."[162]

The Strategic Bombing Survey's *Summary Report* drew a number of lessons for the postwar period. A modernized air force would have to be maintained during peacetime, and it would be necessary to establish a large-scale program of research and development to ensure American technological superiority. Air intelligence would also have to be kept up to date, so that the United States would not again be in the position of hastily assembling a targeting operation from fragmentary information. Inevitably, the Survey also turned to the vulnerability of the United States in a future war, a problem that would become a central preoccupation of postwar strategic thought (see chapter 3). Noting the density and type of construction in American cities, the Survey's report on the atomic bombings of Hiroshima and Nagasaki argued that devastation similar to that witnessed in Japan—or much worse, given new weapons technology—could be expected. But the report nonetheless argued that US vulnerability could be meaningfully reduced if proper measures were taken. "The almost unprotected, completely surprised cities of Japan," the report observed, "suffered maximum losses from atomic bomb attack. If we recognize in advance the possible danger and act to forestall it we shall at worst suffer minimum casualties and disruption." The Survey suggested that measures to reduce American vulnerability should focus on vital production systems. "An enemy viewing our national economy," it contended, "must not find bottlenecks which use of the atomic bomb could choke off to throttle our productive capacity." Consequently, it advocated practical steps toward "a reshaping and partial dispersal of the national centers of activity" as well as measures to ensure that "production of essential manufactured goods" not be "confined to a few or to geographically centralized plants." The report also advocated measures to increase what EOU theorists had called "cushion"—such as the accumulation of "reserve stocks of critical materials and of such products as medical supplies" and capital equipment—to make the future military-industrial economy more resilient.[163]

After World War II, all these measures came to the center of strategic discussion. Among the advocates of such measures were veterans of war-time mobilization planning and air targeting who moved into postwar emergency resource management agencies such as the National Security Resources Board and the Office of Defense Mobilization, the descendants of offices such as NRPB and the War Production Board. We examine this postwar work on vulnerability reduction and preparedness in chapters 3 and 4. Presently, however, we turn back to the early 1930s, to pass through the tumultuous years of the Great Depression and World War II from another perspective. Over this period, the problem of securing vital systems inter-sected with another problem: how to govern emergencies within liberal constitutional government.

2

Emergency Government

On May 21, 1940, following the German invasion of France, President Roosevelt issued an administrative order that proclaimed the existence of a "threatened national emergency" and created an Office for Emergency Management (OEM) within the Executive Office of the President. He charged this new office with managing information about the evolving crisis, coordinating with agencies across the federal government, and performing undefined "additional duties as the President may direct."[1] According to Herbert Emmerich, a specialist in public administration who was working on government reform at this time, OEM served as "a kind of tent and service agency for defense and war establishments and gave immense flexibility, as conditions changed, to constitute and reconstitute units of government."[2] Among the elements of the "defense and war establishments" gathered in OEM were the myriad offices in which statisticians, economists, engineers, and other technical experts worked to manage the massive government effort to ramp up military-industrial production for World War II. These offices included the National Defense Advisory Committee, the Office of Price Administration, the Supply Priorities and Allocations Board, and, after the United States entered the war in December 1941, the War Production Board, whose staff swelled to more than twenty-two thousand at the height of wartime mobilization.[3] As the mobilization planners David Novick, Melvin Anshen, and William Truppner recounted in 1949, these agencies wielded exceptional regulatory powers, which they used to organize "the bulk of the industrial resources of the country into a single integrated production unit."[4]

While officials in the Roosevelt administration created OEM to address the immediate tasks of war mobilization, they understood the office as a response to broader dilemmas of American emergency government. William McReynolds, a special assistant to the president who served as OEM's director, described these dilemmas in 1941. During national emergencies, many problems that under normal circumstances were "delegable to subordinate officers in the regular departments or agencies" became "so acutely significant and so supremely important that the President must exercise more direct and immediate control over them." Such expanded executive authority challenged basic tenets of the US constitutional system, such as the diffused character of sovereignty. And yet, McReynolds observed, both the Constitution and specific laws "vested emergency authority in the President confirming his high responsibilities to furnish leadership in times of crisis."[5]

In formulating their plans for OEM, McReynolds explained, Roosevelt administration officials drew on a "theory of organization" that addressed these competing constitutional and statutory demands. On the one hand, the establishment of OEM would enable the president to "exercise immediate supervision and control" over future emergency situations whose contours could not be anticipated in advance. To this end, the executive order that established OEM did not provide "a detailed statement regarding [its] precise duties and functions." It was written to be "as flexible as possible" so that the authority wielded by the offices within it could be "easily and quickly altered to meet the new or changing demands of the crisis." The president was left "completely free, by administrative order, to determine the character of the Office as conditions might change from time to time, or as new emergencies might arise." On the other hand, OEM's powers were circumscribed. The office did not have a large staff but was "the place in which the Chief Executive can locate liaison, coordinating, and necessary operating activities relating to the emergency." In carrying out OEM's functions, the "facilities of existing organizations"—in particular the federal departments, which were subject to congressional oversight—would be "utilized to the greatest possible extent," and existing administrative procedures would be "disturbed to a minimum degree." In sum, OEM served as what McReynolds referred to as an "administrative device" for the flexible exercise of executive power within the framework of American democratic government. It was designed to allow the president "to discharge his high responsibilities as Chief Executive during an emergency period as required by common sense and as contemplated by the statutes and the Constitution."[6]

Roosevelt administration officials anticipated that, beyond the wartime emergency, this "administrative device" would become a permanent feature of the US governmental system. "National emergencies," McReynolds explained, were "not confined to periods of war or intense preparation for defense." Emergency situations might also arise from any event whose effects were of "more than local importance and become matters of national concern because their impact is seriously felt throughout the country."[7] An executive order of January 7, 1941, thus instructed that OEM would "advise and assist the President in the discharge of extraordinary responsibilities imposed upon him by any emergency arising out of war, the threat of war, immanence of war, flood, drought, or other conditions threatening the public peace and safety."[8] Clinton Rossiter, whose 1948 study *Constitutional Dictatorship* made him the leading contemporary scholar of American emergency government, explained that Roosevelt administration officials imagined that OEM would remain as a "permanent trouble shooter" in the executive branch. It would take responsibility for managing a range of emergency situations as they unfolded and would continuously prepare "plans for future emergencies."[9]

This vision of OEM as a permanent fixture of the federal government was not realized. The office was dissolved in the sweeping demobilization at the end of World War II. But the model of emergency government first assembled in OEM was resurrected in early Cold War offices such as the National Security Resources Board and the Office of Defense Mobilization. In these and subsequent offices, the organizational pattern first established in OEM persisted, as the task of emergency government shifted from resource management for war to preparedness planning to ensure the continuous operation of vital systems in future emergencies.

Taking the "administrative device" of OEM as a starting point, this chapter examines the formation of a distinctive political technology of emergency government. As described in the introduction, the German jurist Carl Schmitt used the term "political technology" to describe the central role of expert rule and executive power in modern states. Schmitt argued that modern states were regularly faced with rapidly unfolding economic, social, and military crises, and that such crises posed particular problems for liberal constitutional governments. Decisive intervention based on strictly rational-technical criteria was required to meet the "dictates" of crisis situations. But such executive power and technical rule was in tension, he claimed, with democratic norms of deliberation and diffused sovereignty. In this chapter, we show that reformers working in and around the Roosevelt administration set up a novel political technology of emergency government to address precisely this tension between the institutions of liberal constitutional government and the demands of the crisis situations of economic depression and war. Throughout both the New Deal and World War II, congressional conservatives and other critics charged that Roosevelt was trying to amass dictatorial powers that were incompatible with American democratic traditions. In response, reformers sought to craft a technical and administrative solution to the problem of crisis government, equipping the executive to manage emergency situations within the framework of the Constitution.[10]

The first sections of this chapter situate the midcentury transformations of emergency government within a broader trajectory of Progressive thought on administrative reform. Initially focused on local government, Progressive reformers emphasized strengthening executive authority and organizing expert knowledge as an input to governmental decision making. In the 1930s, these reformers migrated to the federal government, where they drew on this theory of organization to address the successive emergencies of the Great Depression and industrial mobilization for World War II. Their work culminated with the 1939 Reorganization Act, which was designed to provide Roosevelt with what he called the "governmental machinery" and "modern types of management" that were required to address wars, economic shocks, and natural disasters.[11]

The chapter then examines how this new authority and governmental equipment were first used to manage military-industrial mobilization for World War II. Drawing on the powers of the 1939 Reorganization Act and subsequent legislative enactments, Roosevelt set up a series of emergency agencies to manage the mobilization effort. New Deal experts in economics, statistics, and public administration worked in these new agencies, where they produced a body of systematic knowledge about the vital flows that made up the American military-industrial production system, and assembled instruments for managing these flows, thus grounding emergency economic administration in what they saw as the impartial authority of technical expertise. The final section of the chapter describes how, following the demobilization of wartime emergency agencies, postwar planners reassembled the elements that made up this apparatus of wartime mobilization for postwar emergency government. We suggest that these elements—legal mechanisms, administrative arrangements, forms of knowledge, regulatory devices, and planning procedures—persist as key features of the American emergency state.

Progressive Reform: Governmental Adjustment and the Positive State

The "theory of organization" that underpinned reformers' response to the Great Depression and mobilization for World War II grew out of a particular strain of Progressive thought.[12] When the Progressive movement first emerged in the late nineteenth century, it was defined by opposition to private monopolies and suspicion of any concentration of power, whether private or public.[13] But a new strand of Progressivism in the early twentieth century sought to expand and reorganize government functions, in order to create a "positive" state that would be deeply involved in shaping social and economic life. These Progressive thinkers argued that the institutions of American government had been designed for a predominantly rural and agricultural society. But with rapid urbanization and industrialization in the decades following the Civil War, government would have to take on new roles in managing the problems of an increasingly complex social and economic order.[14]

In the first decades of the twentieth century, a group of Progressive thinkers developed a model of governmental reform on the local level that they would later apply to federal reform during the "emergencies" of depression and war.[15] For these thinkers, the dominance of legislatures, whether at the city or state level, produced a "spoils" politics that was corrupt and narrowly concerned with the interests of local constituents. They sought to strengthen executive authority and create institutions (such as the new position of the city manager) through which professional staff could manage urban problems, independent from local or factional political interests. The Progressives also sought to develop and disseminate expert knowledge and practical experience that could be used by municipal governments, often through the formation of independent research organizations. Examples of such organizations include the Public Administration Clearing House, created by Charles Merriam and directed by Louis Brownlow (and, later, by his protégé, Herbert Emmerich); Luther Gulick's Bureau of Municipal Research; and the Social Science Research Council.[16]

Progressive reformers saw these initiatives for restructuring government and developing social scientific expertise relevant to public administration as part of a broader vision of a modernized American liberalism.[17] Among them, Merriam, a political scientist who was at the center of governmental reform for four decades, was most explicit in articulating such a vision, first in relation to city government and later as a leading "organizer and theorist of

New Deal planning."[18] In his early work, Merriam wrote admiringly of local government in Germany. While he found German militarism "absurd," he admired the application of a "science of administration" to local government and concluded that the German arrangement of "a trained expert as mayor and a permanent staff of employees" produced "efficient administration of a type I had not known before." The question, in Merriam's view, was whether the "democratic spirit of Iowa could be combined with the administrative efficiency of Germany."[19]

A key concept for Merriam in framing the challenge that an industrializing and urbanizing society presented to American democratic institutions was how to "adjust" the US system of government to new conditions. Merriam argued that explosive developments in technology, patterns of settlement, and economic activity had outrun conscious organization and planning. "In many ways," Merriam wrote in the 1925 study *New Aspects of Politics*, "politics has been outstripped in the race for modern equipment supplying the rapid, comprehensive, and systematic assembly and analysis of pertinent facts."[20] "Have we not reached the point," he asked, "when it is necessary to adjust and adapt more intelligently, to apply the categories of science to the vastly important forces of social and political control?" Merriam was convinced that such adjustment did not require abandoning basic tenets of US constitutional government. Rather, as historian Patrick Reagan argues, Merriam maintained a deep "faith in the progress of American democracy," which, he believed, "could be completed by new methods and tools of modern social sciences."[21]

In the late 1920s, Merriam and other Progressive reformers who had been engaged in local government reform turned their attention to the national scale.[22] In 1929, Merriam cochaired the Research Committee on Social Trends, which had been formed by President Herbert Hoover to bring the knowledge of the social sciences to bear on the problems American government faced in the early years of the Great Depression. In his contribution to the Committee's report, Merriam reprised his argument about "adjustment," now in relation to the organization of the federal government.[23] "The political order," Merriam wrote, "can be understood only in its relations to the social and economic forces fundamentally conditioning its activities." He identified several "social and economic forces" as particularly significant: urbanization and industrialization, increasing interdependence through trade, a revolution in public health, and sharp "inflation and deflation of business and agriculture." These developments, Merriam argued, had "obliged the government to undertake new activities," such as "services to business, agriculture, and labor"; public welfare, including education, recreation, health, and social relief; highways and "aid to communications"; the management of national defense in an era of total war; and the "central control over social and economic forces." In part, these new activities raised a question of scale. "Industrial and social relations," Merriam wrote, "overflowed the banks of the states and swept out over the nation," thus implying the need for federal action in areas that had previously been the prerogative of subnational governments. These activities also required that the national government draw on the specialized knowledge of technical experts to ensure "the performance of an endless variety of services now demanded by the community under the new conditions." The key questions were how "far and in what ways" government would "make use of these new techniques?" And how would the administrator make the "wisest use of the technical expert"?[24]

Turning to recent developments in US governmental institutions, Merriam observed that the "outstanding fact" was "the rapid extension of governmental activities and costs on the one hand, and on the other the relatively slight change in governmental units, organization, methods and personnel." The rise of despotic regimes in Europe weighed heavily in Merriam's assessment of this "maladjustment" of American government. "When contrasted with the European situation," he wrote, "it is clear that there has been relatively little shift in fundamental theories and attitudes in America during this period." Fascism, sovietism, and socialism offered radically new governmental forms; democracy and the system of free enterprise had been "subjected to severe analysis on the part of friends and foes." But Merriam expressed confidence that it was possible to "reorganize and reconstruct a type of government and administration in which the factors of modern science and economics were adequately recognized and reconciled with democratic control."

After President Roosevelt took office in 1933, a number of Progressive reformers moved into the federal government. Merriam, along with Wesley Mitchell and Frederic Delano, assumed a position on the National Planning Board, which had been established based on a recommendation from the Committee on Social Trends.[25] Merriam's work focused on "governmental planning," which was understood, according to Reagan, as a matter of bringing "public administrative theory and practice to national planning."[26] By this time, with economic depression and political upheaval in Europe clearly in view, Merriam was explicitly grappling with the challenge that democratic governments faced in addressing crisis situations.

Merriam's evolving thinking about these issues took center stage in the National Planning Board's final report.[27] The report observed that the "financial crisis of 1933, with the closing of every bank in the land, with thirteen millions of unemployed, and with the general prostration of industry and agriculture" had "brought the Nation face to face with stern realities" that required "prompt and bold action to prevent complete collapse." Pointing to the situation in Germany and Austria, where "the parliamentary and democratic balance of authority" had been "violently overthrown, and an entirely different system substituted," the report warned that if US democratic institutions could not "meet the challenge of general discontent and suffering" it would have to give way to "some other system more promising in its prospects of providing security and contentment." This "grim situation" gave new urgency to the problem of governmental adjustment. It was impossible to proceed "as if nothing had happened in recent years, indifferent to the sweeping changes going on among us and elsewhere in the world." But the report insisted that the American system could be adapted to these new circumstances without fundamental changes. "Our presidential-congressional form of government," it affirmed, was "fortunately flexible enough to meet crises within the prescribed limits of the fundamental law." The executive, with congressional approval, was able to act as "swiftly in war or peace as the tension of the time demands." No "modification of the broad lines of our governmental system" were required to "meet the demands of social and economic crises."[28]

Notwithstanding the Board's admonishments, governmental reform and administrative rationalization were marginal concerns in the early New Deal. "The high priority need," as Herbert Emmerich later recounted, "was for reversing the catastrophic downtrend rather than for stopping to work out a tidy organizational pattern."[29] But Progressive reformers soon found their opportunity to push for governmental "adjustment" at the national level in responding to the problem that, as Carl Schmitt had warned, presented a particular challenge to liberal constitutional regimes: the status of emergency executive powers.

Reorganization and the Government of Emergency

Prior to the New Deal, the US government had largely exercised emergency powers only in war, when the president—who during times of peace had only a small personal staff—had broad discretionary authority to intervene in national life. In the aftermath of wars, these powers were rapidly rolled back, upon restoration of the "normalcy" of a minimal executive and leg-islative dominance in the conduct of domestic affairs.[30] This pattern con-tinued through World War I. During the war, temporary offices such as the War Industries Board, the War Trade Board, and the War Finance Cor-poration exercised price and trade controls, regulated consumption and production, and suspended normal laws.[31] As historian William Leuchten-berg has observed, these measures implied a substantial concentration of power, raising "the federal government to director, even dictator, of the economy," whether based on powers that Congress handed to the president through so-called delegatory statutes, or on inherent constitutional powers of the presidency.[32] At the end of the war, Congress shut down the wartime emergency agencies, and political reaction set in against expanded wartime executive power.[33]

President Roosevelt challenged this pattern of emergency govern-ment with his initial response to the Great Depression. The mechanisms of intervention he employed were not in themselves new; indeed, they were overwhelmingly based on precedents from World War I. Rather, what was novel about the interventions of the early New Deal was Roosevelt's use of emergency powers to address a peacetime crisis. Roosevelt portrayed the economic downturn as an emergency situation, calling for "broad Executive power to wage a war" against the Great Depression "as great as the power that would be given . . . if we were in fact invaded by a foreign foe." Accord-ing to Leuchtenberg, this reference to war was meant, in part, to explain "the meaning of the depression" by describing it as "a calamity like war, or, more specifically, like the menace of a foreign enemy who had to be defeated in combat." But it was also meant to justify the use of mechanisms of wartime government as "instrumentalities to combat hard times."[34]

Initially, Congress supported Roosevelt's use of wartime emergency powers to address the peacetime crisis, passing legislation in early 1933 that delegated broad authority to the president to manage the economic down-turn.[35] These legislative acts, as one contemporary observer noted, were "heavily garlanded with 'emergency clauses' describing the dire national peril," affirming that Congress accepted Roosevelt's claim that the Great Depression was a national emergency.[36] But support for this use of emer-gency powers soon collapsed. A bloc of congressional Republicans and conservative Democrats accused Roosevelt of trying to amass dictatorial powers.[37] The Supreme Court also struck down key elements of the New Deal, finding that, as historian Barry Karl puts it, the "difference between the physical threat of war and the large-scale social threat of a depression" was too great to justify the extension of wartime measures to a peacetime emergency.[38] By 1935, according to political scientist Matthew Dickinson, Roosevelt became convinced that the "administrative chaos engendered by the New Deal"—a product of both the proliferation of emergency agencies and ongoing struggles over presidential authority—"threatened his political support." Thus, with the economy stabilizing, Roosevelt "turned his efforts toward rationalizing his institutional house and putting it on a more permanent footing."[39]

"THE BETTER SUPERVISION OF THINGS"

In early 1936, Roosevelt formed a Committee on Administrative Management to address "the management of the executive branch as it is established under the constitution."[40] The Committee was led by Charles Merriam and two other Progressive reformers, Luther Gulick and Louis Brownlow. Its staff included leading specialists in administrative reform such as Herbert Emmerich, Robert Howe Connery, William Yandall Elliott, and James Fesler.[41] Roosevelt asked the Committee to study the relation of the existing organization of the executive branch to the "many new agencies which have been created during the emergency," noting that "some of these agencies doubtless will be dropped or greatly curtailed, while others may have to be fitted into the permanent organization of the Executive Branch."[42]

The Committee's framing of its task reflected both the longstanding concerns of Progressive reform and Merriam's assessment of the challenge that "social and economic crises" posed to democracy. Could the executive as defined by the Constitution, and as constrained by the legislature, be equipped to address both peacetime and wartime emergencies without losing its democratic character? In addressing this question, the members of the Committee occupied a precarious middle ground. Many congressional conservatives resisted reforms to increase the power and autonomy of the executive, fiercely defending congressional prerogatives. At the same time, the Committee's staff shared "an uneasy awareness," as Emmerich later wrote, "that throughout the world democratic government itself was on trial": Could democratic governments manage crisis situations?[43] This challenge was driven home for Brownlow and Merriam on a trip to Europe as the Committee was conducting its studies. Brownlow reported that they encountered a "gulf, wide and deep, between the democratic and authoritarian modes of thought." Political movements "to the right and to the left" professed the doctrine that "governments require an all-powerful man to tell them and their people what to do and to govern them without obtaining the assent or tolerating the dissent of popular elections." The Committee felt that it had to formulate a response to the challenge of these radical alternatives, and to provide, as Karl writes, "an alternative in practical efficient democracy."[44]

Such concerns framed the Committee's report, which was submitted to President Roosevelt in January 1937. "From time to time," the Committee intoned, "the decay, destruction, and death of democracy has been gloomily predicted by false prophets who mocked at us." But the "American system" had matched "its massive strength successfully against all the forces of destruction through parts of three centuries." Facing one of "the most troubled periods in all the troubled history of mankind," it was necessary to "set our affairs in the very best possible order to make the best use of all of our national resources and to make good our democratic claims" to meet the "stern situations we are bound to meet, both at home and elsewhere." The report urged that "our governmental machinery" be adjusted "to meet new conditions"—both the immediate circumstances of the Great Depression and international tensions and the broader development of an urban and industrial society. "Now we face again," the Committee argued, "the problem of governmental readjustment, in part as the result of the activities of the Nation during the desperate years of the industrial depression, in part because of the very growth of the Nation, and in part because of the vexing social problems of our times."[45]

The Committee's discussion of governmental adjustment focused on the office of the presidency itself. Although its functions and responsibilities had expanded dramatically, the national executive remained a small office, limited to the president and a personal staff. This "maladjustment" between functions and capacities had grown particularly acute during the early New Deal. The federal government was engaged in vast new fields of activity that the Committee grouped into five "great categories" of public works, public welfare, public lending, conservation, and business controls. These activities were being taken on by a welter of new offices, created without any plan or theory of administrative organization.[46] More than "100 separately organized establishments and agencies presumably report to the President," but executive control over these entities was ambiguous (see figure 2.1). A dozen more entirely independent agencies constituted "a new and headless 'fourth branch' of the government." These developments, the Committee concluded, had produced an ineffective and overburdened "multiple executive."[47] The problem did not lie in the basic institutions of US government.

FEDERAL AGENCIES DIRECTLY RESPONSIBLE TO THE PRESIDENT

→ ADMINISTRATIVE AGENCIES ←

DEPARTMENTS

SPECIAL STAFF

PRESIDENT

PLANNING COMMITTEES & AGENCIES

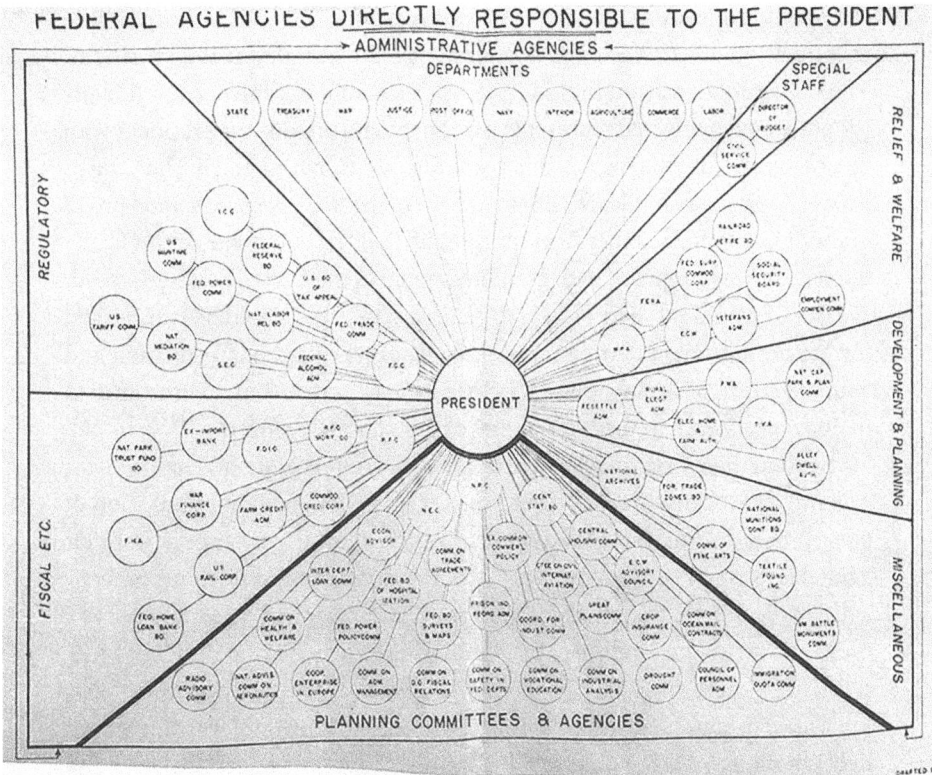

FIGURE 2.1. Federal agencies directly responsible to the president. The Committee on Administrative Management found in 1937 that over one hundred different offices formally reported to the president but that executive control over these entities was ambiguous. The committee advocated reform to rationalize executive administration and to better equip the president to deal with crisis situations. *Source*: Luther Halsey Gulick III Papers, Baruch College Archives, NY

The Constitution had provided the president with ample powers that had been supplemented by acts of Congress. "The American Executive," the report proclaimed, "stands across the path of those who mistakenly assert that democracy must fail because it can neither decide promptly nor act rigorously." The issue, rather, was that the president's "administrative equipment" was far "less developed than his responsibilities." American presidents lacked the "governmental machinery" and the "modern types of management" required to fulfill the constitutional and statutory duties of the office.[48]

The Committee's report laid out various proposals for creating a better equipped national executive. It proposed to provide the president with several special assistants, and to place the Bureau of the Budget and a reconstituted National Resources Planning Board (NRPB)—both offices that performed expert advisory functions—directly under the president. The Committee insisted that such measures would not infringe on powers properly held by elected representatives but would provide the president with means to carry out the responsibilities defined by Congress and by the Constitution. Thus, for example, NRPB would not make "final decisions upon broad questions of national policy," a responsibility that "rests and should rest firmly upon the elected representatives of the people of the United States." Instead, the Board would serve as a kind of "general staff" that would be responsible for "gathering and analyzing relevant facts, observing the interrelation and administration of broad policies, [and] proposing from time to time lines of national procedure in the husbanding of our national resources."[49]

Beyond these organizational proposals, the report argued for permanent executive authority over "the continuous administrative reorganization of government."[50] Such so-called reorganization authority was not new. But it gained new significance—for many observers, constitutional significance—at this time.[51] Traditionally, reorganization authority had been justified as a means to reduce expenditures by eliminating unneeded or redundant functions, or by more efficiently organizing the executive branch of government. For the Committee, however, the purpose of reorganization was "improved management, which would make administration more responsive to the national interest and better able to serve that interest."[52] As Roosevelt put it, the issue was not so much "the financial savings as it is the better supervision of things."[53] Reorganization was thus an instrument of what Progressive reformers had called "positive government." It would allow the president to employ executive administration for the rational-technical management of ever-changing situations.

THE REORGANIZATION ACT

The Committee on Administrative Management transmitted its report to Congress in January 1937. Roosevelt, anticipating that the Committee's proposals would reignite resistance to expanded executive authority, wrote in a message accompanying the report that "in placing this program before you, I realize that it will be said that I am recommending the increase of the powers of the Presidency." But he insisted that this was not the case. "The Presidency as established in the Constitution of the United States," Roosevelt wrote, "has all the powers that are required." Rather, he argued, the recommendation was that the executive be given "tools of management and the authority to distribute the work so that the President can effectively discharge those powers" provided by the Constitution.[54]

Despite these assurances, the report's proposals—along with a Reorganization Act based on them—faced strong resistance in the wake of fights over Roosevelt's "court-packing" plan. Emmerich recounted that the "boldface type used to spell out 'dictatorship' in headlines across the nation's front pages" was "resurrected for the attack on reorganization"; some local papers referred to it as the "dictatorship legislation."[55] Following a protracted struggle, Congress passed a scaled-back Reorganization Act in April 1939.[56] Using the Act's powers, Roosevelt submitted Reorganization Plan No. 1 to Congress. The plan proposed to move several offices and agencies into a new Executive Office of the President, which would remain as a permanent fixture of a vastly expanded and more powerful presidency. By this time, both Roosevelt and many officials in his administration understood reorganization as above all a matter of preparing for US participation in a global war. In his message accompanying the plan, Roosevelt wrote, "In these days of ruthless attempts to destroy democratic governments it is baldly asserted that democracies must always be weak in order to be democratic at all and that, therefore, it will be easy to crush all free states out of existence." Expressing confidence in "our Republic's 150 years of successful resistance to all subversive attempts upon it," Roosevelt proclaimed the importance of "keeping the tools of American democracy up to date." The plan had "one supreme purpose—to make democracy work—to strengthen the arms of democracy in peace and war." It would "let others know that we are strong; that we can be tough as well as tender-hearted" by showing that "what the American people decide to do can and will be done."[57]

Following the German invasion of Poland in September 1939, Roosevelt again used reorganization authority to issue Executive Order 8248, which established the "divisions of the Executive Office of the President" and defined their "functions and duties."[58] The order consolidated a number of existing offices within the Executive Office of the President, including the Office of Management and Budget, the Liaison Office for Personnel Management, the Office of Government Reports, the White House Office, and the National Resources Planning Board. It also provided for what Rossiter later described as an "office-in-embryo": in the "event of a national emergency, or threat of a national emergency," the order stated, the president could create "such office for emergency management as the President shall determine."[59] It was this provision that would be invoked to establish the Office for Emergency Management in 1940.

In sum, Progressive reformers found in the struggles over executive emergency power of the late 1930s an opportunity to apply their theory of organization to reform of the federal government. They clearly perceived a tension between emergency powers and American governmental institutions, and sought to address this tension by creating "administrative devices" and "governmental machinery" that would make possible strong executive authority and rational-technical administration in times of emergency within the framework of the Constitution. This answer to the problem of crisis government in modern constitutional democracy entailed a major revision both to traditional wartime emergency government and to the approach that had been tried out—but rebuked—in the early New Deal.[60] Roosevelt's emergency measures in 1933 had rested on the proposition that *wartime* emergency powers could be used to deal with peacetime crises. The Reorganization Act of 1939, along with the executive and administrative orders that followed it, were designed to make possible a different configuration of emergency, rational-technical rule, and liberal constitutionalism. Still during peacetime, the new law established a governmental machinery that would enable the executive branch to discharge its expanding functions, to continuously prepare for future emergencies, and to manage ongoing emergencies.

Advocates for governmental reform saw this succession of administrative changes as a "nearly unnoticed but none the less epoch-making event in the history of American institutions," as Luther Gulick put it.[61] The creation of the Executive Office of the President, he proclaimed, was "America's answer to the taunt of the dictators that democracies cannot meet the demands of the modern world and still remain democratic." A decade later Clinton Rossiter echoed Gulick's judgment, referring to the "constitutional significance" of Executive Order 8428. "It assures us," Rossiter wrote, "that the Presidency will survive the advent of the positive state" and "may yet be judged to have saved the Presidency from paralysis and the Constitution from radical amendment."[62]

Preparedness for World War II (1940–1941)

During 1940 and 1941, the Roosevelt administration employed these new mechanisms of executive authority to prepare for the massive military-industrial mobilization that would be required for total war: creating and reorganizing government offices, reassigning delegated powers, and coordinating industrial mobilization from new centers of administrative control.[63] Working within these offices, statisticians and economists drew on the New

LEGAL-ADMINISTRATIVE MECHANISMS	KNOWLEDGE PRACTICES	REGULATORY DEVICES
Executive branch emergency agencies created through Presidential reorganization authority	Industrial surveys	Allocation controls (priorities ratings lists of critical resources)
Assignment of delegated powers	Bills of materials	Stockpiling
Coordination between "emergency" agencies and "normal" agencies	Balance sheet analyses of resource requirements and supplies	Plant and production expansion through rapid amortization, supply contracts, etc.
	Feasibility tests	
	Chain of production/flow of materials analyses	

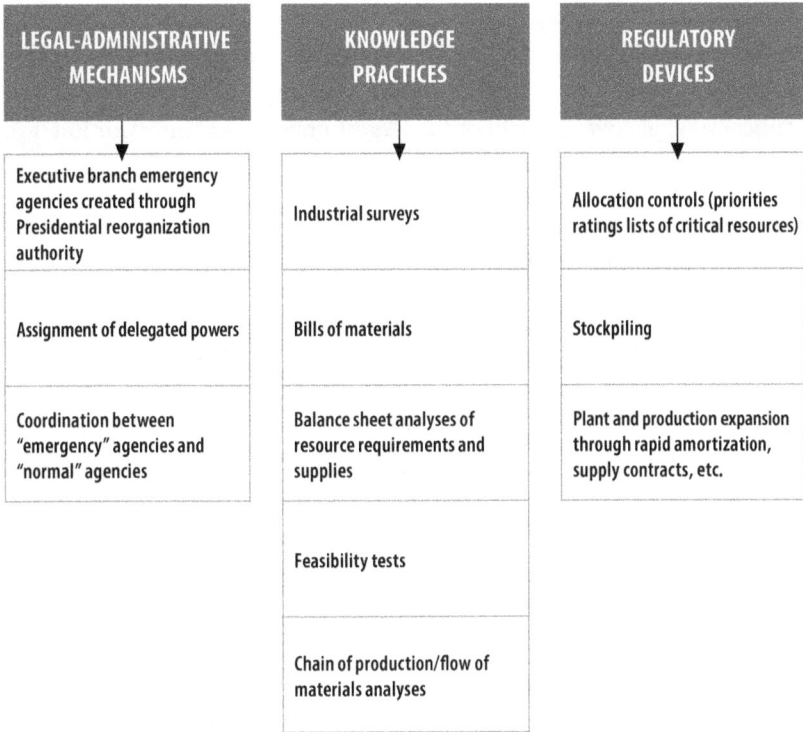

FIGURE 2.2. A political technology of emergency. *Credit*: Janice Yamanaka-Lew.

Deal "science of flows" to specify and manage the otherwise unwieldy problem of mobilization for a total war. Through this process, the "administrative device" of emergency government was linked to new forms of systematic knowledge about the vital flows that made up the American military-industrial production system and to regulatory mechanisms for managing those flows. This combination of an administrative machinery with forms of knowledge and regulatory devices would persist as a template for American emergency government after World War II, as its aim shifted from economic mobilization to nuclear preparedness (see figure 2.2).

The Roosevelt administration's efforts to set up an apparatus of emergency government faced resistance from military and business leaders. World War I mobilization had established a pattern in which these elites controlled wartime industrial production. Along with congressional conservatives—who continued to argue that Roosevelt's arrogation of executive authority undermined American political institutions—military and business leaders sought to maintain their established prerogatives and resisted the concentration of power in civilian planning agencies. Thus, in the run-up to the US entry into World War II and during the first year of the war, the New Dealers' technical assessments and tools of rational management served as weapons in a political struggle over control of the largest government intervention into economic life in American history.[64]

FROM THE GREAT DEPRESSION TO MOBILIZATION FOR WAR

For New Deal economists, mobilization for World War II was above all a matter of determining how much military equipment the American industrial economy could produce: How much expansion of production capacity could be expected? What were the material limits on production? And what kinds of disruptions might result from a massive military production program?[65] Initially working in peacetime agencies and planning bodies, the economists engaged in two kinds of analysis that characterized the New Deal science of flows. On the one hand, they analyzed economic aggregates (total output, employment, etc.) to estimate the military production potential of the American economy as a whole. For example, a group of economists in the Industrial Economics Division of the Department of Commerce investigated the total resources that would be available for a war effort at full employment, estimating that full employment would be achieved only with a 40 percent expansion of gross national production and "would probably require deficits as large as $10 to $15 billion a year."[66] On the other hand, the economists began to study the specific material flows that would constitute a wartime economy. For example, the Energy Resources Committee of the National Resources Planning Board conducted studies of coal, petroleum, hydropower, and electricity that traced the production chains through which these resources were extracted and delivered to energy users. Based on the findings of these studies, the Committee proposed "methods for the elimination of bottlenecks in the production and movement of energy resources." The coal study, thus, identified a shortage of railroad cars as a key potential bottleneck in delivering coal to end users and urged that this shortage "be corrected as rapidly as defense priorities permitted."[67]

In 1940, this form of economic analysis was taken up in the new offices that Roosevelt established to mobilize for war. Referring to a "threatened national emergency," and to Executive Order 8248, in May Roosevelt created the Office for Emergency Management (OEM) within the Executive Office of the President.[68] As Koistinen notes, this use of reorganization power was "thoroughly consistent" with the intention of the 1939 Reorganization Act to create a "structure for the president to preserve his authority over an expanding government and to carry out more effectively his duties as Washington's top, overall manager." By providing a mechanism for asserting and maintaining executive control over industrial production, OEM proved to be "of the greatest significance to economic mobilization," particularly prior to US entry into the war.[69]

Within OEM, Roosevelt created the National Defense Advisory Committee, which served as the center for civilian mobilization planning from May 1940 to January 1941. The Committee was the first destination for many New Deal economists and statisticians who migrated to the emergency agencies to work on mobilization. Some moved on to work on air targeting, while others remained in mobilization planning agencies, where they assembled a knowledge infrastructure and a set of regulatory devices for managing military-industrial production. The Industrial Economics Division of the Department of Commerce was set up as a staff for the head of the Committee's Office of Price Administration, Leon Henderson. Henderson, who had worked in various economic advisory positions in the New Deal, emerged as one of the most influential experts in organizing industrial mobilization.[70] Stacy May, the head of social science programs at the Rockefeller Foundation, directed a Bureau of Statistics.[71] May recruited a number of prominent statisticians and economists to the bureau to work on mobilization, including Ernest Tupper, Edward Mason (who later directed the OSS Economic Division), and Robert Nathan.[72] Nathan, along with Simon Kuznets, had pioneered national accounting in the Department of Commerce and would play a central role in economic mobilization, chairing the powerful Planning Committee of the War Production Board.[73] These statisticians and economists were leading practitioners in a "statistical revolution that was just a few years old," as historian Jim Lacey notes. In recruiting them, May assembled a significant portion of the still-small number of technical experts—fewer than a dozen, in Lacey's assessment—who "actually understood these tools and their capability for directing a total economic mobilization."[74] This group of experts also provided May with connections to "most of the Government departments having statistical facilities and information" that were relevant to war mobilization.[75]

The economists and statisticians working in various offices of the National Defense Advisory Committee were part of a small cadre of "all-outers" within the federal government. Members of this group were convinced that US entry into the war was inevitable, and that a US-led invasion of Europe would ultimately be necessary to defeat the Axis powers. They thus argued for immediate increases in military production and for expansion of industrial capacity in anticipation of much greater future increases.[76] As Donald Nelson, a leading all-outer on the Committee and future head of the War Production Board later recounted, the all-outers were advocates for "the quick conversion of industry, a longer range policy for the accumulation of raw-material stockpiles, [and] a firmer and deeper organization of the economy for war."[77] In pushing for rapid mobilization, the group faced resistance from what Koistinen calls the "conservative mobilization alliance" of military officers, industrial elites, and conservative members of Congress. At this early stage of the war mobilization, many military planners remained isolationist, and the Army and Navy officers who controlled military procurement were not willing to plan for troop levels beyond those already funded by Congress. Business elites, meanwhile, remained most concerned about the danger of building excess production capacity and did not want to contemplate new investment to meet the uncertain prospect of a future war. Finally, congressional opponents of the New Deal argued that a large-scale mobilization would trigger inflation and plunge the United States dangerously into debt. According to Nelson, these forces arrayed against a rapid mobilization "thought that we could avoid a shooting war and that there was no need to shake our economy in anticipation of an emergency which would probably not occur."[78]

Confronted with this resistance to efforts to mobilize the industrial economy—and lacking detailed information about what the mobilization economy would look like—the New Dealers worked to understand both the overall production potential of the US economy and the resource flows that would be needed to produce critical military end products. As Nathan later recalled, in 1940 there was still "considerable slack in the economy" and a high level of unemployment. The economists' initial question, then, was how much military equipment the economy would be able to produce when "all who were able to work, willing to work and actively seeking work had jobs."[79] On this basis, they sought to estimate the overall volume of resources that could be marshaled for mobilization without causing severe economic disruption. These economists also investigated what resources would be needed to produce specific equipment and materiel for war. Drawing on analyses of materials flows conducted by New Deal agencies such as NRPB, they began to identify the inputs that were essential to the production of key military end items and to anticipate what scarcities in such inputs might arise during a mobilization drive. "For airplanes," Nathan recalled, "the key supply was aluminum, ammunition's key supply was copper, and for ships, guns, and tanks, it was steel."[80] Analyses of these metals showed that in the case of an all-out mobilization, resource shortages would rapidly emerge in key inputs and production lines. Accordingly, the all-outers argued for expanding production capacity for both critical materials and military equipment—including a 10-million-ton expansion in steel production and a program to produce eighty-three thousand airplanes by mid-1943—and for accumulating stocks of critical supplies.

In their effort to accelerate preparedness for an all-out mobilization, the New Deal economists and statisticians employed a number of regulatory devices that would continue to be important instruments of emergency government, in the context of both economic mobilization for war and beyond.[81] Here, the new tools and powers created by the Reorganization Act and Executive Order 8248 proved critical, as Roosevelt used these capacities to reassign authorities that Congress had delegated to the president. Some of these regulatory devices were designed to bolster the production of essential war materials and equipment. For example, a program of rapid amortization provided incentives for industrial producers to invest in war production facilities. Meanwhile, special supply contracts incorporated the cost of constructing or converting buildings and of purchasing equipment that was required for the specific demands of war production. According to Koistinen, these production expansion programs proved to be the "most significant accomplishment" of the National Defense Advisory Committee, with the rapid amortization program alone accounting for over half of production expansion prior to the bombing of Pearl Harbor.[82]

Another regulatory device employed by officials in the Advisory Committee was the accumulation of stocks of critical materials to allow for a ramp-up in future production. Authority to undertake such a program was established with the passage of the Strategic War Materials Act in June 1939 and the creation of the Reconstruction Finance Corporation in June 1940. Initially advocated by the Army-Navy Munitions Board, the program became the Committee's responsibility as it assumed greater control over strategic materials planning.[83] But congressional appropriations for the stockpile were limited, and the Committee struggled to procure critical materials with the appropriated funds, due to increasing domestic demand and disruptions in international supply chains.[84] Thus, its expansion programs initially served to meet existing American military procurement plans, which were oriented toward reequipping the armed services for a defensive war, and until the end of 1940, production capacity was adequate to meet demand in a still depressed economy.[85]

PRIORITIES

This situation changed in late 1940 and early 1941. US government purchases of war equipment and materiel rose rapidly. Meanwhile, foreign demand for American-made military equipment increased with the establishment of the British Supply Council and, in March 1941, the passage of the Lend-Lease Act, which financed British (and later Russian) purchases of American military products.[86] This surge in demand led to shortages of industrial materials and production capacity, which soon caused disruptive bottlenecks in critical areas. For example, in late 1941 and early 1942, the shipbuilding program was crippled by shortages of skilled labor, steel, and specific components and equipment, such as machine tools, turbines, gears, and steel valves.[87] As John E. Brigante observed in a postwar account of the mobilization effort, the proliferating bottlenecks and shortages in production "implied a challenge to existing attitudes, facilities, and machinery for coordinating production."[88] How could conflicting demands on scarce resources be reconciled? How might some uses of these materials be prioritized over others?

In June 1940, Congress had given the president legal authority to establish a system of "priorities" to assure that the mobilization needs of the Army and the Navy would not be limited by civilian consumption.[89] In principle, the power to set priorities gave the government the ability to determine how resources in the economy would be used. Roosevelt initially assigned this authority to the National Defense Advisory Committee, where civilian officials negotiated a system for managing priorities with military leadership.[90] When particular materials or manufactured goods became difficult to procure, they were added to the "Priorities Critical List." The first such list, issued in August 1940, included both end items and manufactured components such as generators, motors, and turbines, as well as metal forms such as aluminum and brass castings. Military procurement officers identified what items on the Priorities Critical List were needed to produce "table of organization" items—the equipment and materiel required for a military unit. They then assessed the relative importance of different uses of scarce materials or production capacity using a system of "preference ratings." These ratings ranged from A-1 (highest) to B-10 (lowest), with an additional AA rating "reserved for emergencies."[91] High-rated uses would be first in line to procure scarce resources.

Initially, industrial producers were not required to participate in the priorities system. But as bottlenecks multiplied with the ramp-up of production in late 1940 and early 1941, officials on the Advisory Committee placed aluminum and machine-tool producers on "mandatory priority status." For any orders from such production facilities to be fulfilled, they had to be certified as high-priority items of military production, as designated by preference ratings.[92] But this system soon produced a cascade of unintended consequences. Each military service developed its priorities based on the production program required to meet the force levels defined by congressional appropriations. When production increased and scarcities arose, each procurement agency responded by assigning higher ratings as a way to lay claim to needed materials. The result was what mobilization economists David Novick, Melvin Anshen, and William Truppner referred to as "priorities inflation." As military procurement officers issued ratings in ever-greater numbers, high ratings were worth less and less since they could no longer ensure access to scarce materials. This decentralized system resulted in a mobilization drive that was "basically planless." The military's "inability to array competing claims in the order of their importance to national security" threatened "to become a significant [impediment] to defense production."[93]

Faced with these breakdowns in systems of resource allocation, New Deal economists argued for a different model of emergency resource management. The model they proposed drew both on their views about executive power in managing crisis situations and on the tools of New Deal economics for constituting the national economy as an object of knowledge and a target of intervention. Echoing the arguments that Progressive reformers had made about local legislatures, the economists charged that military procurement agencies were motivated by narrow departmental concerns and were under the sway of private industry. Moreover, they argued, military procurement officers did not have the technical expertise to assess priorities in relation to the total requirements for American military-industrial production, lacking even an understanding of the resources required to produce items on military tables of organization. The army, for example, did not know "how much aluminum went into the manufacture of airplanes," and the Navy "had no bills of materials at all."[94] Further, military bills of materials did not account for the chains of production and for services such as electricity and transport that were needed to deliver those inputs to the manufacturers of military end items. Nor, finally, did these bills of materials assess requirements for "quartermaster items"—such as food and clothing, which were assumed to be readily procured—or for the civilian economy, on which military production was dependent.[95]

The solution proposed by the New Dealers working around Leon Henderson was to place authority for overall resource allocation in the hands of impartial civilian officials and technical experts, who could take an overarching view of both military and civilian needs in a total mobilization for war. This system would effectively replace both the peacetime system of supply and demand in a free enterprise economy and the complex balance of corporate interests (the military, business, etc.) through which the mobilization effort had been organized during World War I.[96] It implied an audacious proposition about technocratic rule and executive power in an emergency: expert determination of priorities would guide the administrative management of resources for the economy as a whole.

In creating such a system for managing the mobilization program, New Dealers faced the challenge of establishing control over the allocation of resources. Initially, they competed for control with both military and industrial leaders, who resisted large-scale industrial expansion in anticipation of a future war. Moreover, the existing system of preference ratings left power in the hands of military procurement officers. During the course of 1941, Roosevelt tried to increase civilian experts' control over resource allocation by repeatedly drawing on executive authority to create a succession of offices in the Office for Emergency Management, installing all-outers in prominent positions in these offices, and delegating to them authorities to address bottlenecks and allocate scarce resources to priority areas (see figure 2.3). Roosevelt's use of reorganization power gave these offices some control over the mobilization, though it was not until the creation of the War Production Board after US entry into the war that the New Dealers' vision of centralized, expert control of mobilization was realized.[97]

In the meantime, however, the New Dealers confronted another problem. Their approach to rationally allocating scarce resources based on expert calculation required an overall statistical picture of available supplies and envisioned requirements, based on a realistic assessment of military strategy. But in 1941 they did not have access to any such information. As we have seen, military requirements were based on current congressional appropriations, which anticipated a defensive war rather than a war that involved global deployments of American troops, culminating in an invasion

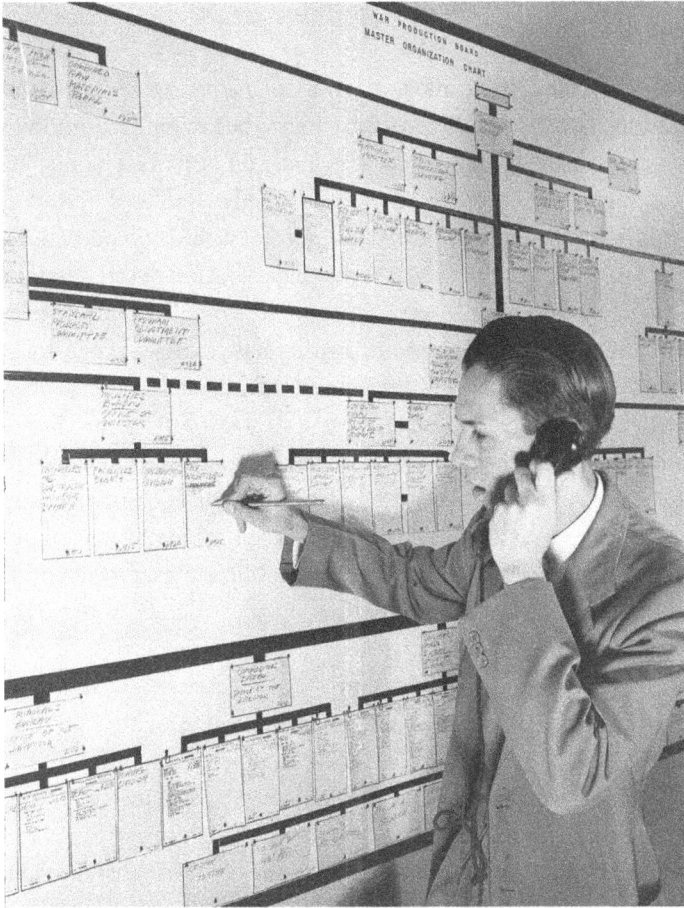

FIGURE 2.3. "Flexible" master organization chart of the War Production Board, Administrative Division. New Deal reformers set up administrative mechanisms through which the executive could flexibly respond to crisis situations as they arose. During World War II, President Roosevelt used one of these mechanisms, "reorganization" authority, to reshuffle the war mobilization agencies.
Source: Photograph by Alfred T. Palmer, December 1942, Library of Congress Prints and Photographs Division, https://www.loc.gov/item/2017698685/.

of Europe. What the New Deal economists and statisticians pushing for an all-out mobilization required was a means to "visualize the relations between aggregate contemplated munitions production, production necessary for defeating the Axis, and maximum United States production possible under an all-out war effort."[98]

THE BALANCE SHEET

To address this lack of basic information about a future war economy, civilian mobilization planners developed what came to be known as the mobilization "balance sheet." The balance sheet was made up of two elements. The first was a statement of resource requirements based on military strategic plans, which implied certain force levels and rates of utilization of materiel. The second was a statement of available supplies—both industrial outputs and existing stocks—that could be used to meet these requirements. In principle, a balance sheet would not present a static picture: requirements and supplies would be scheduled according to rates of industrial production (which might increase as productive facilities were converted to war production and new facilities were built) and strategic military plans (which would imply a certain buildup of forces and use rates of materiel and equipment in combat). The balance sheet was a key instrument through which New Deal planners instituted their program for managing the mobilization effort. It constituted the entire military-industrial economy as a collection of resources that had to be "in balance" with priority uses.

The idea for creating a military-industrial balance sheet came from Jean Monnet, a French production planner who became a leading architect of the European Union after World War II.[99] Monnet moved to the United States in late 1940 and worked to convince high-level officials in the Roosevelt administration that "there had to be a complete exchange of information between the United States and Great Britain" and that Britain would fall without American industrial support.[100] The New Deal all-outers—Leon Henderson, Stacy May, and the economists and statisticians working around them—were receptive to this idea, as they were convinced that a massive increase in production was required to adequately prepare for war, and that an overall picture of the present and future war economy was necessary to make rational decisions about the allocation of scarce resources. In late 1940, May began working with industrial production specialists on his staff—including Robert Nathan and the statistician Edwin B. George—on a balance sheet for the combined war economies of the United States and the United Kingdom.[101] By March 1941, the group had assembled requirements for 1942 from the Army, Navy, and Maritime Commission, as well as from the British Supply Council. After the Lend-Lease Act was passed, its totals were added to the overall list of requirements, which were expressed in terms of troop units: as Nelson put it, "how many tanks, how many small arms, how much artillery, how much ammunition, how much food and clothing and other supplies would be necessary for so many troops of every kind of classification."[102] These requirements were then translated into the total quantities of materials—steel, copper, aluminum, and so forth—that would be needed for a certain overall number of troops.[103]

Having compiled this rough estimate of requirements, May and his colleagues then turned to the calculation of available supplies: "what both Britain and the United States had in the way of resources, materials, and facilities to meet these needs." As Nelson recalled, May and his staff "built a huge book, with comprehensive categories and tabs." Drawing on data that had been collected by the Department of Commerce, the National Resources Planning Board, and other agencies, they filled out the columns for American stocks and production capacity. May then flew to London to collect equivalent figures for the British side, working with the British War Cabinet and the Chiefs of Staff of the Armed Forces. The result was a "composite record of British and American resources and production potentials."[104]

May's initial compendium of requirements and supplies was the first attempt to assess the national military production complexes of the United States and Britain as a whole. When May returned from London, he carried with him what Nelson referred to as "the most precious and important portfolio in the world." May's "encyclopedic work," Nelson explained, resulted in "a body of information which, in a degree, revolutionized our production." It gave military planners a broad understanding of "what we would have to take out of the economy in order to proceed with either a defense or a war production program," and indicated the need to "expand the production of aluminum, steel, copper" and other scarce items. Moreover, by translating troop numbers into requirements for materials and production capacity, it offered a "measuring stick for use against Army and Navy orders." As the military expanded, American production requirements would also "expand in a systematic ratio."[105]

Yet May's balance sheet was limited in a crucial way: its statement of American military requirements was still based on existing congressional appropriations for the armed forces, which assumed a defensive war to protect the continental United States. It did not, Brigante notes, represent a "direct translation into materiel of any grand military strategy then in the mind of either the Commander-in-Chief or the Chiefs of Staff of the Armed Services," which by mid-1941 included plans for an offensive war in Europe and the Pacific.[106] The all-outers had long pushed military leaders to provide estimates of what would be required for such a war, so that it would be possible to lay plans for a total mobilization. Such estimates were not forthcoming. The military had limited information about production chains and was unwilling to plan for a war that implied force levels that vastly exceeded what was implied by congressional appropriations. But given mounting British and Russian requirements, according to Nathan, the all-outers "finally sold Roosevelt and Churchill on the idea of getting the military committed to what they would need to win the war if there were an all-out war."[107] In July 1941, Roosevelt ordered the military to produce estimates of "munitions and mechanical equipment of all types which . . . would be required to exceed by an appropriate amount that available to our potential enemies." Separately, he requested requirements from Russia and the UK. These requirements, compiled and revised over the next several months, ultimately resulted in the so-called Victory Program, which outlined the industrial production that would be necessary for the United States and its allies to defeat the Axis powers.

As the Civilian Production Administration noted in a postwar study, the initial requirements information compiled following Roosevelt's order "were better than any previous requirements statement had been." They were also "enormous by any previous standard."[108] For 1942 and 1943, the estimated outlay to cover the requirements in the program was $150 billion—$90 billion more than had been previously authorized by Congress—at a time when US GDP (in 1941) was $129 billion.[109] In late 1941, first Nathan and then May conducted studies to determine whether such a massive expenditure could be accommodated given the productive capacity of the American economy. Here, as Brigante observes, the "unique value" of the balance sheet became clear. It could be used to translate troop levels into requirements for equipment, materiel, industrial inputs, and raw materials, and to weigh these requirements against the production potential of the economy, in terms of both GDP and specific materials.[110]

Nathan and May concluded that to realize the envisioned production program, national income would have to reach $130–$140 billion by 1943, of which half would have to be allocated to the defense program. In other words, the program was feasible, but only if the United States devoted the maximum possible resources to war production.[111] Based on these findings, Nathan, May, and other all-outers, such as Leon Henderson and Edwin B. George, urged Donald Nelson to push for a doubling of military expenditures to "achieve a full-scale military effort in the sense of utilizing and absorbing all the productive capacity available to support it."[112] In December 1941, the established production schedules for all defense items represented financial commitments amounting to $27 billion for 1942 and $34 billion in 1943. May and Nathan proposed expenditures of \$40–\$45 billion in 1942 and \$60–\$65 billion in 1943.[113]

Wartime Mobilization: The Control of Vital Flows

Just after Stacy May delivered the final report on the feasibility of the Victory Program, the Japanese attack on Pearl Harbor and the subsequent entry of the United States into World War II transformed the politics of mobilization. Roosevelt accepted the all-outers' proposal for an enormous industrial expansion—including forty thousand tanks, fifty thousand aircraft, and 7 million tons of merchant shipping—and added massive additional requirements.[114] Military leaders' resistance to all-out mobilization vanished, as they recognized that all available national resources would now be directed to military-industrial production. The proliferation of requirements led to renewed pressure on the production program. The Army-Navy Munitions Board, writes Koistinen, began to issue priority ratings in an "indiscriminate manner," while blocking preferences for "essential civilian goods."[115] Manufacturers, according to Brigante, found that their priorities certificates were "only 'hunting licenses'" rather than assurances that scarce materials could be secured, leading to wasteful imbalances. "Almost all top-priority producers were obtaining some of the materials they required," and "almost none had enough of the right kinds in the proper quantities to make possible the completion of contracts on schedule."[116] In this context, the New Deal economists and statisticians working on the mobilization program found themselves in a new position. They were no longer arguing for higher production targets. Instead, they were struggling to manage and reconcile multiplying claims on resources.

Over the course of 1942, the New Dealers addressed two key problems in their effort to develop what Brigante referred to as "organizational arrangements and systematic procedures for keeping the [mobilization] programs up to date and in balance."[117] On the one hand, they sought to limit the size of the overall program and the claims on specific scarce and critical materials. Here, the key mechanism of control was the "feasibility test," which the New Dealers used to establish centralized authority over industrial mobilization based on expert evaluations of the total resources available in the economy. On the other hand, they developed tools for controlling the mobilization effort through "detailed and firmly scheduled production programs based on realistic requirements."[118] The latter efforts culminated in the Controlled Materials Plan, which historian Robert Cuff refers to as the "most ambitious attempt ever mounted in the United States to allocate, coordinate, and monitor the flow of scarce materials for the entire industrial economy."[119]

FEASIBILITY

Following US entry into the war, Roosevelt again reorganized the offices lodged in the Office for Emergency Management. In an executive order on January 16, 1942, to "define further the functions and duties of the Office for Emergency Management," he established a War Production Board (WPB) to "exercise general direction over the war procurement and production program."[120] Roosevelt installed Donald Nelson as the chair of the Board. New Dealers such as Luther Gulick, Sidney Hillman, and Leon Henderson ran many of its divisions. Over the next year, WPB evolved into the powerful center of civilian control over industrial mobilization planning that the New Dealers had long advocated. By December, the Board had a total staff of 22,591 working in divisions responsible for the major resources—copper, aluminum, steel, tin, lead, zinc, cork, gas, power, and so on—that were essential to both civilian and military supply.[121]

The activities of New Deal statisticians and economists scattered across WPB were coordinated through a Planning Committee, modeled on the Progressive ideal of the expert advisory body. Formally, the Committee had no operating responsibility or authority. Its role was to provide the Board's chairman with "such plans, procedures, and information as may be helpful to the planned development and realization of the war production program."[122] In practice, the Planning Committee, working with economists in other WPB offices, generated the analyses that civilian officials used to assert control over the production program during the critical first year of all-out mobilization. The members of the Committee were Fred Searls Jr., a mining executive and geologist who had developed copper deposits in Africa;[123] Thomas Blaisdell, an economist who also served on the National Resources Planning Board; and Robert Nathan, the Committee's chair. James Fesler, an expert in public administration who had worked on the Committee on Administrative Management, was the Planning Committee's secretary, and another economist, Edward T. Dickinson, served as executive director. The Planning Committee worked closely with Stacy May, who led WPB's Bureau of Planning and Statistics.[124] May's staff included the economist Simon Kuznets as well as several economists who would play key roles in postwar emergency government, such as Edwin B. George and Shaw Livermore.

Nelson initially charged the Planning Committee with investigating three issues that were critical to the mobilization effort. First, the Committee was to review "the total munitions program in terms of balance among its constituent parts, and the relation of direct military requirements to indirect military needs, essential civilian needs, and the total national economy." Second, it was to examine the situation of shipping. And third, it was to monitor the status of the airplane program, particularly in light of Roosevelt's revised production goal of building sixty thousand planes in 1942. In its initial meetings, the Planning Committee focused on specific bottlenecks in vital industrial production systems that could impede rapid mobilization. For example, it considered various issues concerning the shipment of critical resources: Could production goals for tanks and equipment for a large ground army be met "if shipping facilities are not going to be adequate to transport the equipment and troops to overseas battle fronts"?[125] Given the vulnerability of shipping facilities on the eastern seaboard to German submarine attacks, would a lack of overland oil pipeline capacity constrain the production of vital petroleum products such as toluene, resulting in a "serious decline in the rate of increase of TNT production"?[126]

But the Planning Committee soon turned to the first problem Nelson outlined, when it decided to "devote full attention to review of the objectives of the war munitions program." In a March 1942 meeting, Searls argued that the Committee was disposed to "consider too wide a range of subjects" and urged it to focus on "appraisal of the war munitions program with a view to arriving at reasonably accurate estimates of military requirements and production possibilities, and developing a feasible program in which the several objectives will be mutually consistent." This problem of developing a "feasible" mobilization program was similar to the issue that Nathan and May had examined under different political circumstances in their work on the Victory Program: Could a given war mobilization plan be carried out given the limited resources available in the American industrial economy? In the context of wartime mobilization, the members of the Planning Committee also saw the study of feasibility as an instrument for establishing control over mobilization. As Nathan argued in the same meeting, the current problems in the production program were due not only to a "healthy reaction to military events" but also "to the fact that traditional autonomy of responsibility for each segment of the program has prevented effective auditing of the program from the standpoint of the total economy and the balance among related objectives."[127] Feasibility was a means through which the New

Dealers sought to establish precisely such an authoritative, expert evaluation of the overall possibilities of American military-industrial production.

The Planning Committee charged Kuznets with testing the feasibility of the production program.[128] Kuznets followed the technique that Nathan and May had developed the prior year, creating a balance sheet that compared available supplies to production requirements. But he extended the technique in a crucial way by incorporating the "requirements" of the civilian economy as a whole. This meant including supporting services (such as transportation and energy systems) as well as the basic resources required to sustain the civilian population in the balance sheet analysis. As Kuznets explained to the Planning Committee, the outer bounds of feasible production could be defined by determining "the total production of finished articles of which the national economy is capable" and then subtracting "the irreducible minimum of production required for civilian requirements." He then checked this "test" based on overall production by analyzing "specific military requirements as compared to the specific resources for production," such as labor power, raw materials, and production capacity for critical manufactured items.[129] In a preliminary analysis, Brigante records, Kuznets found that the requirements of the current mobilization program vastly exceeded total US industrial capacity. The planned program for 1942 would run into "definite shortages in a number of critical materials," while the program for 1943 was "completely out of line" with the "expected supply of almost all basic materials." The picture was even more dire when "civilian and indirect military needs"—including such crucial items as "steel for the upkeep of railroad facilities, indispensable to the movement of troops, raw materials, and finished military goods"—were taken into account.[130] Based on these findings, the Planning Committee argued that the 1942 munitions program had to be dramatically cut back.[131] Nathan and May sent a letter to WPB chair Donald Nelson, warning of the economic chaos that would be unleashed by an unbalanced program. "Any attempt to attain objectives which are far out of line with what is feasible," they wrote, "will result in the construction of new plants without materials to keep them operating; vast quantities of semi-fabricated items which cannot be completed; production without adequate storage facilities; idle existing plants due to lack of materials; and similar disrupting situations." To press their point, the Planning Committee formed a Committee on Feasibility, consisting of the heads of all the WPB operating divisions, that oversaw a full-scale feasibility analysis, which Kuznets and his assistants worked on for four months in mid-1942.[132]

This push to set limits on the size of the production program based on feasibility tests set the stage for a struggle between the New Dealers centered in the Planning Committee and the industrial and military leaders who were running production and procurement. Military leadership wanted to maintain control over military supply, while industrial leaders resisted government intervention into their production plans.[133] The calculation of feasibility was particularly contentious, according to Koistinen, because the "analytic tool originated with economists, statisticians, and academics, who were resented, perhaps even feared, by industry and the military."[134] A series of increasingly bitter confrontations with military officials in charge of procurement were resolved only when—likely following Roosevelt's intervention[135]—the military relented and cut back its requirements to levels that approximated those deemed feasible by the civilian experts. As Koistinen observes, the resolution to the struggle over feasibility "set a critically important precedent." Civilian mobilization planners had established the right to determine "whether claimants' demand was realistic" based on expert assessments of the industrial economy's potential.[136] Although leading New Dealers would soon be ousted from positions of power, the authority of expert evaluation in determining the limits of mobilization—as well as the specific techniques of the balance sheet and the feasibility test—would endure. After World War II, feasibility testing and balance sheet planning were employed to test mobilization plans for future conflicts with the Soviet Union. By the late 1950s, as we show in chapter 6, mobilization planners had redeployed these techniques for a very different purpose: planning for the survival of the population following a devastating nuclear attack on the United States.

PROGRAMMING PRODUCTION

In parallel to their efforts to establish control over the total size of the mobilization program, New Deal economists working in and around the Planning Committee also set up a system for controlling the allocation of scarce materials. They understood this problem of resource allocation in terms of the diminishing marginal utility of particular items of production to the war effort. As a summary of a Planning Committee discussion of the president's objective to produce forty-five thousand tanks put it, "while the first several thousand tanks are absolutely essential, and must take precedence over most alternative uses of our economic resources, the last thousand tanks . . . may be less important to the war effort than the building of some of the planned ammonia and synthetic rubber facilities."[137] "At some point," explained the economist and mobilization planner David Novick, "the 1000th tank of a certain type was less important than the stainless steel milk pails essential for milk to be supplied to either soldiers or civilians."[138] The system of priorities that had been initially worked out in 1940, which decentralized determination of the relative importance of alternate uses of scarce materials, offered no means to bring such disparate requirements for the war program into a common calculus. "What was imperatively called for," as Novick, Anshen, and Truppner described the New Dealers' understanding of this problem, "was the formulation of a policy and a program which would encompass, in common units, both materials and end products, and fit the job in order of importance of its component parts within the resources of the economy."[139]

Prior to US entry into the war, in 1941, the civilian mobilization agencies had set up a system for production programming called the Production Requirements Plan (PRP).[140] Although the PRP was not mandatory until mid-1942 (and then only briefly), it established a knowledge infrastructure and a set of technical practices that remained essential to emergency resource management, both during and after the war. In the PRP, procurement agencies—which generally meant military agencies purchasing equipment and supplies—submitted their resource requirements to civilian officials. In principle, these requirements would be based on detailed bills of materials that described the inputs required to manufacture a specific end item (steel for a tank, for instance). Civilian officials would then allocate available materials among competing requirements, and issue priorities certificates directly to industrial producers, who could use these to acquire controlled materials. Thus, for example, a producer of airplanes and airplane components that made use of aluminum, which was a controlled material before and during the war, would have to secure priorities certificates for the quantity required for each unit. Through this system of production programming, civilian officials sought to limit the number of such certificates to the available supply of aluminum or specific aluminum forms. Here, balance sheet analysis was used not only to test the feasibility of plans but also to control the actual flow of resources in the economy.

After the United States entered the war, a Requirements Committee within WPB was given responsibility for compiling data on industrial requirements and allocating scarce materials. The Committee was to serve as a "central tribunal" for the production program, adjudicating conflicting claims on resources. Initially, however, the civilian officials based in WPB had limited tools to implement such a plan. Edwin B. George, who directed the Requirements Committee's economic and research office, explained in a March 1942 memorandum that the Committee struggled to assert authority and to gain "security access" to military information about production. Moreover, members of the Committee lacked the analytic tools necessary for "reorganizing requirements data into comparable categories of materials by shapes and forms and subjecting them to common tests of urgency"— which would involve developing comprehensive and standardized knowledge about the entire military-industrial economy.[141] According to Novick and Steiner, notwithstanding the significant work in data collection that had been performed during the New Deal and the period of war preparation after the bombing of Pearl Harbor, "facts simply were not available for the most fundamental type of policy decision."[142]

Analysts working in WPB addressed this demand for information about the wartime economy by launching a massive program of data collection. "More statistical information on the industrial aspects of the American economy was collected by the War Production Board," Novick later wrote, "than had ever before been assembled in the history of this country." This "mass of statistical information" covered basic materials, components, and subassemblies; productive capacity; and finished end items, as well as the "the purposes for which the products are used, and the kinds of ultimate consumers who use them." A key device that WPB used to compile this "mass of statistical information" was the familiar tool of the industrial survey, which had been developed in the Department of Commerce during the New Deal. Over the course of the war, more than five thousand different survey forms were used to collect information about various categories of materials, each often circulated to tens of thousands of producers or users of scarce materials.[143] The most comprehensive survey of this type examined metal usage. WPB officials sent over twenty thousand forms to industrial producers, ultimately covering more than 90 percent of the metal requirements in the American economy. The survey, which collected data on requirements, receipts, inventory, and actual use for "all critical metals in mill shapes" offered, according to Novick, Anshen, and Truppner, the "first overall measure of metal consumption and requirements segregated by significant product classifications." For the first time, Novick and Steiner emphasized, the "entire structure of input-output relationships in American industry . . . was silhouetted for study, analysis, and application to the problems of directing the flow of materials through industry."[144]

WPB analysts used a revised version of this metals survey to administer the Production Requirements Plan.[145] Manufacturers reported to WPB what military orders had been placed with them and what scarce metals—such as steel, aluminum, and copper—were required to complete the orders. On this basis, WPB calculated the overall demands for metals and balanced it against available supplies, including inventories held by producers. Policy bodies within WPB's materials branches then worked with the Requirements Committee to allocate metals among industries, and "assigned a preference rating on a plant-by-plant basis to these requirements and authorized firms to obtain specific quantities."[146]

The Production Requirements Plan forced all participants in the industrial mobilization effort to adopt standard procedures for accounting and data handling. This presented a particularly significant challenge for the military branches, which did not have systematic or automated processes for procurement planning.[147] To manage this vast flow of information, military procurement agencies employed tabulating machines and punched cards, on which they stored resource data. These technologies would soon be made obsolete by the digital computer (see chapter 3) but were used on a massive scale for managing war mobilization. Thus, for example, the Army—which accounted for a large portion of overall military procurement—had to hire and place "many hundreds of additional employees" and to acquire "batteries of complicated tabulating machines" to handle its requirements submissions to WPB (see figure 2.4).[148]

The accounting process for the Production Requirements Program began with the formulation of accurate bills of materials that identified "thousands of material ingredients of finished items of military equipment." Requirements for individual end products were then recorded on punched cards and processed on tabulating machines to generate summary requirements for controlled materials. These summary requirements were then forwarded to civilian officials working in WPB. As R. Elberton Smith, a historian of Army economic mobilization, described this process, "Beginning with manually punched cards and progressing through various stages of electrical punching, machine operators converted the raw data from individual, detail[ed] bills of materials up through various summary stages." These data were collated with production schedules to produce a "mechanically calculated and recorded summary of total material requirements by material classes for time periods and by programs." To meet deadlines for requirements submissions, "whole batteries of IBM machines would be in operation twenty-four hours per day behind locked doors."[149] The job of processing these

FIGURE 2.4. "Computers" processing forms for the Production Requirements Plan. The World War II–era Production Requirements Plan required the accumulation and management of a vast amount of information about American military-industrial production. Despite the large-scale use of automated techniques, such as punched cards and tabulating machines, human "computers" played a significant role in this massive task of information processing. *Source*: Photograph by Howard Lieberman, October 1942, Office of War Information, Library of Congress Prints and Photographs Division, https://www.loc.gov/item/2017694093/.

aggregated requirements was passed on to the Bureau of the Census, which tabulated data on requirements and inventories for all claimants. Based on these results, which were reported for more than two hundred separate product groups, the Requirements Committee "cut pies for all critical metals and issued directives allocating critical metals to industries." In a final step, priorities certificates were distributed to each producer in the military-industrial production chain.

Civilian officials in WPB saw the Production Requirements Plan as a significant step in establishing control over production programming. The centralized system of allocation control brought supply and requirements into balance for key materials. Moreover, the program succeeded in establishing a knowledge infrastructure for future versions of the control program.[150] But the Production Requirements Plan also introduced several problems. In allocating priorities certificates directly to producers, civilian planners had no

way to ensure that the materials acquired with these certificates were actually used to fulfill military requirements or that allocations for various inputs
to a given product would be in balance. "It was entirely possible," economist
Charles Hitch explained in 1943, "for Ordnance to schedule the production
of, say, 500 tanks per month in an arsenal; for the arsenal to apply . . . for this
material and be given 90% of the steel and 80% of copper . . . and for some
subcontractor making a vital part to be given only 60% of his material requirements."[151] The Requirements Plan also created perverse incentives for producers and procurement agencies to inflate estimates of requirements and to
hoard scarce metals. Finally, the complexity of the program—which required
allocating scarce materials directly to thousands of individual producers—
plunged WPB into what Koistinen calls "administrative tumult."[152]

THE CONTROLLED MATERIALS PLAN: VERTICAL CONTROL

As troubles mounted with the production program in the first half of 1942,
Nathan's Committee on Feasibility concluded that "without the prompt
institution of a comprehensive and unified set of controls," the war production program was "in immediate danger of disastrous consequences." It
urged that a "top authoritative group" in WPB be formed to "immediately
put aside all other work and give full attention to the development of a
control system."[153] Nelson responded by reconstituting the Feasibility Committee as the Committee on Control of the Flow of Materials and charged
it with studying alternative systems for controlling production. This Committee solicited input on alternative designs for a materials control program
from the research office of the Requirements Committee. These alternatives were presented in a memorandum written by Edwin B. George, who
directed the office, and two economists on his staff, Lincoln Gordon and
Melvin de Chazeau.[154]

One approach that George and his colleagues presented in the memo
was what came to be referred to as a "horizontal" system of control. Allocations would be made "directly to all major plants, whatever their level in
the industrial process." This system, which was a variant on the Production
Requirements Plan, would put civilian planners in the position of determining allocations on a plant-by-plant basis for the entire economy. A second
option was referred to as a system of "vertical" controls, which focused
not on individual producers but on production chains. This system worked
through "warrants"—authorizations to purchase controlled materials—that
would be issued only to the manufacturers of end products, such as tanks or

airplanes.[155] In practical terms, vertical controls would enable civilian planners working in the WPB to focus on end products and the inputs required to produce them, allowing prime contractors to manage relationships with subcontractors and distribute warrants throughout the economic system. This proposal for vertical controls likely provided a template for the approach to "vertical bombing" developed a few months later by New Deal economists who moved to London in mid-1942 to work on strategic intelligence for air war (see chapter 1). In both mobilization and air targeting, the focus was on end items needed for military operations; the task was to assess the material flows of resources essential to end-item production. But where air-targeting specialists worked to identify how, and when, the disruption of production chains would affect the enemy's deployment of forces on the front, production planners sought ways to manage resource flows to maximize domestic production.

Based on George's memo, the Committee on Control of the Flow of Materials concluded that a vertical system of control would be preferable and began work on what would eventually be called the Controlled Materials Plan (CMP).[156] The details of the CMP were developed by a consultant to the Committee, economist Charles Hitch, who would later work on systems analysis at the RAND Corporation with the mobilization planner David Novick.[157] The Plan was implemented by Ferdinand Eberstadt, a former head of the Army-Navy Munitions Board who was appointed as WPB's vice chair for program determination in late September 1942.

The key feature of the CMP, following the proposal for warrants, or "vertical" controls, originally suggested by Edwin George, was "its emphasis on programs rather than products." The CMP controlled three metals—aluminum, copper, and steel. These metals were chosen, as Novick later explained, because of their near-ubiquitous use in the production program, which meant that they could be treated as "addable currency common to virtually all programs." In effect, they were money equivalents at a time when material resources were the key constraint in military production. CMP technicians broke these three metal "currencies" down into more than a hundred specific forms and types, each of which was separately controlled. Thus, copper products were divided into four categories, each including many distinct controlled shapes. Steel was broken down into thirteen categories of carbon steel and ten of alloy. "The objective of this reporting detail," as Novick, Anshen, and Truppner explained after the war, "was the proper balancing of supply and demand, giving full consideration to the limits imposed by facilities available for producing each reported shape and alloy."[158]

In operation, the CMP consisted of two streams of paper flow that effectively constituted an accounting and control system for the entire American industrial economy. The first stream, which built directly on the knowledge and accounting infrastructure of the Production Requirements Plan, led from subcontractors through prime contractors and procurement agencies "up to the supply-demand balance for the total economy."[159] First, subcontractors prepared detailed bills of materials that specified "the amounts of each contained material required to make one unit of a fabricated product." These bills of materials were then passed to the "prime contractors" that produced military end items, such as tanks, airplanes, and munitions. Prime contractors combined material requirements for their own and their subcontractors' requirements and passed them on to military procurement agencies. These agencies prepared "estimates of controlled-materials requirements in total and by program detail," which they submitted to branches of WPB that planned allocations for a given material. In a final step, the Requirements Committee reconciled any disputes or imbalances, based on assumptions about the end products required for both civilian and military economies, whether war equipment such as planes and tanks, or essential items for the civilian economy such as combines and tractors.

The second stream of paper ran down from WPB through prime contractors to every producer that used controlled metals in the American industrial economy. Based on WPB's determinations, the program provided claimant agencies with credits—equivalent to deposits in a bank account—that consisted not of money but of specific amounts of metals in particular forms. Prime contractors could "draw" from this bank account, distributing these credits to subcontractors as they were required. The aim of the program was to balance centralized control with minimal disruptions of the existing relationships between suppliers and producers that made up the nation's industrial production system.

According to Novick, Anshen, and Truppner, the CMP was "the most complex piece of administrative machinery created during the period of the war emergency." It also marked a kind of apotheosis of a particular vision of expert control in governing emergencies—one that had been first formulated by Progressive reformers facing the recurring crises of economic depression and war. A body of civilian experts provided advice to officials in the executive branch who, ultimately at the direction of the president, managed industrial production for war. Employing the analytical tools of the "science of flows" and a comprehensive knowledge of the American economy as a complex of interdependent processes, their interventions targeted what Merriam had called the "strategic points of a working system." Their aim was to accommodate a powerful new instrument of executive authority with American political and economic traditions, as they understood them. They sought to leave in place the established supply relationships in the American economic system. Moreover, instead of consolidating control in a massive new administrative apparatus, WPB worked with procurement agencies in the military to minimize disturbance of existing administrative relationships.[160]

These efforts did not assuage conservative critics of the Roosevelt administration, or military and industrial leaders, who were eager to reestablish control over the mobilization program. Even as the CMP was being implemented, these groups worked to marginalize the New Dealers. By early 1943, the New Dealers working on mobilization were being replaced by military and industrial elites. The Planning Committee was shut down, and, according to Koistinen, central figures like Stacy May, Leon Henderson, and Donald Nelson were "soon off the scene."[161] But the knowledge practices and patterns of administration that they had invented—feasibility tests, balance sheet analyses, systems of control over materials flow, priorities ratings, and authority for reorganization and delegation—persisted. These remained key techniques of American emergency government for the rest of the war, in mobilization preparedness during the early Cold War, and in emergency preparedness for domestic catastrophes to the present day.

A "State of Preparedness": From World War to Cold War

As World War II drew to a close, the wartime apparatus of emergency government was rapidly demobilized. Emergency agencies shut down, wartime powers were withdrawn, and military procurement declined precipitously. A familiar pattern of retrenchment against big government and executive power set in.[162] The abolition of the National Resources Planning Board in 1943 was one indication, as President Harry Truman's advisor Clark Clifford wrote, of a broader "resentment and opposition of the Congress to strong or highly centralized control over the Executive Branch by the Chief Executive, and the even stronger resentment against 'planning' by the president to improve his executive position."[163]

In a key respect, however, the World War II demobilization was different. Even as emergency agencies were shutting down, military strategists, government officials, and planners, both inside and outside government, argued that the United States faced a new and threatening strategic reality.[164]

In the prior two world wars, the United States had significant time to ramp up military-industrial production while its allies fought overseas. But given the experience of World War II—in which American industrial production had proved crucial to Allied victory—and given the development of long-range bombers and atomic weapons, military strategists assumed that at the outset of the next war, there would be no such time for mobilization. A 1945 report by the Joint Chiefs of Staff thus argued that in light of the "decisive role the United States had played in the first two World Wars," any future enemy "would not give America time to mobilize our forces and productive capacity; the United States would be attacked first." Hap Arnold, the Air Force general who had championed strategic bombing during World War II, similarly warned of "a sudden, unpredictable and paralyzing blow from the air." In the next war, there would be no "opportunity for our gradual mobilization." Industrial mobilization plans "could be buried and lost in rubble."[165]

The implication was that the United States now had to accept what James Fesler referred to as a "new kind of normalcy."[166] Emergency government could no longer be limited to the exceptional measures taken during—or in anticipation of—a specific national emergency. Instead, ongoing emergency preparedness had to become a permanent activity in the federal government. As an influential 1945 report on the organization of the postwar armed forces put it, the "interval of time for preparation and mobilization after the outbreak of hostilities has reached the vanishing point, thus nullifying the capacity for hasty planning and emphasizing the importance of maintaining a state of preparedness."[167] Such a permanent "state of preparedness" implied not only "immediate readiness of weapons and troops" but also the "smooth and efficient functioning" of a governmental machinery to manage industrial mobilization.[168] In particular, the federal government had to be ready to rapidly set up emergency government to manage the national economy for total war. Postwar planners modeled such mobilization preparedness on the wartime arrangement for governing an emergency: concentrated executive power and expert controls derived from the Progressive tradition of governmental reform, as well as techniques of wartime resource management, based on the New Deal "science of flows."

"INDUSTRIAL READINESS": THE EBERSTADT REPORT

The elements of postwar mobilization preparedness were laid out in a 1945 study of the national security establishment by a committee chaired by Ferdinand Eberstadt, who had overseen the Controlled Materials Plan during World War II. The report of this committee is best known for its recommendations on the organization of the armed forces and as a template for the unification of the military that was instituted with the passage of the National Security Act in 1947. But the report also addressed broader problems of postwar organization of national security. Based on his wartime experience, Eberstadt was convinced that a new security structure would have to link "strategic plans with their conversion into national resources."[169]

Eberstadt recruited historian Robert Howe Connery to oversee work on the report. Connery, who had been a member of the Committee on Administrative Management, was in close contact with the social scientists who circulated among such institutions as the Public Administration Clearing House, the Brookings Institution, and the Social Science Research Council (SSRC). Connery convinced Eberstadt to expand the study to include industrial and resource mobilization and recruited a number of experts in administrative reform to work on it. Most influential among these experts was E. Pendleton Herring, a political scientist who would later become director of the SSRC and a major architect, along with Eberstadt, of the 1947 National Security Act. In 1941, Herring had written an influential book, *The Impact of War: Our Democracy under Arms*, which took up the arguments of social scientists, such as Carl Friedrich and Harold Lasswell. Herring argued that the advent of total war posed a fundamental challenge to American democratic institutions.[170] "We face a world," he wrote, "where discipline, organization, and the concentration of authority are placed before freedom for the individual and restraints on government." Echoing Charles Merriam's concern with "readjustment and maladjustment," he observed that "our government was originally designed for no such complex necessities."[171] Moreover, the distinction between war and peace had broken down, and the country faced a "constant state of crisis." The approach to addressing this "constant state of crisis" proposed by Herring and other experts in administration who consulted on the study followed the now-familiar prescriptions of Progressive reform. "All of the individuals surveyed," according to Dorwart, "wanted an organizational elite that stood above local interests and politics to run this permanent war machine," whether this elite body comprised "eminent scientists," a "business and labor advisory council," or, on the suggestion of Herbert Emmerich, experts in public administration.[172]

Herring's arguments shaped the study committee's final report, *Unification of the War and Navy Departments and Postwar Organization for National Security*, which outlined the rationale for ongoing mobilization preparedness.[173] "Modern wars," the report argued, "bring the total resources of the combatant countries into conflict." Military forces were only "the apex of the pyramid of national strength." The pyramid's base was to be found in "fields, forests, and mines"; in "factories, power plants, and transportation systems"; and in "managerial, technical, and labor skills." Therefore, the "vast military program" required to prepare for a future war would have to be "fitted into the civilian economy" by translating military requirements into "the basic elements of their production." In short, "military preparedness" would have to include "industrial readiness as an integral element."[174]

Here, the World War II mobilization experience—which the report reconstructed in minute detail—provided crucial lessons. Pointing to the "complicated and sometimes stormy evolution" of wartime materials control, the report's authors identified several principles and practices of mobilization that had been established during the war, following drawn-out struggles among the military, industrial interests, and civilian experts. These practices included the formulation of requirements and the calculation of "the limits of feasibility in production," as well as the institution of a control program that would "assure the actual flow of materials in accordance with programs for distribution." The wartime experience of "gradual progress toward success—finally achieved, but tardy in many instances" offered a cautionary tale, whose lessons had to be applied "vigorously and promptly." The government would have to institute measures in advance to ensure the "development of balanced and realistic over-all production programs."[175]

This program of mobilization preparedness, according to the report, required a continuous process of assembling "adequate information on the manpower, material resources, and productive facilities of the Nation" as well as the development, testing, and adjustment of "plans and proce-dures for implementing the military program in the civilian field."[176] It also implied ensuring that the proper authority would be in place to run a materi-als control program to bring the economy "under central direction." These functions, the report argued, should be placed in the "highly trained and experienced hands" of civilian experts, who were in "a position to make a comprehensive and objective appraisal of all the competing demands in the interest of the successful prosecution of the war."[177] The report proposed that these functions could be assumed by a National Security Resources Board. The explicit model for this new organization was the War Production Board, but it also resembled the resource planning boards of the New Deal. "One of the first tasks of the National Security Resources Board," the report suggested, "should be to take an inventory of our resources" by "develop-ing and maintaining adequate information on the manpower, resources, and productive facilities of the nation." The proposed Board would also be responsible "for the translation of current military plans into plans for an integrated and balanced civilian mobilization," taking into account "essential nonmilitary requirements." More broadly, its aim would be to "keep our national balance sheet of resources solvent" by developing policies to ensure the adequate supply of strategic materials in a future emergency.[178]

THE NATIONAL SECURITY RESOURCES BOARD

After a contentious congressional debate over the organization of the armed forces, the main recommendations of the unification study were instituted with the passage of the National Security Act in July 1947.[179] The Act created a new National Military Establishment[180] as well as two civilian planning agencies: the National Security Council—which became the preeminent advisory body on national security in the executive branch—and the National Security Resources Board (NSRB). Its clauses on NSRB included many of the functions related to national resource management and emergency government organization that were carried out by the wartime mobilization agencies. Thus, NSRB was responsible for developing policies to "assure the most effective mobilization and maximum utilization of the Nation's manpower," as well as the "effective use in time of war of the Nation's natural and industrial resources for military and civilian needs." The Act specifically referred to the Board's role in "maintaining the national balance sheet" by continuously assessing "the relationship between potential supplies of, and potential requirements for, manpower, resources, and productive facilities in time of war." The Board was further charged with maintaining "adequate reserves of strategic and critical materials" in a national stockpile, using powers that had been put in place by the Strategic and Critical Materials Stockpiling Act (1946).[181] Finally, NSRB was to develop plans for emergency government organization by establishing "policies for unifying, in time of war," the work of the various federal agencies involved in "production, procurement, distribution, or transportation of military or civilian supplies, materials, and products."

A cryptic final clause on NSRB in the National Security Act pointed to a quite different sphere of activities: the "strategic relocation of industries, services, Government and economic activities, the continuous operation of which is essential to the Nation's security."[182] This clause pointed to a novel problem for postwar mobilization planning—the vulnerability of American industrial production systems to enemy attack—that would become central to NSRB's work by the early 1950s (see chapter 4). But in 1947, this clause went virtually unnoticed in congressional debates about the National Security Act, and the functions it suggested were marginal to NSRB's early work.[183] In the Board's first years, "the bulk of NSRB's staff and effort" was occupied with mobilization preparedness for a future war.[184]

NSRB was staffed with veterans of wartime mobilization. Indeed, as historian Robert Cuff notes, the Board served as a "significant holding environment for personnel in both governmental and nongovernmental institutions concerned with industrial and economic mobilization" between the end of World War II and the creation of new emergency mobilization agencies during the Korean War.[185] In its first years, these veterans of wartime mobilization followed the Eberstadt report's recommendations and set about reconstructing elements of wartime mobilization, or planning for their future reconstruction in the event of war. As historian Harry Yoshpe writes, this meant, above all, occupying the "central position in the development and continuous refinement of the overall appraisal of resources and requirements."[186] As during the war, the "central tool" in this process of assessment was the "testing of military plans for economic feasibility."[187] NSRB's first feasibility studies assessed a strategic plan completed by the Joint Chiefs of Staff in summer 1948. To evaluate the feasibility of the plan, the Board undertook a survey of national resources in coordination with twenty-one federal agencies and departments, constructing balance sheets for thirty-five critical resources, as well as for manpower, transportation, electric power, and fuels.[188] Ralph Watkins, the director of NSRB's Office of Plans and Programs, explained in 1948 that in "the aggregate" these "balance sheets would embrace all the major or critical requirements for a mobilization planning period." Using these balance sheets, NSRB tested the plan in terms of particular resources, including manpower, steel, copper, aluminum, petroleum, merchant shipbuilding, construction, and 240 key items of military equipment. Such studies, Watkins explained, provided the "factual basis on which you determine either policy recommendations in time of peace for economic readiness steps or planning for administrative action in time of war."[189] Thus, Board staff used balance sheet studies to advocate policies to address potential bottlenecks and shortfalls, such as maintaining standby production facilities, entering into voluntary allocation agreements with industry (since the Board initially had no power to compel allocation decisions), and pushing for "substantially increased" appropriations for the national stockpile of critical materials.[190]

In another area of its work on mobilization preparedness, the Board formulated plans to rapidly set up an emergency government organization in the event of a future war. As Watkins explained, the Board developed a concept of "continuous mobilization planning," formulating "blueprints for mobilization" that could be regularly tested and adjusted. One focal point of such "blueprint" planning was the development of standby legislation, including a War Powers Act. The Board's general counsel, Charles Kendall, who had previously served on the legal staff of the War Production Board, led work on this standby legislation. The draft legislation laid out provisions for "re-creat[ing] in the Executive, in wartime, the basic authority found necessary in World War II" to manage the mobilization economy, including regulatory devices such as priorities and allocations powers, production loan guarantees, and rapid amortization. The law provided broad authority for the president to "organize the government for war by redistributing functions among existing agencies" or creating new agencies.[191] The Board also developed techniques to prepare government officials to employ these powers in a future wartime emergency. Beginning in 1947, it formulated plans for the organization of emergency government, resulting in a September 1950 "Organization Plan for Executive Branch of Federal Government for Full Mobilization." This plan outlined the structure of emergency agencies and the essential functions that they would have to perform.[192] Finally, NSRB staff developed "action programs" that outlined the steps federal offices and agencies would have to take in the event of the sudden outbreak of war, such as the issuance of emergency orders, the establishment of production controls, and the creation of emergency operating units within existing agencies.[193] These actions were compiled in checklists that indicated emergency actions to be taken by each government agency and office, the authority on which each action would be based, and the current state of preparedness of the office or agency that had to carry it out. The aim, according to Yoshpe, was both to lay plans for necessary emergency actions and, through recursive testing, to identify "gaps in existing plans" that could be addressed through further preparedness measures.[194]

Resurrecting the Emergency State: The Korean War

As NSRB continued its work on plans for government in a future emergency, the tensions of the Cold War mounted in 1949, with, among other developments, the Soviet A-bomb test, the Communist victory in China, and the division of Germany. President Truman, who was averse to large military expenditures, faced increasing pressure to invest in rearmament. In April 1950, the National Security Council presented him with its influential study NSC 68, which called for the "containment" of global communism and for a dramatic increase in military spending. Truman did not initially act on these recommendations, insisting, according to historian Paul Pierpaoli, that "the actual costs of the plan be calculated." But he changed his position after North Korean forces crossed into South Korea in June 1950. Truman committed troops to a UN force and began a rearmament program along the lines that had been proposed in NSC 68. In August, Congress appropriated $12 billion for rearmament, the beginning of a massive increase in military spending, from a post–World War II low of $13.5 billion to an annual $50 billion by the end of 1951.[195]

With the rapid growth in military spending and the looming prospect of a broader war, mobilization planners working in NSRB and advocates of a global confrontation with communism pressured Truman to reestablish the emergency economic controls that would be needed for a large-scale rearmament program. These advocates of renewed mobilization faced resistance from conservative critics, who argued that a permanent state of preparedness for war would "transform the country from a democratic state into a garrison state," as historian Michael Hogan puts it. But given the unfolding conflict in Korea, as well as rising prices caused by a surge in military spending, both Truman and conservative opponents of increased military spending came to accept measures that "went further than they thought desirable."[196] In September, Truman signed the Defense Production Act (DPA), which was based on the National Security Resources Board's standby War Powers Act.[197] The Act provided the executive branch with authority to exercise many of the powers—including reorganization authority and emergency economic controls—that had been used to manage mobilization during World War II.

The passage of the DPA was a signal moment in the postwar development of US emergency government. Along with the 1947 National Security Act and the 1950 Federal Civil Defense Act (discussed in chapter 4), the DPA has formed the statutory bases for much of the post–World War II emergency state. As Cuff puts it, the Act "sustained a succession of executive branch agencies," from the Cold War Office of Defense Mobilization to the present-day Federal Emergency Management Agency.[198] These offices oversaw the transformation of American emergency government that we trace in this

book: from resource management for war mobilization to ongoing preparedness for a range of future domestic catastrophes.

At the time of its passage, however, the Truman administration employed
the DPA to manage industrial mobilization for the Korean War. In exercising DPA authorities, Truman was initially unwilling to create a mobilization
"superagency" along the lines of the War Production Board, as NSRB staff
suggested. Instead, he set up a decentralized structure for administering the
growing mobilization program in which functions were distributed among
existing departments and agencies.[199] Truman delegated some functions to
NSRB, including responsibility for production expansion through accelerated amortization, loan guarantees, and purchase commitments. A Business
Expansion Office within NSRB developed a priority rating system for the
allocation of these subsidies. While the Board was officially responsible for
coordinating the mobilization effort across the government, operational
responsibility remained, Yoshpe notes, "in the hands of existing departments
and agencies." Thus, in a reprisal of the early period of mobilization prior to
US entry into World War II, NSRB's role of coordination had "little meaning
in the absence of a central programming of the defense effort."[200]

This decentralized system for coordinating mobilization broke down following the Chinese intervention in Korea in fall 1950, which, according to
Pierpaoli, "caused military as well as economic havoc." A dramatic acceleration in rearmament, along with panic buying and hoarding by consumers
who feared a broader war, sparked renewed inflation and shortages.[201] In
the face of these mounting pressures, Hogan recounts, concerns "about a
garrison state began to recede," and Truman faced growing calls to abandon
his decentralized approach in favor of a centralized mobilization program.[202]
Leon Keyserling, the head of the Council of Economic Advisers, thus argued
that the "need of our economy for steel and other vital commodities cannot
be appraised realistically or comprehensively" by the loosely coordinated
efforts of many agencies and in the absence of definitive information about
"the likely future trend of defense requirements." Performing a balance
sheet analysis—what Keyserling referred to as the "basic task of measuring
resources against requirements"—had to be centralized as a "precedent to
overall programming."[203]

Truman acceded to these demands in December, declaring a state of
emergency and, on that basis, drawing on Defense Production Act powers to
issue an executive order that established the Office of Defense Mobilization
(ODM) as a "super mobilization agency." The role of ODM, according to the

order, was to "direct, control, and coordinate all mobilization activities of the Executive Branch of the Government" on behalf of the president. ODM would assume the delegated authorities that had previously been dispersed across the federal government. Truman installed the president of General Electric, Charles E. Wilson, as director, a powerful position that was referred to at the time as a "co-presidency."[204] Truman also created several other emergency offices to direct mobilization, including the Defense Production Administration, which was responsible for requirements planning. The staff of the new mobilization offices included large contingents of specialists in resource mobilization who transferred from NSRB (311 staff members moved to the Defense Production Administration, for example), as well as other veterans of World War II mobilization.[205] In July 1951, the administration set up a new Controlled Materials Plan—this time, according to a postwar report, without "trial and error, and far more rapidly and with less initial friction than was the case in [the wartime] WPB."[206]

NSRB officials viewed the Korean War mobilization as a validation of their work on mobilization preparedness, as Edward Dickinson, who had been a member of the WPB Planning Committee and became vice director of NSRB in October 1951, argued in a 1952 lecture. The "emergency machinery" of the Korean War, Dickinson explained, was based on "stand-by legislation and plans," including "executive orders, and everything else" that had been prepared by NSRB. "In the first place," he noted, "the Defense Production Act of 1950 itself was based on draft legislation we already had prepared." At the same time, the "background and knowledge" of NSRB staff enabled them to "take the plans and programs developed in the Resources Board and make such adjustments as were necessary to meet the new situation."[207] Rearmament for the Korean War provided, for these mobilizers, a proof of concept of their approach to mobilization preparedness. As we will see, many of the same mobilization specialists would later apply this template to a different problem: preparing for a nuclear attack on the United States.

Warfare, Welfare, and the Emergency State

Writing in 1950, political scientist Clinton Rossiter took stock of the astonishing change in American government over the prior two decades. "The years since the inauguration of Franklin D. Roosevelt," he wrote, "constitute one of the critical periods of American constitutional development." Most important were "changes in the presidential office" that followed two longstanding elements of Progressive proposals for governmental "adjustment." The first was the expansion of executive authority to address rapidly unfolding social, economic, and military crises. The postwar executive had recourse to what Rossiter described as a "well-stocked arsenal" of "permanently delegated statutory authority" that could be "call[ed] into action in the event of a 'sufficient emergency' or 'in time of war or similar emergency.'" The second important change was the creation of an "administrative machinery" to assist in the technocratic management of rapidly unfolding crisis situations. This administrative machinery was housed in "several hundred offices, at least twenty of them top-flight positions by any standard," that served the president as "eyes, ears, arms, mouth, and even brain." This transformation in American government, Rossiter wrote, had been precipitated by a "startling succession of major emergencies": "depression, recession, threat of war, war, inflation, industrial war, cold war, and still another threat of war." Its underlying cause, however, was the formation of what Rossiter referred to as the "positive state." In particular, Rossiter pointed to the still-unfinished "process of building an American 'welfare state,'" which had "piled so many added duties and decisions upon the President."[208]

In the aftermath of World War II, the politics of executive power, emergency government, and the positive state shifted. The construction of a social welfare state had been interrupted by conservative attacks on the early New Deal, and then by World War II. New Deal planners anticipated that the effort could be resumed after the war and laid out ambitious plans to do so. For instance, the National Resources Planning Board's (NRPB) 1942 volume *Security, Work, and Relief Policies* linked generous institutions of social insurance to fiscal and monetary interventions for maintaining full employment. But these proposals were "engulfed in a storm of conservative attacks," as historian David Brinkley writes. In part due to the role played by Merriam and other prominent New Dealers, critics charged that the proposals were a vehicle for the permanent peacetime expansion of executive power.[209] In the wake of these attacks, according to Rossiter, the NRPB was "senselessly murdered" in 1943 by congressional conservatives, who "insisted on regarding Charles E. Merriam as a dangerous Red," or who saw an opportunity for budget cutting and undermining the New Deal.[210] The 1945 Full Employment Act represented a final push to institute an expansive welfare state that would combine the provision of social insurance with government intervention to manage economic growth. This effort was similarly rebuffed amid charges by congressional conservatives—both Democratic and Republican—that it was "another vehicle for inflating the president's power at the expense of their own."[211]

Events of the early Cold War dramatically changed the political dynamics around American state building. At the time, many national security strategists and policymakers argued that the United States faced a period of "semi-war"—a time of "prolonged struggle and constant preparedness," requiring massive ongoing expenditure on national defense, internal security, and readiness for industrial mobilization. As Hogan has documented, conservatives resisted the vast expansion of the national security state just as they had resisted the welfare state, essentially opening up a second front in their campaign to return to a prewar "normalcy" of a small federal budget and a limited presidency. They failed to "derail the national security state," however, and in the run-up to US entry into the Korean War, they accepted unprecedented levels of peacetime military expenditure and a permanent expansion of presidential power. But they learned, Hogan argues, to "decouple it from the economic and social policies of the New Deal." The broad result, he writes, was a "transformation of the nation's post-war identity from that of a welfare state to that of a warfare state."[212]

The story we trace in this book adds an additional dimension to such accounts of postwar American state building. By following the practices and practitioners of emergency government, we bring into focus a different legacy of New Deal planning, and a different dimension of the rise of the "positive state" after World War II. As public administration specialist Edward Henry Hobbs noted in 1954, if legislators resoundingly defeated the New Deal project of building a social welfare state, then in another sense, their attitudes about the New Deal project of resource planning were ambivalent. This ambivalence was exemplified, he argued, by "the kiss of death [Congress] gave the National Resources Planning Board in 1943 only to fill the void within four years" when it created the National Security Resources Board. The latter office, according to Hobbs, was in effect "a mutational revival in the genealogy of Presidential planning and coordinative agencies" that congressional conservatives tried to vanquish with the abolition of the National Resources Planning Board.[213] Indeed, as we have shown, many mobilization planners who had worked in the NRPB and in wartime mobilization agencies moved to the NSRB, where they continued practices of New Deal planning that they had adapted for wartime mobilization—such as resource studies, interdependency analyses, and balance sheet analyses. These were not the leading New Dealers, who had been ousted from their positions during the war. Rather, they were members of a younger generation of mobilization specialists, which included Shaw Livermore, Edwin

George, Melvin Anshen, William Truppner, David Novick, Lincoln Gordon, Walter Skuce, and Bertrand Fox. As we will see, these actors would play a crucial role in the transformation of the mobilization apparatus over the following decade to address the problem of preparing for a nuclear attack on the United States.

The persistence of these practices of New Deal planning in the face of conservative attacks was made possible, speculates Cuff, by the political cover offered by the rubric of "national security"—most concretely manifest in the subtle shift from "national resources planning" in NRPB to "national security resources" in NSRB. If Congress had "been aware that [the NSRB] not only contained holdover personnel from the New Deal's National Resources Planning Board, but also pursued similar studies of the country's natural resources," Cuff writes, "the board might not have escaped scrutiny."[214] The association of NSRB's work on resource planning with national security was not, however, mere political cover. In the first years of the Cold War, the planning and management of specific resources, interventions to regulate the broader economy, military-industrial production, and military strategy were tightly linked, as they had been during the "total wars" of the twentieth century.

But gradually, at first on the margins of NSRB's work on preparedness planning for total war, this particular formation of emergency government began to mutate. Mobilization planners continued to employ many of the same knowledge practices (resource studies, interdependency analyses, balance sheet studies) and governmental tools (stockpiling, emergency planning, priorities ratings, and production controls), and drew on many of the same statutory and nonstatutory authorities, that they had employed in New Deal planning and in both wartime and postwar mobilization. But with the rise of more powerful weapons and delivery systems, their attention turned to a new problem: the prospect of a devastating enemy attack on American industrial systems, essential services, and metropolitan areas. Working in NSRB and, after 1953, in the Office of Defense Mobilization—which Hobbs also locates in the "line of descent" of New Deal planning agencies—mobilization planners began to focus on anticipating and preparing to manage the damage such an attack would cause.[215] By the mid- to late 1950s, with the proliferation of thermonuclear weapons and the advent of long-range delivery systems, the prospect of a long war with sustained military-industrial production seemed increasingly implausible. Emergency planners gradually decoupled the task of anticipating and preparing for the devastation of a future war from military-industrial production. Mobilization preparedness instead began to focus on what, through the 1950s, was increasingly treated as a distinct aim and problem: ensuring the continued functioning of vital systems in a future emergency. Alongside and intertwined with the evolving warfare and welfare states, this problem of vital systems security came to the center of what we now recognize as the American emergency state.

PART II

Demobilization and Remobilization

3

Vulnerability

The prior two chapters have examined the forms of knowledge, techniques of intervention, and patterns of governmental organization that were assembled to manage the emergencies of the Great Depression and World War II. In these contexts, the problem of emergency government referred specifically to the management of *economic* emergencies. Chapters 3 and 4 focus on the war and the early years of the postwar. They describe a transitional moment in the broad shift in emergency government that we trace in this book: from managing ongoing economic depression and wartime mobilization to preparing for a future catastrophe. During this period, mobilization planners continued to prepare for military-industrial production during a long military conflict. But they were increasingly concerned with a new problem: assessing American vulnerability to an atomic attack and preparing for emergency government operations in its wake.

———

In 1954, Harvard economist Carl Kaysen published an article entitled "The Vulnerability of the United States to Enemy Attack." Writing shortly after the Soviet Union detonated its first hydrogen bomb, Kaysen lamented that "wars and rumors of wars are always with us." The prospect of "heavy enemy attack on the cities of the United States" implied "problems in public policy never before faced," such as identifying the most critical American vulnerabilities and devising measures to mitigate them. To address these problems, Kaysen posited, policymakers would have to understand "the character of the United States as a target today." Beyond military bases, an enemy in a future war would likely focus its attacks on American cities to meet various strategic aims: "to cause casualties, to destroy housing, to destroy industrial capital in general, to disrupt centers of leadership and communication, [and] to attack symbolic images of national power."[1]

Kaysen described two kinds of analysis that could be employed to evaluate US vulnerability. The first, industrial analysis, examined "specific industrial installations as targets."[2] Drawing on comprehensive data about American industrial structure, this type of analysis could identify specific facilities that an adversary in a future war would target for destruction.[3] The second kind of vulnerability analysis considered a different problem: How would the effects of an atomic attack unfold in urban space? Such "area analysis" would view a city "as a collection of economic resources." Graphical techniques and engineering assessments could be used to assess likely damage "in terms of destruction of resources—demolition of buildings and machinery, wounding and killing of workers." "For each city studied," Kaysen explained, "maps of population, housing, industrial capital, utilities and services, commercial building, and government centers in terms of, say, a quarter-mile grid could be made." From such data, it would be a "straight computation problem to estimate the impact of damage caused by weapons of different sizes exploding at different ground zeros."[4]

Kaysen situated his discussion of American vulnerability in relation to the history of strategic bombing theory. From the birth of air war through World War II, area analysis and industrial analysis had been formulated in the service of different and opposed approaches to air warfare. The early "prophets of air warfare," such as Giulio Douhet and Billy Mitchell, Kaysen noted, had argued that the main function of air war would be to attack "the 'morale' or 'will to resist' of the enemy"; for these strategists, "social organization" was the "target par excellence." They imagined that in a future war, air attacks would be unleashed on dense urban centers, where people and governmental machinery were concentrated.[5] Reacting against this vision of the indiscriminate bombing of cities, a different group of airpower theorists formulated an alternative strategy. In the 1920s and 1930s, officers at the US Army's Air Corps Tactical School, arguing that area attacks would be both immoral and ineffective, developed a theory of strategic airpower based on precision bombing of an enemy's vital industrial production systems. Air-targeting organizations like the Enemy Objectives Unit, where Kaysen worked during World War II, took up and systematized this approach. Advocates of precision bombing continued their criticism of urban area attacks after World War II, most prominently through the US Strategic Bombing Survey, the exhaustive overview of the air war in Germany and Japan. "The major lesson of strategic bombing in World War II," Kaysen argued, citing the Survey's findings, was that the prophets of air war—such as Mitchell and Douhet—had "prophesized falsely." With the "weights of attack and the types of weapons used in that conflict, social organization appeared not to suffer any but immediate shock damage, from which it quickly recovered."[6] The German industrial economy had proven surprisingly resilient to massive area attacks.

But the introduction of atomic weapons, Kaysen went on to suggest, had seemingly broken down the distinction between area and precision bombing. No matter how precisely it was targeted, an atomic detonation would have catastrophic effects both on "area" targets, such as the population and government organization, and on vital industrial production systems. Thus, in the late 1940s and early 1950s, in devising plans for a future air war, military targeting specialists worked to combine the economistic approach of industrial analysis with the engineering assessments and spatial modeling of area analysis in a single method of vulnerability assessment. By 1954, when Kaysen wrote his article, a new kind of expertise and a new kind of expert had emerged—what an Air Force report on computerized bomb damage assessment referred to as a "vulnerability specialist."[7]

Vulnerable America

This chapter describes how, in the aftermath of World War II, the United States became "vulnerable" in a new way. It does not focus on advances in weapons technology or on growing international tensions that convinced strategists that the country might be subject to a crippling attack at any moment. Rather, the chapter examines how the vulnerability of collective existence was constituted as an object of specialized knowledge. It focuses on the work of technical experts in largely forgotten government organizations who undertook a massive and meticulous labor of assembling standardized and comprehensive data on national resources and devised techniques for managing these data. The resulting knowledge infrastructure was used for various purposes: to identify American facilities that required protection from sabotage; to select foreign targets of aerial bombing; and to prepare for a future enemy attack.[8]

The first two sections of this chapter trace the evolution of what Kaysen called "industrial vulnerability" analysis. Our account begins with the practice of "resource evaluation" conducted by economists in the War Production Board during World War II. Using tools drawn from mobilization planning—such as interdependency analyses, punched cards, and tabulating machines—these economists identified industrial facilities and "essential services" whose destruction or damage would cripple vital military-industrial production. We then show how air intelligence specialists adapted these techniques to analyze the vulnerability of an enemy nation's vital industrial production systems, both during and after World War II. In this context, vulnerability expertise was used to identify "precision" targets for aerial bombardment.

The chapter then examines the development of another technique of vulnerability expertise—"area analysis"—which modeled the spatial distribution of damage-causing effects of a future air attack, such as fire, blast, or radiation. Area analysis was initially developed by wartime air intelligence specialists, who used maps and transparent overlays both to identify targets for incendiary bombing and to assess the damage caused by attacks on Japanese cities. After the war, air intelligence specialists combined area analysis with industrial analysis in an integrated "system of methods and procedures" for target selection in a future atomic war. This set of procedures included the basic elements of vulnerability expertise that Kaysen identified in his 1954 article, integrating a model of the "event" of a future atomic detonation as it unfolded in space with assessment of the damage to population centers, facilities, and systems that would be caused by such an event.

The final two sections describe how these methods were transposed from the analysis of air bombing targets to the assessment of domestic vulnerability, as vulnerability specialists addressed the novel problem of anticipating the effects of an atomic attack on American cities. These sections trace a significant technical development in the early history of vulnerability expertise: the "digitization" of vulnerability analysis using computer simulation. Initially, planners in civil defense offices used "analog" techniques—such as maps, transparent overlays, and information about bomb effects—to produce deterministic models of an atomic detonation at a particular "aiming point" in a given city. These techniques were subsequently adapted by defense mobilization specialists for the new platform of the digital computer. Computerization made it possible to simulate large numbers of randomized attack patterns and to assess the impact of each simulated attack on vital systems, whether industrial production or essential services. In the process, the basic elements of vulnerability expertise as we know it today were assembled: interdependency analyses that identify vulnerability as the emergent product of complex interlinked systems; tools of spatial analysis that anticipate the geographic extent of future catastrophic events; and computer-based simulations for calculating the probable effects of such events.

Resource Evaluation

Although military strategists had long been concerned about American vulnerability to enemy attack, the focus of their attention shifted in the early decades of the twentieth century. Prior to this time, strategists had concentrated on vulnerability to naval attack or blockade, and therefore on coastal areas and major waterways.[9] But with the advent of the airplane as an instrument of war, they came to conceptualize strategic vulnerability in much broader terms, to include urban and industrial areas across the country. In the 1920s, for example, Army Air Corps General Billy Mitchell warned that the concentrated centers of industry and population in the United States were highly vulnerable to attack. The development of long-range bombers in the 1930s heightened this concern with the vulnerability of vital production systems. In response to this threat, an Army directive in 1935 called for both "active" measures, such as warning and air interception, and "passive measures for protecting population and industry."[10]

During World War II, despite anxieties about an imminent attack on the United States, measures to protect population and industry never became a national priority (see chapter 4).[11] But outside of public view, the Roosevelt administration did take steps to mitigate the vulnerability of American vital systems to disruption. Following the Japanese attack on Pearl Harbor, in December 1941 Roosevelt issued an executive order that warned of the "serious and immediate potential danger of sabotage to national defense material, national defense premises, and national defense utilities which may menace our maximum production effort." The executive order instructed the War Department to institute "special precautionary measures" such as "establishing and maintaining military guards and patrols" as well as other "appropriate measures" to protect "national-defense material, national-defense premises, and national-defense utilities" from "injury or destruction."[12] It assigned responsibility for instituting these measures to the US Army's Office of the Provost Marshal General.[13] In response to the order, the Provost Marshal General set up an Internal Security Division whose "principal objective" was to ensure the "continuity of production and delivery of materials required by the armed forces" against "all hazards, except those in a combat zone." "Internal security" was thus concerned with potential events that might disrupt military-industrial production, such as "sabotage, and other destructive acts and omissions, including natural disasters, accidental fires, and industrial accidents to war plant labor."[14]

The Internal Security Division carried out some of the most infamous measures taken by the US government during World War II. Most notorious among these was the internment of American citizens of Japanese ancestry, which the Roosevelt administration justified as a preemptive measure to prevent industrial sabotage. The Provost Marshal General from 1941 to 1944, Major General Allen W. Gullion, was a prime mover in the internment.[15] Another area of Internal Security addressed a different problem: not the interdiction of "enemies within" but the mitigation of vulnerabilities in the nation's vital systems. In this area, officers worked with mobilization planners in the civilian War Production Board to develop a practice of "resource evaluation," which generated the first comprehensive analysis of the United States as a complex of vital and vulnerable systems.[16]

VULNERABILITY METRICS: FACILITY RATINGS

During World War II, the Internal Security Division addressed its responsibility for the continuity of industrial production by inspecting military and civilian facilities to ensure that their proprietors were instituting protective measures, such as posting armed guards to prevent saboteurs and making emergency plans for fires or industrial accidents. To carry out this function, the division maintained a Master Inspection Responsibility List of industrial facilities considered vital to military production. Inspectors rated these facilities according to their "criticality," so that the limited resources allocated for internal security could be directed to the most vital and vulnerable parts of the military-industrial production complex.

Initially, officials in the Internal Security Division based decisions about which facilities to include on the Master Inspection Responsibility List on recommendations made by military procurement officers. Their assumption was that these procurement officers had the most intimate knowledge of production processes. But staff in the division soon concluded that procurement officers did not have the economic understanding needed to make such assessments. As Colonel George Engelhart wrote in 1943, procurement officers were concerned "almost exclusively" with items from the Army's "table of organization and equipment," which defined the needs of combat units. The most dangerous vulnerabilities, however, were actually to be found in industrial facilities that were "remote from War Department contacts."[17] To identify these facilities, it was necessary to analyze the contribution of a particular facility to overall war production by tracing the flow of resources through production chains—from raw materials to military end items. As Engelhart put it, individual facilities could "be revealed in their true values" only by examining the "relative importance of facilities, as to total national available production, and the relative importance of facilities producing vital raw materials, parts, assemblies, and tools."[18]

In seeking out the information and expertise needed to conduct such analyses, Engelhart explained, the Internal Security Division looked beyond the military. In part, they drew on data that had been produced by industry organizations such as the American Petroleum Institute and the Association of American Railroads. They also drew on the "expert assistance of other federal agencies." For example, the Federal Power Administration had "administrative jurisdiction over electric power and gas utilities, and very complex information on them"; and the Bureau of Mines had "invaluable information on mining and related facilities." But the most valuable information about the military-industrial production complex was held by the civilian mobilization planning agencies, such as the War Production Board (WPB). Under its "broad wartime control over production," Engelhart reported, the Board "has available to it very complete statistical information as to all classes of production, including the availability of raw materials."[19] The same information and analytical tools that were being used to manage war production, he suggested, could also be employed to assess the vulnerability of the nation's vital industrial systems.

To make use of this body of knowledge, the War Department asked WPB chair Donald Nelson to create a "board of war production experts . . . for the purpose of evaluating facilities of all categories"—that is, assessing their criticality to the mobilization effort.[20] In response, in May 1942, Nelson created the Resources Protection Board, with members representing the War Department, the Navy Department, the Office of Civilian Defense, and the WPB. The Resources Protection Board was charged with determining "the relative importance of plants, facilities, installations, utilities, materials, and other economic resources to the war effort."[21] It carried out this function by working with civilian mobilization planning specialists to determine whether a particular facility would be included on its Master Inspection Responsibility List.

In making these determinations, the Resources Production Board drew on "factual analyses of the importance of facilities" that were conducted by a Resources Analysis Branch within the WPB.[22] This branch, in turn, was divided into a number of different sections. Some focused on the specialized products of the war economy, such as aircraft, shipbuilding, and ordnance. Others focused on the evaluation of raw materials, on facilities producing electrical and other machinery, or on chemicals. Still another section evaluated "non-production facilities" or "essential services": electric power, gas, communication, highways, petroleum pipelines, railroads, water systems, waterways, research laboratories, and stockpiles.[23]

Resource evaluations were used to assign "criticality" ratings to every industrial facility that was deemed to be "of such outstanding importance that if destroyed its loss would have a substantial adverse effect upon the war effort." Drawing on the same knowledge infrastructure that was used for mobilization requirements planning, resource evaluation specialists recorded information about tens of thousands of industrial facilities and products on punched cards.[24] These cards were then processed on tabulating machines to assess two factors on which the ratings were based: first, the importance of the output of a given facility for overall military production based on the "priority" ratings issued by WPB; second, the percentage of national production of a given output that a particular plant accounted for—the "concentration" of production in that facility. The Board assigned the highest ratings (AA) to facilities that accounted for a large percentage of total national production of a particular output that was considered vital to war mobilization (see figure 3.1). By the end of the war, the Board had rated "more than 1,200 individual products and services," encompassing tens of thousands of individual facilities.[25] This practice of resource evaluation made it possible to understand the American industrial economy as a complex of vital but vulnerable production systems.

In a series of memorandums composed in 1944, Engelhart described the practice of resource evaluation and documented the most significant vulnerabilities of the military-industrial production complex.[26] Engelhart noted that each facility on the Provost Marshal General's Master Inspection Responsibility List could potentially become a "critical production capacity bottleneck." But he singled out a smaller number of "facilities producing vital commodities which would be of utmost urgency in the event of loss or damage." Among these critical facilities were the Buffalo Electro Chemical Company in Tonawanda, New York; the plants of the DuPont Company in Niagara Falls, New York; the American Cyanamid Company in Bridgeville,

Percent of Total Output or Capacity	Classification of Product or Service		
	I	II	III
25% ⬦	AA	AB	AC
10 - 25%	AB	AC	–
5 - 10%	AC	–	–

FIGURE 3.1. Wartime resource evaluation: Facility ratings. Wartime resource evaluation determined the "criticality" of facilities, based on their importance for war production and the percentage of total national output of a particular good or service accounted for by the facility in question. The most critical plants (rated AA) were those whose destruction would seriously impair war production. *Source*: Office of the Provost Marshal General, *Defense against Enemy Action Directed at Civilians*, Study 3B-1, Historical Manuscripts Collection, Office of Chief of Military History Manuscripts, US Army Center of Military History, Washington, DC.

Pennsylvania; plants of the Monsanto Chemical Company in St. Louis, Missouri, and in Everett, Massachusetts; the Timken Roller Bearing Company of Canton, Ohio; the Hudson Motor Car Company of Detroit, Michigan; and the Boeing Company of Richmond, British Columbia. These "bottleneck plants" accounted for a high percentage of the total national production of items that were vital to military production—whether phthalic anhydride, tetraethyl lead, roller bearings, or wing assemblies for B-29 bombers. These plants would, Engelhart argued, "be ideal targets for sabotage, in that their destruction or damage would interrupt or seriously affect the conduct of war."[27]

The major source of "dangerous vulnerability" identified by resource evaluation was industrial concentration: the more vital production was concentrated, the more widespread the disruption that could be caused by a single natural disaster, industrial accident, or act of sabotage. One kind of concentration was the spatial clustering of vital facilities. Engelhart cited as an "outstanding example" the complex of plants operated by the Dow Chemical Company in Midland, Michigan, which produced twenty-two different items that were vital for military production.[28] Within the chemical industry, "similar geographical concentrations" could be found "in Niagara Falls, NY, in West Virginia, New Jersey, and elsewhere." A second form of concentration—more important, according to Engelhart, but also less immediately evident to the nonexpert—involved cases in which a small number of plants accounted for a large portion of total national production of a vital product. Resource evaluations indicated that the most dangerous vulnerabilities of this kind were not to be found in the production of "table of organization" items—the tanks, planes, or finished munitions directly provided to equip military units.[29] Rather, analysts would find such vulnerabilities in the production of often-obscure inputs to these end items, typically buried deep in production chains.

Engelhart illustrated this vulnerability in a description of the concentration of "Vital Chemical Intermediates," referring specifically to Buna-S—a synthetic rubber that was used to replace natural rubber when Southeast Asian suppliers were cut off by Japanese occupation. The production of Buna-S itself was reassuringly deconcentrated. Sixteen different plants produced the synthetic rubber; the largest accounted for only 12.1 percent of US production capacity. But an investigation of the inputs required to produce Buna-S revealed a much more troubling picture. Potassium persulfate, which was "essential to the manufacture of Buna-S," was produced by only two plants, one of which accounted for 91 percent of total American production capacity. The destruction or disruption of this particular plant could bring Buna-S production to a halt. Engelhart cautioned that potassium persulfate was not the only "Achilles' heel" of the Buna-S industry. Tertiary butyl catechol, used in certain technical processes, was "prepared by a single producer." The production of dodecyl mercaptan—a "genuine sine qua non" for Buna-S production—was also highly concentrated: three facilities accounted for 96 percent of American output. And Buna-S itself was only the tip of the iceberg. "When one considers that the finished tire is the real interest of military procurement," Engelhart wrote, "the pyramid expands in all directions." The resource evaluation specialist had to "take into consideration the many industries contributing to high tenacity rayon, to nylon, to tire cord weaving, to the manufacture of complicated anti-oxidents, accelerators, carbon black, tire molding, etc., etc."[30]

Buna-S was only one example among the "scores" of cases in which "small industries . . . are the vulnerable spots of large ones, even though one does not think of them as being within the large ones." As further illustration, Engelhart turned to metallic sodium, on which "depend some of the most important industries contributing to our war machine," including tetraethyl lead, used in aviation gasoline; sodium cyanide, used for hardening metals; sodium peroxide, used for "secret Navy interests"; airplane engine valves; sulfa drugs; stainless steel; and sodium hydrosulfide, used to produce mercaptans, which were, in turn, an essential element in the manufacture of Buna-S. This list of essential military products dependent on metallic sodium, he concluded, "is a self-explanatory indication of its importance to war-time American industry." Yet only two production facilities accounted for the entirety of American production. "<u>It can be stated conservatively,</u>" he wrote, underlining for emphasis, "<u>that upon these two plants rests the possibility of prosecuting a modern war.</u>"[31] As Engelhart traced out the web of industrial interdependencies through the chain of supply, more and more critical vulnerabilities became visible. The American war economy was revealed to rest on fragile foundations indeed.

ESSENTIAL SERVICES

Wartime resource evaluation also analyzed the vulnerability of "essential services"—the nonindustrial facilities that were "necessary to the maintenance and operation of production facilities, military installations, and production communities and to the delivery of raw materials, supplies, and end-products."[32] The category of essential services thus included rail transport, electricity generation and distribution, water systems, communications, and petroleum production and distribution. A report by the Office of the Provost Marshal General examined the significance of essential services for the functioning of cities and industry, analyzing the systems of interconnected flows that underpinned modern urban and industrial life as critical vulnerabilities. Electric power, for instance, was "so inextricably woven into everyday life in the United States that a study of power systems literally becomes a study of modern industry and urban existence," the report argued. "Factories, railroads, locks, utilities, offices, and homes are dependent upon the continued flow of electrical energy." Water systems were "immediately indispensable to the maintenance of life in any metropolitan area." Cutting off the water supply of New York City "would force its evacuation in short order."[33]

After World War II, the vulnerability of these essential services would become a central concern for mobilization planners as they turned from wartime industrial production to the survival of the population after an atomic attack (see chapters 5 and 6). But wartime resource evaluation specialists assessed essential services primarily as inputs into military-industrial production processes that were strategically significant for the war effort. Thus, the Provost Marshal General's report noted that alongside its indispensability for the "maintenance of life," water must be "available in sufficient quantity and purity for industrial purposes . . . as an ingredient, reagent, solvent, cooling agent, and a source of power."[34] Railroads were "vital both to logistics and to the movement of supplies and materials for industrial production." Oil was the "vital life blood" of the American war machine: "Not a wheel in industry can turn without lubrication," and "essential automatic transportation, many railroads, and much of industry are totally dependent upon petroleum for power." The report highlighted the importance of essential services for producing the items that the Resources Protection Board rated as most critical, such as synthetic rubber and metallic sodium. For example, the petroleum industry, "besides providing those products usually associated with its name," was among "the principal sources of butadiene and other components of synthetic rubber."[35] The two plants that accounted for all American production of metallic sodium—identified as one of the most critical vulnerabilities in American war production—were themselves perilously dependent on the uninterrupted provision of essential services. Disruption of power supply to a single plant in the Niagara Falls area that produced 57 percent of American metallic sodium by electrolysis "would have 'frozen' certain cells, requiring nine months to replace." Just a few minutes without power at this plant would deal "the nation's war machine a crippling blow."[36]

As with the assessment of production facilities, the Resources Protection Board's evaluation of essential services drew on a vast amount of data about each sector. For example, each American railroad submitted "a list of their critical structures"—more than 2,500 in total—ranked "in order of relative importance to the system." By September 1943, the Board had compiled a list of 2,041 rated facilities, which included 649 rated A (vital), 465 rated B (critical), and 927 rated C (important).[37] Of 115 major petroleum pipelines identified in the survey, 65 were rated high enough to receive wartime protection in the form of security guards or funds for physical protections; 85 high pressure wells and 66 petroleum storage and distribution terminals also received such ratings.[38]

In some cases, the Board assigned criticality ratings to individual components of essential services that were vulnerable to disruption and difficult to repair, replace, or circumvent. For railroads, the most vulnerable points were "switchpoints, frogs, crossings, curves, and control towers" as well as "bridges, tunnels, viaducts, fills, and cuts."[39] For oil pipelines, critical vulnerabilities included "key pumping stations," "strategic river and other water-crossings," and "important interconnections with other pipe lines" as well as "originating, intermediate, and terminal storage facilities upon which continuous operation of the line was dependent."[40] In other cases, the Board evaluated not individual facilities but the "systems" of which they were a part.[41] Thus, the report's discussion of railroads referred to the "methods of evaluation of [railroad] systems"; and its study of electric power noted that "the problems of interconnections, surplus potentials, voltage and cyclic requirements are so complex that the study necessary to the evaluation of any particular facilities becomes a study of the system of which it is a part."[42] The analysis of petroleum also described a "system" comprising a vast and dispersed range of facilities that tied together virtually the entire American military-industrial production complex: "All of the most important refineries and storage and distribution terminals on the East Coast," the report explained, "are dependent in time of war on the integrated system of numerous connected pipe lines transporting petroleum supplies from inland oil fields and refineries of Texas, Oklahoma, Ohio and other far distant areas." One component of this "vast net-work of eastern flowing pipelines," the Buckeye Pipe Line Company, "not only acts as a supplier of crude oil to other pipe lines connecting directly with Eastern refineries but also directly supplies refineries in the Toledo, Detroit, Cleveland, and Sarnia (Ontario) areas."[43]

The evaluation of these vital systems emphasized their spatial and material structure: the availability or absence of alternative routes that could be taken if a critical connection was destroyed or disrupted; interconnections that would allow one source of supply to be substituted for another, or whose absence would make a given system particularly vulnerable. Thus, evaluations of railroads accounted for the length of different lines and the availability of alternative routes connecting specific points. The highest rated railroad bridges were those that carried vital materials for military production, and for which no alternate routes were available. The New York, New Haven, and Hartford Railroad Bridge over the Hudson at Poughkeepsie, New York, which carried 2,377 cars per day, was one such critical span. "The damaging or destruction of this bridge," the Board's report noted, "would have necessitated a major reconstruction project during which a large portion of this vital traffic would have been paralyzed." The Pennsylvania-Southern Railroad Bridge over the Potomac at Washington, DC—which handled "all north-south traffic along the Atlantic seaboard"—presented another critical vulnerability. The destruction of this bridge would force a "long circuitous detour of all such traffic," resulting in "paralyzing traffic congestion and serious delay in the delivery of vital supplies."[44] The evaluation of electricity facilities also considered patterns of interconnection, taking into account the "emergency potential capacity available through interconnections of power plants and power systems." The Long Island Lighting Company's Glenwood No. 2 Steam and Electric Station in Long Island, New York, received an "Aa" rating because "this 223,000 kilowatt station was the principal source of power for

'Aa' rated war industries on Long Island" and was "inadequately interconnected with the Consolidated Edison System." The Boulder Hydro-Electric
Station (later renamed the Hoover Dam) at Boulder City, Nevada, received
an Aa rating as "the sole source of power supply for Basic Magnesium Inc.,
Las Vegas, Nevada, and the principal source for the vital aluminum, shipbuilding, aircraft, and oil industries in the vicinity of Los Angeles."[45]

Some apparently vital facilities were not rated as "critical" because the
systems of which they were part could adaptively respond to disruption.
Thus, in the case of electricity, surplus generation capacity or interconnections might allow diversion from nonessential uses that could be curtailed
to compensate for a facility that had been knocked offline. In Chicago, for
example, power production capacity provided a "10 per cent margin over
the peak operational load," and approximately 200,000 kilowatts could be
saved by restricting nonessential consumption. Thanks to the power system's
extensive interconnections, this surplus could be routed to critical needs.
Resource evaluation specialists reached similar conclusions in their analysis
of electricity in Detroit: surplus potentials plus interconnections meant that
"the loss of the largest power station would have forced only a ten per cent
reduction of consumption." Because of the "availability of alternate interconnecting systems," special protection was not required for "thousands of
power facilities which must otherwise be deemed vital installations."[46] In
the terms of World War II–era target analysts (see chapter 1), these systems
had both "cushion" and a capacity to replace damaged facilities through
"substitution," making them resilient to enemy attack.

In sum, wartime resource evaluation specialists built on practices of
mobilization planning to conduct the first comprehensive analysis of American industrial vulnerability. They drew on the War Production Board's "priorities" ratings, which indicated the importance of specific products to the
total national military-industrial production effort, as well as on data about
industrial facilities that had been gathered by mobilization planners. But
they used this knowledge for a novel purpose: identifying the facilities or
systems whose disruption would cripple the American war effort.

Modeling Future Attacks: Air War Planning

During World War II, techniques similar to those used in resource evaluation were employed for a different purpose: selecting targets for strategic
bombing attacks. Initially, the Army Air Forces asked resource evaluation
specialists to identify "certain vital products to the war economy of Axis

countries, in order that facilities producing those products in those countries might be selected as targets for Allied bombing."[47] Air war planners working in the Pacific theater adapted these techniques to model the effects of aerial bombing on Japanese industrial production systems. Air intelligence specialists recorded standardized information about Japanese industrial facilities on punched cards and used "runs" on tabulating machines to assess the effects of various attack patterns on specific target systems, such as oil distribution or airplane manufacturing. After World War II, air war planners consolidated these techniques into a comprehensive method for modeling future attacks. The vulnerability of vital systems was thus constituted as an object of anticipatory knowledge.

"A PERFECT AVALANCHE OF INFORMATION"

At the outset of World War II, Allied air war planners did not possess the detailed economic intelligence necessary to identify the most vital and vulnerable targets in the enemy's military-industrial production systems. As a result, they had to painstakingly piece together information from disparate sources to plan for air war (see chapter 1). As the war proceeded, military leadership set up programs to gather and maintain intelligence about enemy production systems. In September 1944, the Joint Chiefs of Staff created the Joint Target Group (JTG) as an interservice organization focused on target planning in the Pacific war, in part to remedy the highly fragmented air intelligence and targeting work in the European theater. Its charge was to identify air targets whose destruction would "injure most seriously the enemy's war-making ability," to prepare information for the military units responsible for planning and carrying out air attacks, and to assess bomb damage after such attacks.[48]

In assessing the vulnerabilities of the Japanese industrial economy, JTG relied mainly on widely available data. James T. Lowe, who had conducted air intelligence analysis early in the war in Europe before joining JTG, noted that information about the Japanese industrial economy came from "open intelligence—facts about industry, location of plants, machinery that is in them, the materials that go into the plants, who uses the output, etc."[49] The challenge was thus not in generating such intelligence but rather in collecting and handling the massive volume of data required to plan air attacks. JTG analysts began by compiling detailed intelligence on potential target systems in the Japanese war economy, such as petroleum or transportation. They then translated this information into numerical codes, which were recorded on IBM punched cards for automated processing using tabulating machines. The target analysts used tabulator "runs," which sorted the data according to various criteria (industrial sector, spatial location, etc.), to identify the components of a particular target system.[50] Analysts could then assess the resources (equipment, personnel, etc.) that would be required to destroy these targets through aerial bombardment. They could also anticipate the damage to industrial production systems that would be caused by a particular attack.

As World War II came to an end, airpower advocates argued that this knowledge infrastructure of air intelligence and target planning should be maintained.[51] The US Strategic Bombing Survey's summary report on the war in Europe, completed in July 1945, described the inadequacy of "information on the German economy available to the United States Air Forces at the onset of the war," noting that there was "no established machinery for coordination between military and other governmental and private organizations." This experience suggested the imperative to maintain "such arrangements on a continuing basis" after the war concluded.[52] The Strategic Bombing Survey's *Summary Report* on the Pacific war observed that following the Japanese attack on Pearl Harbor, "the obtaining and analysis of economic and industrial information necessary to the planning of an attack on Japan's sustaining resources required several years of the most strenuous effort." A similar situation at the beginning of a future war, argued the report, "might prove disastrous."[53] The basis for adequate air intelligence could "only be laid in peacetime" through coordinated intelligence gathering and analysis across the military establishment. The current lack of "recognized responsibility for intelligence work by the various operating organizations" gave "cause for alarm" and required "correction."[54]

Responding to these concerns, in November 1945, the Joint Chiefs of Staff established the Strategic Vulnerability Branch within the Air Intelligence Division of the Army Air Corps.[55] The new unit, which grew out of and was modeled on the Joint Target Group,[56] was charged with determining the "strategic vulnerability to air attack of all the countries of the world," and with recommending "targets for destruction in the event of war."[57] Like JTG, the Strategic Vulnerability Branch was jointly staffed by officers from the Army Air Corps, Marines, and Navy, as well as by civilian experts.[58] Several officers who had served in JTG took up positions in the new unit, including Joseph Coker, James C. Pettee, and James T. Lowe, who served as director of research. The Strategic Vulnerability Branch was divided into multiple sections. A section on General Analysis examined the sociological, political, and psychological aspects of target selection. A Physical Vulnerability Section—staffed by scientists and engineers—investigated the effects of munitions on various kinds of structures. Finally, two "geographic" sections, staffed by economists, worked to identify potential targets in a future war.[59] The Central Section investigated the Soviet economy, as well as those parts of Europe and Asia that were likely to be captured by the Soviet Union in a future war. The East-West Section, meanwhile, was concerned with the strategic vulnerability of the United States itself.[60]

Following the approach developed in wartime air intelligence units, analysts in the Strategic Vulnerability Branch sought to identify targets whose destruction would cripple the enemy's capacity to wage war. Lowe described the branch's "philosophy of target analysis" in lectures to the Air War College.[61] "War in its final analysis," he explained, "is nothing more nor less than a contest of national energies." Thus, in a future war, the first mission of US strategic air forces would be to "find a target system supporting the offensive capability of the enemy against our own industrial heartland." The job of air intelligence was to identify targets whose destruction would disrupt that system's functioning, taking into account the same factors that had been used for target selection in World War II: importance, depth, dispersion, cushion, reserves, location, vulnerability, and recuperability (see chapter 1).[62]

Lowe began with the problem that had been identified in the Strategic Bombing Survey: how to maintain, in peacetime, the knowledge base and analytic tools that would enable air war planners to rapidly identify targets in a range of possible future military emergencies. Following the widely held view that a future war would "begin with an attempted Pearl Harbor on the industrial heart land of the United States," Lowe argued that the US military would not have sufficient time "to determine what the targets are or should be, and to collect the necessary operational data to lay bombs on the target." For this reason, target intelligence had to be "ready on the trigger with the decision to wage the war."[63] As was the case during World War II, the difficulty in maintaining such intelligence was not that relevant information about enemy economies was inaccessible or secret: "90 percent of the information is freely available to anyone who wants it," Lowe observed. "You can get it from libraries, bookshops, newsstands." The challenge, rather, was one of data collection, management, and analysis. The relevant information was "scattered everywhere—in military and government files, and in the files of banks, insurance companies, engineering offices, and religious organizations." Moreover, the volume of data to be collected and processed was overwhelming. "Within the military orbit of practically every major power,"

Lowe explained, there were "some 70,000 or more potential objectives of air attack." The task for vulnerability analysts was to screen this "great mass of data" through a "fine mesh" to get to "the 70 odd or 7 that are within the capabilities of the attacking air force," and whose destruction "would make the maximum contribution to the attainment of the mission."[64] What was required, then, was a "plan of operations to handle this great mass of data," which he described as a "perfect avalanche of information."[65]

THE BOMBING ENCYCLOPEDIA OF THE WORLD

The centerpiece of the Strategic Vulnerability Branch's "plan of operations" was a vast compendium of data called the Bombing Encyclopedia of the World. The Bombing Encyclopedia—which exists to this day, though its name has changed—was a "huge index of basic, factual information" relating to "all the potential objectives of air attack throughout the world," including the targets that enemy air forces might attack in the United States.[66] It enabled military strategists to rapidly sort through data on potential target systems. The contents of the Bombing Encyclopedia were summarized in a set of large books. But in its "physical aspects," Lowe explained, the Encyclopedia consisted of thousands of IBM punched cards. Each card contained information about a particular industrial facility that might be selected for targeting, including the facility's longitude, latitude, and elevation; its purpose, indicated by a seven-digit product code; and its importance as a target, defined by the volume of its output, in the case of industrial facilities, or, in the case of cities, the size of its population.[67] By June 1947, the Encyclopedia included four thousand industrial targets in the Soviet Union.[68] A decade later, it contained records on tens of thousands of industrial facilities and other critical installations worldwide.[69]

The information on US industrial facilities contained in the Bombing Encyclopedia came from the same sources that federal agencies had used for economic management during the Great Depression and for mobilization planning and resource evaluation during World War II.[70] For example, the Encyclopedia drew on the Department of Commerce's Survey of Manufacturers for information about the function and volume of industrial facilities. By 1952, this survey included twenty-eight thousand US manufacturing establishments with more than one hundred employees and annual added value in excess of $500,000. In some cases, air intelligence specialists supplemented survey data for the purpose of target planning. For instance, the survey's information on the location of manufacturers was imprecise—limited to street addresses—so in the early 1950s, the Directorates of Air Intelligence and Management Analysis launched Project Congreve with the RAND Corporation to establish precise location information for key US industrial facilities. By 1954, the Air Intelligence Directorate had added geographic coordinates to records for nineteen thousand vital facilities, including all of those that were located in major metropolitan areas.[71]

Lowe noted several advantages of the Bombing Encyclopedia's automated procedure for data storage and processing. The use of machine methods for handling information meant that air intelligence organizations were "able to operate with a small fraction of the number of people in the target business" than had previously been necessary.[72] Moreover, the system of punched cards and tabulating machines made it possible to automatically sort data to generate reports about the elements of a given target system, based on a range of criteria such as industry and location. The Bombing Encyclopedia thus offered a "degree of flexibility." The mission of American air forces in a future war, he argued, "is not something that can be predicted with accuracy, and it is not something that is usually repetitive, as it is responsive to many factors all of which are very fluid in nature."[73] "By punching these cards," Lowe explained, "you can get a run of all fighter aircraft plants" near New York or Moscow. "Or you can punch the cards again and get a list of all the plants within a geographical area." Indeed, he continued, "Pretty much any combination of industrial target information that is required can be obtained, and can be obtained without error."[74] Using the Bombing Encyclopedia, a target analyst could rapidly survey the vulnerability and essentiality of disparate enemy target systems, comparing, for example, "the electric power plants in a given country and the aluminum plants, the transportation system as against the steel industry, or the aircraft industry as against small arms and ammunition."[75] Air intelligence specialists could then recommend specific target systems to the Strategic Air Command, which would "prepare the operational target charts [to be] ready for any national emergency."[76]

In the late 1940s and early 1950s, Strategic Vulnerability Branch analysts used the Bombing Encyclopedia to conduct studies of potential target systems. Employing concepts from the economics of target selection such as cushion, depth, substitution, and recuperability, these studies analyzed system vulnerability by modeling how a specific event—damage to a given facility or target system—would propagate through the interdependent systems that constituted a war economy. Based on such analyses, they determined the most valuable targets for attack. A 1951 assessment of Soviet industrial vulnerability by the chief analyst of the Central Section, Joseph Coker, illustrates this approach to target selection.

Coker offered a panoramic survey of the vulnerability of Soviet military-industrial production, considering the suitability of various potential target systems for strategic air attack. He noted that, at first glance, the Soviet economy's dependence on rail transportation, which handled 88 percent of all the nation's travel, appeared to be a critical vulnerability. Rail transport was essential to "the vital arteries of coal, ores, steel, other essential material as well as components, finished products and foodstuffs." If these lines were cut, "little of the traffic could be transferred to other types of transport facilities" due to the lack of interconnections and limited capacity in other transportation modes. But because Soviet transportation facilities were widely distributed and could be repaired quickly, they were not suitable targets. Soviet electrical power also initially appeared to offer a promising target. The system was concentrated—Soviet industry was dependent on just 150 power plants—and the effects of a campaign to interrupt the electric power system would be felt quickly, since electricity was essential at all depth levels, from basic industries (such as steel) to military end products. Moreover, the system was physically vulnerable to attack and, given the absence of "interconnecting transmission lines," the power supplied by the 150 plants that might be targeted could not be replaced by electricity from the other 1,900 power plants in the Soviet Union. Recovery "would require at least six months."[77] Nonetheless, given the large number of targets, the cost of disrupting this system through air attacks would be high. Coker's survey proceeded through a range of other target systems, ultimately settling on petroleum refining and coking as particularly acute Soviet vulnerabilities.

The Branch's East-West Section, which was responsible for assessing American strategic vulnerability, used the Bombing Encyclopedia to conduct similar studies—analyzing US military-industrial production systems from the perspective of an enemy target planner. In a January 1953 lecture to the Air War College, Colonel John G. Fowler, Deputy Director of Targets, summarized the section's work.[78] Fowler's presentation followed the same template as Coker's, surveying the "specific target systems in the United States that would appear extremely attractive to the enemy" in a future war. "High concentrations of capacity in the earlier stages of processing or in the production of components," according to Fowler, presented an "Achilles' heel" whose destruction would "stop production of the entire industry." Facilities in these areas of concentrated production were the "most probable targets which the enemy might attack." One example was the production of high-octane gasoline. While 350 refineries in the United States were capable of producing this fuel, only two plants produced tetraethyl lead, an essential input to its production. A successful attack on these two plants would reduce annual production of aviation fuel by 76 percent. The stockpile of aviation gasoline and tetraethyl lead was sufficient for only ninety days' use, and at least that amount of time would be necessary to rebuild the plants. A much longer time would be required before full production could be reached. The antifriction bearing industry presented another critical vulnerability. "An attack on 19 plants," Fowler reported, "may be expected to reduce the overall output by 75 per cent." Given the lack of stockpiles of critical supplies, the effects of this attack would be felt by producers of military end items within three months. Reconstruction of the plants would take twelve to eighteen months. Fowler concluded his survey by arguing that vital American industrial production could be successfully disrupted by an attack that focused on just 154 targets in total. Many of the vulnerable target systems that Fowler identified were, like tetraethyl lead and antifriction bearings, familiar from wartime resource evaluation. Among these were the aircraft industry (particularly engine production), arms and ammunition (production of synthetic anhydrous ammonia), electronics (tubes), and the steel industry (open hearth furnaces). Others, however, were new to the postwar world. For example, Fowler identified the atomic energy industry as a particularly lucrative target system. Sixteen installations constituted the most "vital and vulnerable targets in this industry."[79] A successful attack on these plants would halt the production of fissionable material. Reconstruction and restoration of production would take up to two years.

Urban Area Analysis: "A Graphic Story of Urban Destruction"

Thus far, this chapter has followed the development of industrial analysis—a method for identifying the vital and vulnerable nodes of production systems—both in wartime resource evaluation and in wartime and postwar air intelligence. We now turn to the development of another key technique of vulnerability expertise: area analysis. Area analysis employed graphical tools such as maps and overlays to model the blast, radiation, and fire damage that an air attack would cause to urban features such as vital industrial facilities and concentrations of population. Air intelligence specialists in the Joint Target Group initially developed area analysis to select urban targets for firebombing attacks in Japan. In the early postwar period, the technique was adapted to a new purpose: target planning for atomic war.

SELECTING URBAN TARGET SYSTEMS

Air intelligence specialists first began to develop area analysis during the latter stages of the war in the Pacific, in the context of a turn away from precision bombing of vital industrial targets. In late 1944, Haywood Hansell, a former Air Corps Tactical School instructor who was serving as the commander of the 21st Bomber Command in the Pacific theater, had led a campaign of high-altitude precision bombing.[80] Hansell's campaign ran into problems due both to operational factors (such as high winds) and to the difficulty of finding high-value precision targets, given the dispersal of Japanese war production. As a result, in January 1945 General Hap Arnold replaced Hansell with Air Corps Major Curtis LeMay. With the support of key Air Corps commanders, LeMay launched the infamous campaign of nighttime incendiary bombing of Japanese cities, beginning with the firebombing of Tokyo on March 8, 1945, which killed an estimated one hundred thousand civilians.[81] LeMay was unapologetically indifferent to the human costs of this strategy, proclaiming (in an interview conducted decades later) that "there are no innocent civilians" and that it therefore did not "bother me so much to be killing the so-called innocent bystander."[82] In response to this shift to an area bombing strategy, Joint Target Group analysts devised a method for assessing the effects of such attacks on Japanese military-industrial systems. This method—"urban area analysis"—modeled the impact of a catastrophic event (here, the firestorm that followed an incendiary raid) as it unfolded in a city.

Given "very poor ground intelligence" in Japan, JTG used aerial photographs to identify different functional zones of a given city—residential, industrial, mixed, transportation, storage. They also identified physical characteristics, such as building density, types of roof cover, and fire breaks, that would determine the effects of an incendiary attack.[83] These features were charted on maps of Japanese urban areas, outlining installations that were part of target systems, such as vital production or logistics facilities. The assessment of damage to industrial facilities then involved plotting blast and fire effects on this map.

JTG analysts also used the area technique to investigate how incendiary bomb attacks on a particular part of a city would affect industrial systems that extended beyond the zone of damage—thus linking area analysis with industrial analysis (described in prior sections).[84] First, they estimated the roof areas of "each industrial activity . . . and of each industrial type (e.g., the aircraft industry)" and recorded this information on punched cards. Then, analysts used punched card "runs" on tabulator machines to calculate the "total pre-attack areas" of a particular target system across all Japanese cities (for example, the total roof area of facilities that were part of the aircraft production "target system"). In a final step, they used damage by roof area as a proxy for damaged production capacity. The result was a "quantitative picture" of the effects that an incendiary attack on a particular part of a given city would have on a target system that was arrayed across multiple regions.[85]

In the aftermath of World War II, Army and Navy intelligence officers produced a joint report on the use of area analysis in selecting targets for incendiary bombing of Japanese cities. The report made the case for the efficacy of urban area attacks in Japan and for the continuation of urban area analysis after the war. "Surveys of strategic bombing in World War II," according to the report, had found that area attacks made an "effective contribution to the decline of the enemy civilian economy, morale, and industrial productivity through casualties, dehousing, absenteeism, administrative disorganization, and destruction of key facilities." Given the effectiveness of such bombing, it maintained, urban area analysis should be adopted as a "standard system of methods and procedures" for the selection of "urban target systems."[86] This standard system would involve both a "general study" of density, construction, and other urban features to assess area effects and an "industrial study," which would examine vital industrial installations and their relationship to broader target systems (see figure 3.2).

The report made several suggestions for improving the method that the Joint Target Group had developed. For example, it proposed to analyze the "structural content" of an urban area by distinguishing among buildings of wood, iron, steel, or concrete frame construction. The report also suggested new graphical techniques for data analysis. By mapping information about cities (density, structural types of buildings, etc.) on "transparent overlay[s] to the basic zone map," target analysts could flexibly juxtapose different types of spatial data, including information about cities and about the effects of bombing, such as firestorms and blast pressures.[87]

URBAN·AREA·ANALYSIS

MAPS | TOURIST GUIDES | INSURANCE PLANS | AGENTS REPORTS | OTHER SOURCES

AERIAL PHOTOGRAPHY

GROUND INTELLIGENCE

PRE-ATTACK ANALYSIS

GENERAL STUDY | INDUSTRIAL STUDY

ZONE MAPS | DENSITY OVERLAY | STRUCTURAL CONTENT OVERLAY | INDIVIDUAL INSTALLATIONS | INDIVIDUAL INDUSTRIES

ZONE GROUND ROOF AREAS | ZONE DENSITY FIGURES | ROOF AREA FUNCTIONAL & STRUCTURAL ANALYSES

WEAPON·DESIGN
FORCE·REQUIREMENTS
WEAPON·SELECTIONS
AIMING·POINT·SELECTIONS

AERIAL PHOTOGRAPHY

GENERAL ASSESSMENT | FUNCTIONAL IMPORTANCE

INDUSTRIAL ASSESSMENT | MAXIMUM INDUSTRIAL CAPACITY | INDEX OF CAPACITY

BASIS FOR TARGET SELECTION

TARGET VULNERABILITY

WEAPON·EFFECTIVENESS
FUTURE
FORCE·REQUIREMENTS
WEAPON·SELECTIONS
AIMING·POINT SELECTIONS

ECONOMIC APPRAISAL

POST-ATTACK ANALYSIS

·GENERAL··STUDY· | INDUSTRIAL STUDY

DAMAGE PLOT | DAMAGE REPORT | INDIVIDUAL INSTALLATIONS | INDIVIDUAL INDUSTRIES

ROOF AREA DAMAGE REPORT PLOT FUNCTIONAL & STRUCTURAL DAMAGE ASSESSMENTS

GENERAL ASSESSMENT | DAMAGE ASSESSMENT | EVALUATION OF UNDAMAGED AREAS

INDUSTRIAL ASSESSMENT | DAMAGE ASSESSMENT | EVALUATION OF UNDAMAGED AREAS

BASIS FOR FURTHER TARGET SELECTION

FIGURE 15

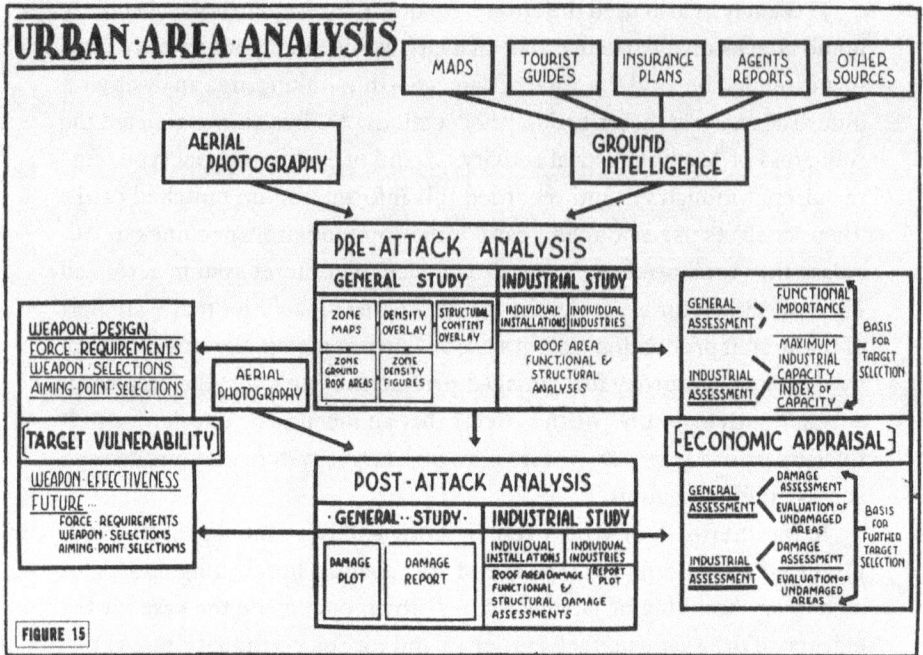

FIGURE 3.2. Urban area analysis. During the latter stages of World War II, air intelligence specialists in the Joint Target Group developed the technique of urban area analysis to select targets in Japan for firebombing raids. The technique drew on maps, overlays, reconnaissance photos, and detailed analysis of industrial structure to develop a "quantitative picture" of the effects that an incendiary attack on a particular part of a given city would have on a target system that was arrayed across multiple regions. *Source*: United States Strategic Bombing Survey, *Evaluation of Photographic Intelligence in the Japanese Homeland, Part Four: Urban Area Analysis* (Washington, DC: Photographic Intelligence Section, June 1946), 4.29.

MODELING ATOMIC WAR

Soon after the conclusion of World War II, US military leadership directed air intelligence units to identify potential enemy targets for a future atomic war. To conduct such assessments, air target analysts incorporated area analysis into their procedures of target selection. Given their background in precision bombing strategy, many Air Force targeting specialists were initially suspicious of thinking in terms of area bombing. For example, Joseph Coker wrote in 1949 that area bombing during World War II had produced "a spectacle of destruction so appalling as to suggest a complete breakdown of all urban activity and to give at least a first impression that the area attacks must have substantially eliminated the industrial capacity of Germany."[88]

But this impression was mistaken, he argued, citing the findings of the Strategic Bombing Survey. Proponents of area attacks had underestimated the German economy's "recuperability," or "resilience"—that is, its ability to adaptively adjust to disruption. It would have been more effective to focus air attacks on "some very small and vulnerable but very significant part of the war economy"—its "Achilles' heel."[89]

But for Coker and other advocates of strategic target selection, the advent of the atomic bomb seemed to undermine the distinction between precision bombing and area bombing. The Strategic Bombing Survey's studies of Hiroshima and Nagasaki—as well as the US government's test program—had demonstrated the vast extent of blast and firestorm damage that would result from an atomic bomb (see figure 3.3). No matter how precisely or imprecisely it was targeted, any atomic bomb functioned as an area bomb. Air intelligence specialists soon concluded that target selection for atomic warfare would have to anticipate the effects of atomic bombing on both vital industrial systems—comprising many facilities that were geographically distant from each other—and spatially concentrated area targets.

Initially, target analysts did not have access to information necessary for anticipating the area effects of atomic bombs.[90] The results of atomic weapons testing were highly classified and closely held by the Atomic Energy Commission. Intelligence about Soviet facilities and cities that was needed for performing a target analysis was sparse. Consequently, the Air Intelligence Directorate's early studies of targeting in a future war employed crude estimations of the area that would be destroyed by an atomic bomb, and then calculated the number of bombs that would be required to destroy all Soviet cities containing vital military-industrial targets. In search of more precise tools for target selection, military leadership looked to the small circle of experts with access to still-classified information about atomic weapons. In 1946, the War Department General Staff commissioned one such expert—nuclear physicist Ralph Lapp—to conduct a study of the damage that an atomic weapon would cause in an urban area.[91] Lapp's study drew on new information that was emerging from atomic tests to model the effects of an atomic detonation at a particular location in a city. This technique of "aiming point analysis" superimposed graphical representations of blast, radiation, and fire effects on city maps. The lines on these maps designated anticipated zones of common blast intensity, radiation exposure, and firestorm spread, determined by the characteristics of the bomb and by weather, topography, and the built environment (see figure 3.4).

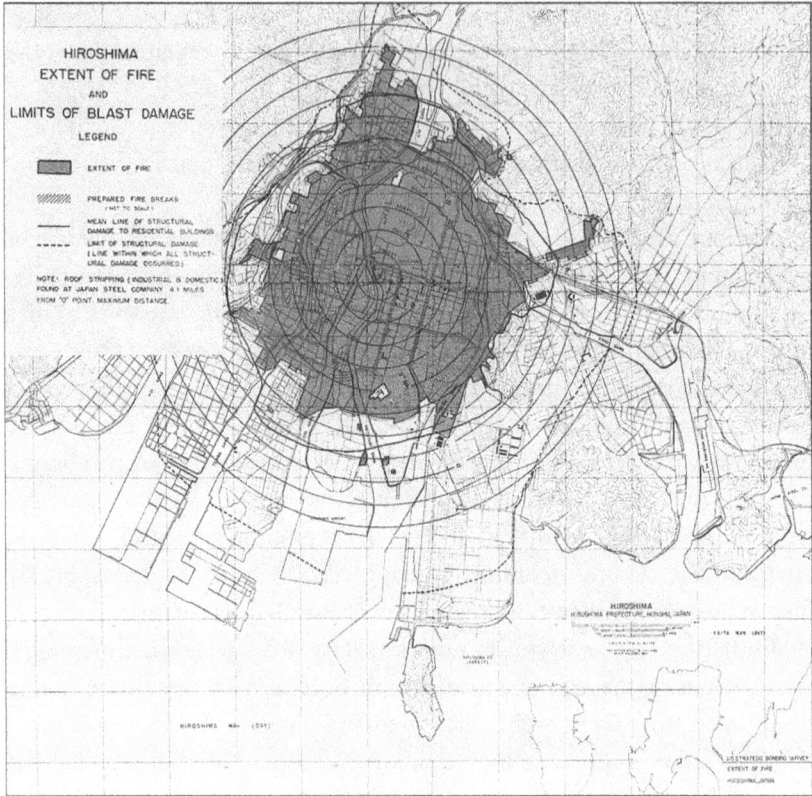

FIGURE 3.3. Hiroshima: Extent of fire and limits of blast damage. After World War II, the US Strategic Bombing Survey conducted detailed analyses of the effects of the atomic bombings of Nagasaki and Hiroshima. The Survey's analysts mapped the extent of firestorm and blast effects around "ground zero." *Source*: United States Strategic Bombing Survey, *The Effects of the Atomic Bombings of Hiroshima and Nagasaki* (Washington, DC: USSBS, June 30, 1946), 48.

Lapp's study indicated that in an atomic bomb attack, blast and shock waves would diminish in a regular pattern with greater distance from ground zero. The map of these effects thus took the form of concentric circles, radiating out from the assumed detonation point. Meanwhile, Lapp reported, the "vast quantity of radiant energy" liberated in the form of "ultraviolet, visible and infra-red radiation" would ignite "hundreds or even thousands" of fires in a three-mile radius from ground zero. These fires would produce a "mass influx of air from outside the region," resulting in an "enormously destructive fire-storm" that would burn out large areas that had been unaffected by the blast. The area consumed by such firestorms would have an

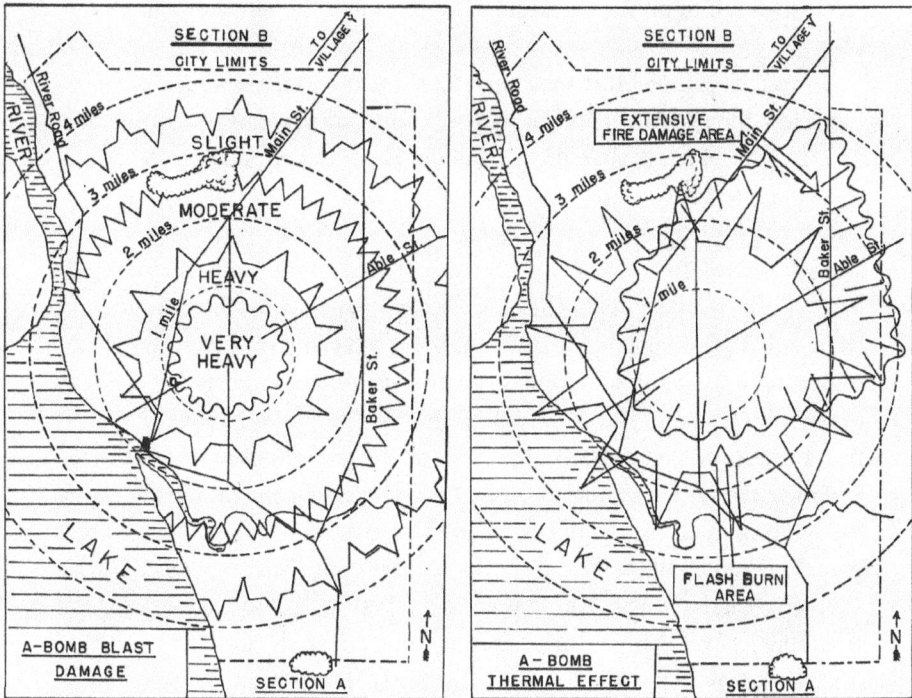

FIGURE 3.4. A-bomb blast damage and thermal effects. The atomic physicist Ralph Lapp conducted classified studies for the War Department to simulate blast, firestorm, and radiation effects of atomic weapon detonations. For these studies, Lapp and other experts drew on mapping techniques that had been developed to plan aerial bombing in Japan—as well as on the results of atomic bomb tests. *Source*: Ralph Lapp, "Atomic Bomb Explosions: Effects on an American City," *Bulletin of the Atomic Scientists* 4, no. 2 (1948): 13–14.

irregular shape, determined by building materials, features of the landscape such as parks or roads, and climate conditions. Finally, Lapp's study considered the effects of radiation: direct exposure to gamma rays from the initial detonation, fission products that would "fall out from the atomic cloud," and "induced" radiation in materials such as concrete or earth that were exposed to the blast, which might last for months or even years.[92]

In the late 1940s, air target planners systematized this "aiming point" method for modeling the effects of atomic weapons. Physical vulnerability specialists drew on information about the effects of atomic weapons to model how, for example, a structure of reinforced concrete or wood frame construction located two miles from an atomic detonation would be affected. As historian Lynn Eden notes, this new information gave them "a great detail of predictive ability" in modeling the effects of an atomic blast.[93]

Industrial vulnerability specialists then integrated the aiming point technique into their assessments of alternative bombing strategies in a future war. The director of the Strategic Vulnerability Branch, James Lowe, described this integrated approach.[94] Using a city map and a set of transparent overlays, Lowe explained, target planners could construct a "graphic story of the strategic vulnerability of a city in an urban area attack." One map overlay, for example, would indicate the construction types—wood, steel frame, and so on—that predominated in different parts of a city. Other overlays would indicate the spatial distribution of various "factors of importance," such as population concentrations, fire and radiation breaks, and industrial densities. Lowe reported that his group was in the process of producing maps and overlays for Soviet urban areas, a job that could be completed "when the 100 principal cities of the USSR are all analyzed in this fashion." Target planners then combined this "graphical story" of urban vulnerability with industrial analysis using the punched cards that had been coded for the Bombing Encyclopedia. By conducting punched card "runs" on tabulating machines, Lowe explained, it would be possible to estimate how an atomic attack in a particular city would affect a Soviet target system such as electricity or aircraft production.[95]

Lowe emphasized that target selection for atomic bombing would require the analysis of multiple target systems. "Within the limitations of any great city or industrial area," he explained, "there will not be one thing, such as the seat of the government, the headquarters of the military establishment, a steel plant, an aluminum plant, or an aircraft factory." Instead, in most industrial concentrations, analysts would find "something of each," and target selection would therefore involve "adding up apples and oranges, in a sense, and determining the value of the composite." In describing this method of target selection, Lowe offered a chilling vision of rationalized annihilation. In place of the search for pinpoint targets, a target planners would take a map of the country under consideration "and put an overlay over it and divide that overlay into grids, little squares." Each square "would represent the area of destruction that could be accomplished by one atom bomb." The task would then be to "weigh these little squares," to choose the one that "would be the best square in that whole country to obliterate from the standpoint of the mission of the attacking Air Forces."[96]

For air intelligence specialists of the late 1940s and early 1950s, urban vulnerability analysis brought into focus the previously unimaginable scope of destruction that would result from large-scale atomic attack. As Lowe observed, whereas the "kill rate" of the average 2,000-pound bomb in World War II was twenty people, for "an atom bomb in an urban area drop," the number exceeded 40,000 people. Faced with the vast destructive power of the new weapons, Air Force target analysts, who had long advocated a strategy of precision bombing, began to consider the possibility that an atomic war would involve primarily the selection of what had previously been considered "area" targets, such as the civilian population and government facilities. "Does that open up a situation," Lowe asked, "whereby in a future war we could forget all about this traditional economic or industrial type of analysis and concentrate upon selecting targets that will destroy the very fabric of society itself?"[97]

In fact, air target planners did not "forget all about" economic or industrial analysis. Rather, they weighed the "value" of attacks on area targets such as concentrations of population or government facilities against attacks on industrial target systems. For example, in his 1951 lecture on Soviet vulnerability, Joseph Coker drew on area analysis to determine whether an attack that focused on urban populations would cripple industrial production. Air force estimates at the time suggested that, given existing technology, a surprise attack on Soviet cities using three hundred atomic weapons would either kill or injure one-third of the 35 million workers that resided in these cities. But even this seemingly incomprehensible holocaust would not impede industrial production for long, Coker argued, since workers who were killed or incapacitated would be quickly replaced by workers from the countryside.[98] Two years later, in an analysis of US vulnerability, and now considering the vast scope of the effects of a thermonuclear attack, Colonel Fowler argued that an enemy in a future war might well select area targets. A Soviet attack that targeted vital military-industrial production facilities launched without warning would generate 2 million casualties. An attack that focused exclusively on population rather than industry would kill 12 million people. Meanwhile, an attack on Washington, DC, using hydrogen bombs that targeted the federal government would cripple "the single agency with the legal authority and administrative machinery to mobilize and direct the employment of the nation's resources for war." A daytime attack with three atomic bombs would produce "estimated casualties amounting to 175,000, or more than three-fourths of all government workers in this area." If the attack were carried out with a hydrogen bomb during daytime, casualties among government workers could approach 98 percent "and only a slightly lower proportion for the entire population."[99]

To summarize, after World War II air target planners devised a method for modeling the effects of atomic weapons that combined graphical techniques for assessing the "area" effects of atomic detonations as they unfolded in space with techniques of industrial analysis, which assessed how the destruction of particular facilities would affect the functioning of vital systems. In the coming years and decades, this method for assessing vulnerability would have significance well beyond the world of air targeting. As we show in the last two sections of this chapter, the method was adopted in civil defense and mobilization planning to generate knowledge about American vulnerability to enemy attack.

Urban Analysis of American Cities

One area in which the Air Force's new method for vulnerability analysis was applied was in planning for civil defense—the protection and relief of civilian populations in a future war.[100] By the late 1940s, the mounting tensions of the early Cold War led to calls to prepare for an enemy attack on US cities (see chapter 4). The Truman administration ordered the National Security Resources Board (NSRB), which had statutory responsibilities related to vulnerability reduction, to formulate a national plan for civil defense and to provide guidance to state and local governments in planning for civil defense.

In late 1950, the chair of NSRB, Stuart Symington, asked the military to prepare instructions on the assessment of urban vulnerability in domestic preparedness planning. Symington had been secretary of the Air Force in the late 1940s, when air intelligence specialists were developing methods of vulnerability analysis for use in target selection. Now, as chair of a civilian agency responsible for nonmilitary defense measures, Symington proposed to apply these methods to civil defense planning. In response to Symington's request, a *Civil Defense Vulnerability Manual* was delivered to the Board in 1951. As one NSRB official noted, the method described in the manual drew on "techniques in use by Air Intelligence in analyzing the vulnerability of foreign urban areas." These techniques, the official emphasized, gave "full consideration to the atomic bomb" as the central weapon in a future war.[101]

Federal officials circulated a handbook summarizing the Air Force's approach, *Civil Defense Urban Analysis* (*CDUA*), to local civil defense agencies in 1953. The handbook described how maps, acetate overlays, and detailed information both about a particular city and about the effects of atomic weapons could be used to assess local vulnerability. This method of urban analysis had three key elements, which had been developed in resource evaluation and target analysis: (1) an inventory of the significant features of a given city that would be arranged on maps; (2) an event model that anticipated the "aiming point" of a future enemy attack and the resulting distribution of weapons effects; and (3) an assessment of the damage that these effects would cause to significant urban features. As part of postwar civil defense planning, these elements were used to model a domestic landscape of destruction—a catastrophic future—that officials would have to navigate following an atomic attack.

INVENTORY

The first step of urban analysis described in *CDUA* was to catalog "all pertinent aspects of the city"—that is, the urban features that were significant in preparing for an atomic attack.[102] The manual listed these "pertinent aspects" in a table of forty-seven "urban features," including land use patterns, building density, industrial plants, population, police stations, the water system, electric power, streets and highways, streetcars, port facilities, the telephone system, hospitals, zoos, and penal institutions.[103] The manual also indicated the significance of these features for understanding the potential effects of an attack. Information about land use, for example, could help local planners in estimating possible damage to urban facilities and in mapping the distribution of population in order to assess likely casualties from a blast. Industrial plants were significant as possible targets of sabotage or bombing. Water distribution systems were critical to fire-control plans and were needed for emergency provision for attack victims and civil defense workers. Streets were potential targets, particularly at vulnerable points such as bridges and tunnels, but would also serve as routes for evacuation and mutual aid. Subway stations could be used as bomb shelters.

Once they had identified significant urban features, civil defense planners were to plot these features on maps of the city. The manual emphasized the importance of this mapping process in clarifying interrelationships that might not otherwise be apparent, pointing to the distinctive insight gained from spatial analysis. "The various features represented are dissimilar," the manual explained, "but are significant because of their interrelationship. For example, one particular street may be important as an emergency route because bordering buildings are not sufficiently high to block the street with rubble in event of their destruction by bomb blast."[104]

EVENT MODEL

The next step in an urban analysis was to model the "event" of an atomic detonation, initially by identifying a ground zero that could be used as a basis for local civil defense planning. Here, the manual drew on the technique of aiming point analysis, instructing local officials to view their city through the eyes of an enemy target planner: at what location in the city would an atomic detonation inflict the "maximum number of casualties" and destroy important industrial and transportation facilities? To find this "assumed aiming point," local planners were to draw a circle on a transparent acetate overlay and place this overlay on a map of industrial facilities and population concentrations (see figure 3.5). The radius of this circle, which indicated the area of extensive blast damage, depended on the size of the bomb to be used for planning purposes. Thus, a bomb 2.5 times the size of the Hiroshima bomb would have a 1.4-mile damage radius; a bomb 8 times the size of the Hiroshima bomb would have a 2-mile damage radius. The manual then instructed planners to "shift the overlay experimentally on the map of industrial plants," making a mark at the center of a blast area that encompassed the maximum number of industrial facilities.[105] Similar marks were to be made at the center of population concentrations. The next step was to transfer these points to a base map and join them by straight lines. Using the transparent overlay designating the radius of bomb damage, planners could determine the location of a detonation that would destroy all the facilities and population centers that had been identified. If some of these targets were beyond the damage radius of a bomb 2.5 times the size of the bomb detonated in Hiroshima—which was used as a reference—the manual suggested using "two smaller bombs" for planning purposes.

Civil Defense Urban Analysis emphasized that the assumed aiming point was not meant to predict the exact location of a future attack. The enemy's intended target could not be precisely known in advance, and, in any case, the enemy might miss. Rather, aiming point analysis was meant to represent a worst-case scenario—a detonation that would cause the heaviest damage—so that planners could estimate the maximum demand that would be placed on emergency services. "In target analysis," the manual explained, "the assumption is made that if bad marksmanship or ballistic error causes the bomb to fall elsewhere . . . the resulting damage will be more easily dealt with." By determining the maximum possible damage from an attack, officials could ensure that response plans were "sufficiently broad and flexible to meet all possible conditions." The manual indicated that this assumption

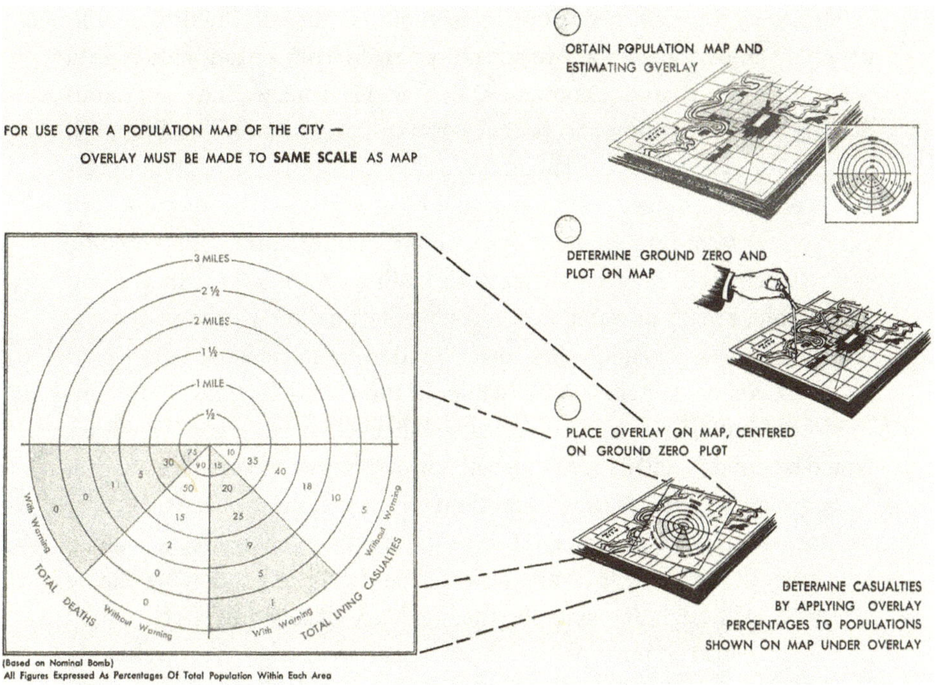
OBTAIN POPULATION MAP AND
ESTIMATING OVERLAY
FOR USE OVER A POPULATION MAP OF THE CITY —
OVERLAY MUST BE MADE TO SAME SCALE AS MAP
DETERMINE GROUND ZERO AND
PLOT ON MAP
3 MILES
2 ½
2 MILES
1 ½
1 MILE
½
PLACE OVERLAY ON MAP, CENTERED
ON GROUND ZERO PLOT
With Warning
Without Warning
TOTAL DEATHS
TOTAL LIVING CASUALTIES
DETERMINE CASUALTIES
BY APPLYING OVERLAY
PERCENTAGES TO POPULATIONS
SHOWN ON MAP UNDER OVERLAY
(Based on Nominal Bomb)
All Figures Expressed As Percentages Of Total Population Within Each Area

FIGURE 3.5. Estimation of civilian casualties. In the early 1950s, federal civil defense planners instructed local officials on how to estimate civilian casualties in a future attack. The first step was to determine an "assumed aiming point" by shifting a transparent overlay with concentric circles designating bomb effect radii over a map of a city. The point where this area covered the largest concentrations of population and industry was assumed to be the most likely enemy target—and was used, therefore, as a basis for preparedness planning. *Source*: Federal Civil Defense Administration, *Health Services and Special Weapons Defense*, vol. 34 (Washington, DC: Government Printing Office, 1950), 16–17.

should be "tested before being adopted" by using another "experimental" process: planners should select "a point or points at random" and then make "an approximate assessment of the resulting damage" to ensure that the scenario identified in the initial analysis was indeed a worst case.[106]

DAMAGE ASSESSMENT

The final step of the procedure described in *Civil Defense Urban Analysis* was to estimate the damage a bomb of a given size, hitting an assumed aiming point, would inflict on significant urban features. Physical damage to facilities would be determined by two factors: blast intensity at a particular location and the extent of ensuing firestorms. Blast damage could be estimated by using an acetate overlay in combination with a table that indicated the amount of damage to structures made from various materials in blast zones, designated as zones of A, B, C, and D damage. The exact radius of these zones would vary according to the size of the bomb (see figure 4.1, p. 183). For example, in the zones of A and B damage (a 1.4-mile radius for a 2.5× bomb), ordinary buildings would be entirely destroyed; reinforced concrete or steel-framed buildings would remain standing but would be severely damaged; highways, streets, and railroads would be damaged and unpassable; breaks in water, gas, and sewers would be widespread; while telephone poles, elevated water towers, and overhead electric lines would be destroyed or damaged. Meanwhile, in the C and D zones, only non-reinforced buildings would be damaged; blockages of streets and highways would be limited; and major infrastructure systems would not be damaged.

The manual also described a method for estimating the likely effects of firestorms and massive conflagrations—the atomic weapon effects that had "the greatest potential as destroyers of life and property following enemy attack."[107] Although, in principle, the incidence of such fire effects would depend on construction type, previous studies had concluded that relatively few buildings in American cities were fire resistant.[108] Therefore, the main determinant of fire spread would be building density. According to the manual, areas in which buildings occupied more than 20 percent of the land area were at the highest risk for conflagrations, which would also be influenced by meteorological factors such as wind and humidity. Finally, the manual described a method for calculating the effects of radiation, instructing planners to plot the distribution of the population in the city at the time of attack on a map, based on estimates of daytime migration patterns.[109] Drawing on information about injuries and fatalities that would result from exposure to different levels of bomb effects—summarized in a series of tables and charts—the planner would then "record the fatal casualties, nonfatal casualties and uninjured" for different zones.[110]

The purpose of this method of anticipatory damage assessment, the manual explained, was to identify gaps in emergency response and to plan preparedness measures. Each emergency service in city government should be "furnished specific maps and information pertinent to their operations." For example, information about anticipated damage to streets and highways, or about the likely spatial distribution of casualties in a future attack, should be incorporated in a "general civil defense transportation map." Evacuation plans could then be formulated based on assumptions about the likely volume of evacuees over certain routes, and civil defense planners could determine areas where fallen buildings or trees might block evacuation and plan alternatives. Casualty maps, meanwhile, would be "especially valuable for estimating shelter needs." These specific maps would also enable planners to identify vulnerabilities in systems of response: a single police station that housed "all of the police broadcasting equipment"; an electric station with the only transformer available to step down voltage from the transmission network to the local distribution grid; or a dangerous concentration of physicians' offices, which were generally located "near the probable target center" of cities.[111] Such information would allow planners to identify high-priority preparedness measures: building redundant facilities, stockpiling supplies, and relocating critical facilities outside likely target areas.

Simulating a Future Attack

At the same time that civilian mobilization planners were using urban area analysis to assess the vulnerability of US cities to atomic attack, a group of Air Force mobilization specialists was developing a method for assessing the damage such an attack would inflict on American industrial production systems. This method used the same basic procedures as vulnerability assessment in civil defense—an inventory of critical resources; a model of a potentially catastrophic event; and an assessment of the damage that such an event would cause to individual facilities and critical systems. The key difference was that the Air Force mobilization specialists proposed to automate vulnerability assessment, using the new tool of the digital computer. This method involved the standardization of heterogeneous kinds of information—about vital resources, production facilities, essential services, spatial location, and attack effects—into a quantitative format for computer processing.

PROJECT SCOOP: FEASIBILITY TESTING THROUGH COMPUTER SIMULATION

The proposal for a computerized bomb damage assessment procedure emerged from the world of postwar Air Force mobilization planning. Specifically, it was developed within the Air Force's Project on the Scientific Computation of Optimum Programs (SCOOP), which was part of a broader effort in the late 1940s by airpower advocates to justify the massive investments (in airplane development and procurement, pilot training, air intelligence, etc.) needed to create and maintain a large peacetime air force. As part of such efforts, Air Force planners developed techniques for analyzing the kinds of wars it would have to fight and the weapons it would use to fight them, and for translating military plans for such wars into programs for procurement and production that implied certain budgetary requests to Congress.[112]

Project SCOOP was conceived in mid-1947 by air force programming specialist Marshall K. Wood and mathematician George Dantzig, whose work on optimization algorithms was a key development in early computing.[113] In its initial stages, the project focused on Air Force requirements planning.[114] Technicians working in SCOOP built complex quantitative models that calculated the total amount of resources required to carry out a given war plan. Following the approach of wartime mobilization planning, this calculation of resource requirements included the immediate inputs to a military operation—such as personnel, equipment, and materiel—as well as the demands that military operations placed on the civilian economy for industrial production, basic materials, and labor power. As Edward Dunaway, a computer technician who worked on SCOOP, later explained, "There is a continuous string that you could put together mathematically to show that, in order to fight this kind of war you've got to have these kinds of resources from the civilian economy."[115] The project's aim was to use such mathematical formalizations and new computational tools to dramatically speed up the formulation of mobilization programs, a process that could take two years using manual methods. "Through the use of modern techniques of numerical analysis and the use of high-speed electronic computers," according to Wood, "we might be able to short-cut these operations to reduce this time span and thereby make it possible to explicitly relate our peacetime requirements to war plans, wartime programs, and the requirements stemming from them."[116]

When Wood and Dantzig began their work on Project SCOOP, Air Force programming specialists did not yet have access to digital computers. Their initial objective, as Dunaway recounted, was to develop the "techniques, the principles, and the concepts" that could ultimately be used in computer processing of mobilization planning data.[117] SCOOP technicians used the existing punched card system that the military had created for wartime requirements planning, incorporating coefficients to account for the ratios among the units of equipment, personnel, materiel, fuel, and provisions that would be needed to carry out battlefield activities in a future war. By 1950, according to Wood, Project SCOOP analysts had reduced the time required to translate a war plan into an industrial mobilization program from two years to a few months using existing punched card technology.[118] As of 1952, SCOOP analysts were working on the Air Force's UNIVAC I, one of the first digital computers.[119] The analysts initially used the UNIVAC to process the existing set of punched cards that had been coded with information about industrial facilities and national resources. Eventually, this inventory was transferred to magnetic tapes, making the vast amount of data about the national economy that had been laboriously collected by civilian and military agencies during the New Deal and World War II available for processing on a high-speed digital computer.

In the early 1950s, Project SCOOP's work on computerized requirements planning came to the attention of officials in three civilian advisory bodies: the Council of Economic Advisers, the Bureau of the Budget, and the National Security Resources Board. These officials were responsible for overseeing the feasibility of military mobilization plans—that is, ensuring that the resource requirements of these plans were in balance with civilian needs and available resources in the national economy. In the context of the ongoing Korean War and a large-scale rearmament program, their concern was that, as in World War II, the military might "seek forces and operations large enough to collapse the civilian economy."[120] To address this concern, the officials formed a Committee on Interindustry Economics to promote the use of more advanced methods of feasibility testing, with a particular interest in the use of digital computers.[121] This Committee channeled funding to Project SCOOP, which had the most advanced computational capabilities in the federal government.[122] Within SCOOP, the feasibility testing program was housed in a new Interindustry Research Program, led by mathematician H. Burke Horton.

Technicians in the Interindustry Research Program devised computerized methods to conduct test runs of industrial mobilization plans. Through such test runs—which were effectively simulations of a future war economy—they hoped to avoid the problems and delays that had hampered mobilization for World War II. As Horton explained in 1952, during World War II "the nation's productive apparatus" had been effectively used as a "gigantic analog" that served to "test ex post facto, the feasibility of a whole series of programs." The results of such "tests" showed up "in the form of insuperable production difficulties, time delays, and constantly changing delivery schedules." In the event of a "new emergency," such "'computation' by trial and error" might "cost the nation more time than it can spare," he argued. The Interindustry Research Program aimed to remove at "least a part of this trial and error process from the nation's industrial establishments and to place it instead in the computation laboratory." Imbalances in a mobilization plan would then show up as "red lights on a control panel" during test runs on a high-speed computer, rather than as actual breakdowns of military production during a future war emergency.[123] By late 1952, program analysts had completed their first computerized feasibility test, using what they called the "Emergency" model to project the economic effects of a four-year war plan.[124]

"THE BOMB DAMAGE PROBLEM"

In 1954 the Interindustry Research Program issued a report, *The Bomb Damage Problem*, which noted that the computerized system of feasibility testing had identified a number of potential problems in existing mobilization plans, and that these plans had therefore been adjusted to "remain within feasible production levels." But the report cautioned that despite this "careful initial planning," the mobilization program was "probably completely unrealistic." Current techniques of feasibility testing had "largely ignored the enormous damage to military installations and to industrial facilities, and the accompanying manpower losses, which would probably occur in the first few weeks" of a future war.[125]

To address this anticipated loss of resources, the report proposed to incorporate bomb damage into mobilization planning by using a digital computer to simulate a future atomic attack. The simulation procedure would make it possible to estimate the effects of such an attack on the resources available for military production. This procedure drew on the same elements of vulnerability analysis that were being employed in civil defense and air target analysis: an inventory of vulnerable facilities; an event model; and an assessment of damage. But in place of the analysis of the effects of a bomb striking a single aiming point, the procedure modeled a nationwide attack and computed damage for each detonation point. The results of these computations were then summed up to create a composite picture of the overall effect of a simulated attack on national resources.

Inventory of vulnerable facilities. The damage assessment procedure began with an inventory of the national resources—mainly industrial facilities important for military production—that might be damaged in a future attack. For each record in this inventory, it was necessary to provide basic information about the facility in question: Where was it located? What was its significance? Here, the procedure drew on existing databases that had been compiled for military and civilian mobilization planning. Information about the product and volume of production of each facility was drawn from the Department of Commerce's Survey of Manufacturers and the Standard Industrial Classification for industrial establishments and military product codes. This information was stored on punched cards or magnetic tapes for processing on a UNIVAC computer.

Location information presented a more difficult problem, since existing federal surveys contained only street addresses of individual facilities. The report described a method developed in Project Congreve (discussed above) for deriving precise location information on any structure in the United States, using a universal grid system that could be readily entered into a digital computer. First, Sanborn maps—which included outlines of buildings—were employed to identify precise center points of critical facilities.[126] The Air Force's Aeronautical Chart Service then transferred these points to maps or photo mosaics. Finally, using an overlay grid based on the Universal Transverse Mercator (UTM) system, analysts determined precise coordinates for each plant for transfer to facility records.[127]

Turning to future improvements in the system, the report recommended that information about essential services also be included in the computerized inventory. For example, given the "importance of electric power to the nation's survival," information should be collected from the Federal Power Commission regarding the coordinates of all major dams, hydroelectric power stations, steam power–generating stations, and central transformers. Since an atomic attack on the United States might "generate huge additional transportation loads needed for the purpose of tying together the jagged remains of US industrial capacity," the report also proposed recording the "coordinates of major railroad freight classification yards, truck terminals, rail and highway bridges, canals, locks, docking facilities . . . and airfields."[128] It suggested methods for identifying the location of "mobile targets" such as the working population. For example, regular patterns of movement could be modeled using information about employment at work sites or about the residential population by time of day.[129]

Event model. The next step in the proposed bomb damage assessment procedure was to construct an event model that would indicate the spatial distribution of blast, fire, and radiation from a series of atomic detonations that corresponded to a given enemy attack pattern. Air intelligence specialists on a "Red Team" would "choose a set of ground zeros" for the enemy attack and "specify the weapon size and time for each designated point of attack."

The location of these ground zeros would be based on estimates of Soviet capabilities ("the ranges of missiles or bombers, and the number and sizes of weapons") and on the most likely targets of enemy attack: centers of population, critical industries such as aircraft production, or essential services, such as electric power or transportation.[130] The report cautioned that simulation of a "perfect" attack under any given set of assumptions—in which every weapon was successfully delivered to its aiming point—would be a "poor reflection of real conditions." To produce a more realistic assessment of likely future damage, each weapon used would have to be "required to run the gamut of interception risks and aiming errors." [131] The report proposed to account for these uncertain factors by incorporating assumptions about the number of bombers that would be shot down and the probability that bombs would miss their targets. These chance factors could be introduced by using "tables of random numbers"—the area of mathematical specialization of the program's director, H. Burke Horton[132]—to generate parameters for "[a] number of independent runs." It would then be possible to "ascertain the influence of chance factors on the success of the attack."[133]

Damage assessment. The final step of the simulation procedure was to assess the damage to key resources and facilities that would be caused by the enemy attack. Using the "general coordinate system" recorded on facility records, it would be a "simple matter," the report explained, to calculate the distance of each facility from every weapon detonated in a particular area.[134] An initial computation of damage could be made for a "standard" weapon, with each facility assigned to one of four damage categories—destroyed, heavily damaged, damaged, or operational—depending on its distance from ground zero. Variations in bomb size would then be accounted for through a "distance adjustment factor" based on the assumption that the outer radii of damage zones would vary proportionately with the cube root of a bomb's explosive power.[135] The procedure would also factor in differences in the physical vulnerability of facilities made up of varying materials and types of construction, drawing on data provided by air intelligence specialists.[136] In sum, the computer simulation technique automated the type of aiming point analysis that had previously been carried out using maps and acetate overlays.

The damage assessment procedure could also model the effects of firestorms and radiation. Firestorms, the report noted, were "a very likely sequel to atomic attack" that would dramatically enlarge the area of total destruction and would also alter the "shape of the various damage zones, changing them from concentric rings to irregular patterns."[137] Given the impossibility of defining a simple quantitative model of firestorms for computer processing, the report proposed a method derived from the approach that wartime air intelligence specialists had used in the Pacific War. A photo interpreter would manually outline the likely boundaries of firestorm damage on a map, taking into account construction materials, firebreaks, and other factors that would affect the spread of firestorms in various parts of a city. The zone would then be mathematically described as a series of trapezoids that could be readily processed through a computing routine that would automatically assign plants to a damage zone, as with blast damage.

How did Air Force planners envision using the "complex experiments" or "war games" described in *The Bomb Damage Problem*? One use, described in an internal Air Force presentation on the project in late 1954, was to assess the feasibility of a given mobilization plan: Would supplies of critical resources be sufficient to meet requirements once the damage caused by an attack was taken into account?[138] Installations in different damage categories, as determined by a particular "run" of the damage assessment procedure, could be sorted and summed up by Industrial Classification Code.[139] The result would be "a measure of damage to each industrial group expressed in absolute units."[140] With these damage assessments in hand, planners could estimate the loss of vital industrial production that would result from a given attack pattern. It would then be possible to construct a postattack balance sheet, assessing whether the nation's requirements could be met with remaining resources that had not been damaged in the attack.

Notably, Air Force vulnerability specialists also considered another set of possible uses for the damage assessment procedure—planning nonmilitary measures to reduce the vulnerability of the civilian population and the civilian economy. It could indicate, for example, whether certain industries were "so concentrated in large urban areas as to create a very high probability that they will suffer disproportionately heavy destruction." By identifying particularly vulnerable industries, the procedure might point to "a need for additional standby capacity in dispersed locations." It could also aid in designing an optimal pattern of such dispersion. For example, simulation runs might show that "the larger weapons [are] 'efficient' only against the very largest and congested urban areas," thus suggesting the value of dispersion in small, spread-out industrial cities.[141]

Planners could also employ the procedure to estimate civilian casualties and to assess whether sufficient resources would be available to sustain the civilian population in the aftermath of an attack. "Fairly reliable answers can probably [be] found," the report posited to the question of whether the "subsistence needs of the surviving population" could be met after an attack by constructing a different kind of postattack balance sheet: of survival items rather than industrial resources.[142] For example, the procedure would make it possible to study whether the transportation system could be "depended upon to supply subsistence needs" or whether it would be so badly damaged as to "necessitate the prior stockpiling of food." Alternatively, it could be used to determine whether losses of hospital and medical personnel in urban centers would be so great that injured victims would be left to "fend for themselves."[143]

———

At the time that *The Bomb Damage Problem* was published, the kind of mobilization planning that Project SCOOP was engaged in was falling out of favor in the Air Force. New military leadership appointed in 1953 by President Eisenhower rejected centralized planning and pushed many civilian experts out of the armed forces. By 1954, the Interindustry Research Program had been slated for closure, and Project SCOOP had been cancelled.[144] This strand of technical research on American industrial vulnerability shifted to civilian mobilization planning offices in the Executive Office of the President. The Office of Defense Mobilization recruited H. Burke Horton, who had overseen work on bomb damage assessment in the Interindustry Research Program, along with a number of other Air Force vulnerability specialists, to continue their work on the simulation of a future attack. Over the next several years these vulnerability specialists worked together in ODM to build the computerized damage assessment system that was proposed in *The Bomb Damage Problem* (see chapter 6).

What is the significance of this obscure technical work on bomb damage assessment, initiated in a now-forgotten planning agency of the early Cold War, for the genealogy of emergency government? The procedure for using a simulated attack to test mobilization plans was built out of the knowledge infrastructure of surveys, resource records, and punched cards that had been assembled over the prior two decades for economic management in a national crisis, whether depression or war. Moreover, the immediate aim of the simulation procedure was to conduct "feasibility tests" of Air Force mobilization plans, in order to assess whether supplies of critical resources would be available to meet requirements in a future war. Thus, in developing the damage simulation procedure, mobilization planners specified the problem of preparedness for nuclear attack as a question of resource management: Could requirements for critical resources in a future war be met by available supplies, particularly if those supplies were severely limited by a large-scale enemy attack?[145]

The technique of computer-based simulation initially proposed by Air Force mobilization planners and then instituted in the Office of Defense Mobilization prefigured many knowledge practices that are central to how the contemporary world is constituted as "vulnerable." For example, in translating graphical analyses using maps and acetate overlays to digitized "layers" of geospatial data (terminology that would only come into use years later), the bomb damage assessment procedure anticipated modern geographic information systems.[146] Similarly, the procedure for modeling the effect of hazards using "test runs" that incorporate randomization techniques is the direct antecedent of what is now called "catastrophe modeling." In this sense, *The Bomb Damage Problem* brings to light one line of connection between midcentury emergency government, focused on economic management during depression and war, and the form of emergency government we know today, oriented to preparedness for future catastrophes.

4

Preparedness

In 1948, atomic physicist Ralph Lapp published an article entitled "Atomic Bomb Explosions—Effects on an American City" in the *Bulletin of the Atomic Scientists*.[1] The article drew on classified research on atomic weapons to argue that domestic vulnerability was a public issue that demanded governmental action. Lapp asked his readers to imagine an attack on a hypothetical American City X with a population of 1 million people. He described a pleasant sunny day in this city, with a "cooling breeze coming in off the lake," the streets in the city's downtown thronged with shoppers. Suddenly, an atomic bomb dropped from an unseen and unheard jet plane explodes at an altitude of 2,500 feet, sending a "vari-colored sphere of flame" rushing toward the earth. Directly under the blast, pedestrians are seared into "unidentifiable charred, grotesque forms." An initial blast wave is followed by a firestorm that "persists for several hours turning the entire area around Ground Zero into a raging holocaust." Lapp took his readers on a tour of the devastated city, describing the journey of a survey party through zones demarcated on a map by concentric circles radiating out from the detonation point. Within 1 mile of ground zero, the survey party finds nearly complete destruction. In many places, the street grid cannot be discerned. In a second damage zone—extending to 1.75 miles—some well-built structures are intact. But the view from the top of a fire escape nonetheless reveals a scene of "utter desolation." Within a circle roughly 2.5 miles in diameter, City X has "ceased to exist" (see figure 4.1).[2]

HYPOTHETICAL ATTACK
CITY "X"

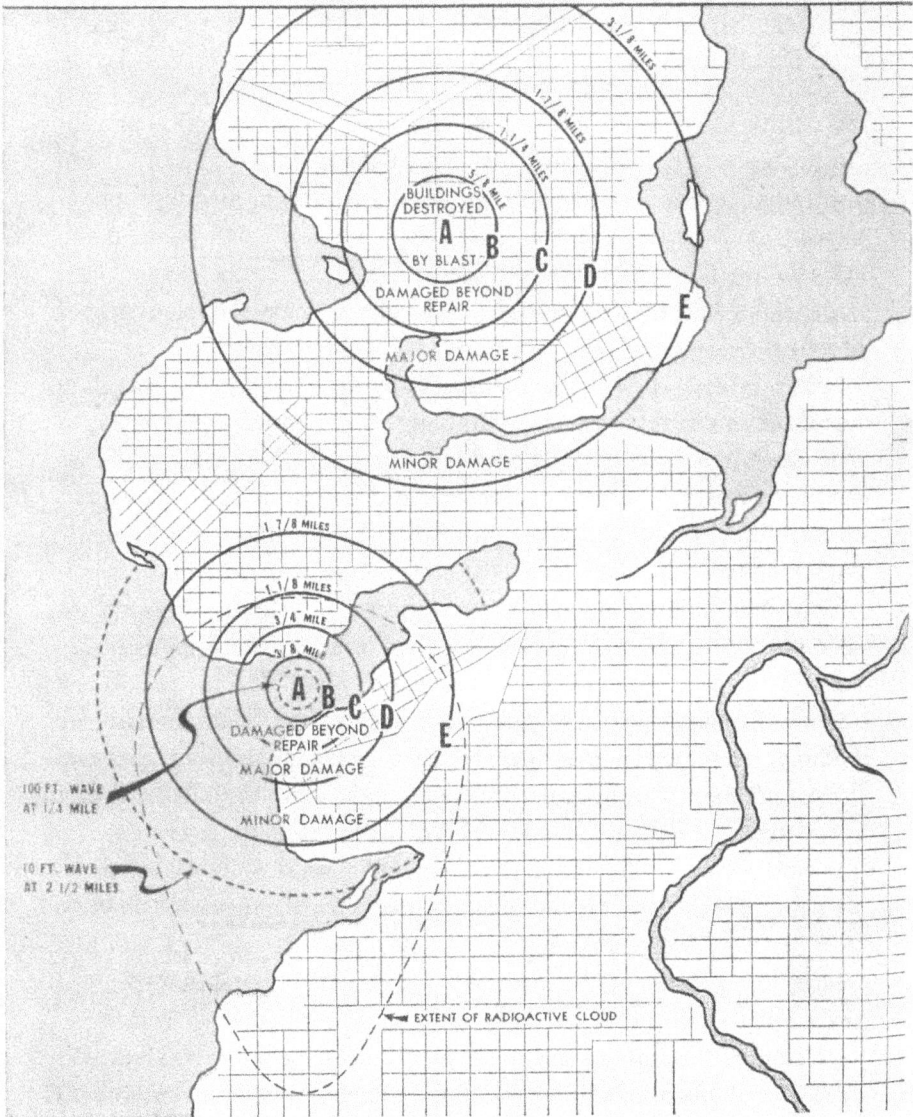

FIGURE 4.1. Hypothetical attack—City X. In the early 1950s, the National Security Resources Board disseminated methods for conducting vulnerability analyses of American cities, often illustrated with a hypothetical City X. These methods were meant to enable local governments to anticipate the effects of a future attack and, on this basis, to institute preparedness measures. *Source*: National Security Resources Board, *United States Civil Defense* (Washington, DC: Government Printing Office, 1950), 118.

Lapp then analyzed the measures that could be taken to mitigate American vulnerability to such a devastating attack. "Active" military defense measures such as the detection and interdiction of enemy bombers using radar systems offered no guarantee of security. "In spite of the best efforts of interceptor forces," some bombs would inevitably "get through to their targets and be detonated successfully." Consequently, it was necessary to undertake "passive defense measures," such as relocating communication and medical facilities outside urban centers, as well as "long range city planning" to prevent "further concentrations of facilities within built up areas of the city." Lapp also underlined the importance of "preparedness planning for catastrophes," which included training for specialized firefighting, evacuation, and medical response. If "a well worked out preparedness plan can be implemented immediately after the detonation," he concluded, "many thousands of lives may be saved."[3]

After World War II, techniques initially developed in the classified worlds of atomic testing and air targeting served both as vivid means to illustrate American vulnerability and as pedagogical tools, as federal officials instructed local governments and enterprise owners to assess their own vulnerability. Images similar to those presented in Lapp's study—with isometric lines designating zones affected by firestorms, blast, and radiation contamination, superimposed on a map of an American city—appeared in numerous articles, reports, and manuals that described "passive defense measures" to prepare for an atomic attack. For example, a 1948 pamphlet by the National Security Resources Board (NSRB), *National Security Factors in Industrial Location*, presented an image with isometric lines superimposed over two urban settlement patterns to demonstrate the greater vulnerability of dense, concentrated cities and the advantages of urban dispersal (see figure 4.2). The same pamphlet directed factory owners to conduct vulnerability analyses of prospective sites for new facilities, to ensure that they were not near likely targets. Two years later, City X itself appeared in another report by NSRB, *United States Civil Defense*, which instructed local governments in conducting aiming point analyses as a basis for preparedness planning.

A number of scholars of the political culture of the early Cold War have argued that urban dispersal and civil defense programs were venues through which forms of knowledge and techniques of governing invented for military purposes were transposed to civilian governance.[4] NSRB often appears in this scholarship as a site of translation from military to civilian government. Thus, historian Jennifer Light argues that NSRB's dispersal program inaugurated "a new approach to city planning and management."

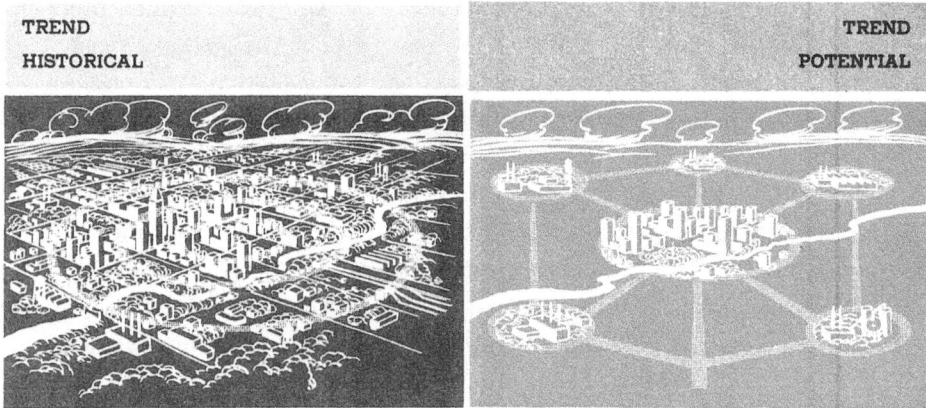

FIGURE 4.2. Trend historical and trend potential. In the late 1940s, many experts argued that the dispersal of industry and populations outside dense cities was the most effective way to reduce American vulnerability to attack, advocating for the restructuring of cities into small satellite settlements. *Source*: National Security Resources Board, *National Security Factors in Industrial Location* (Washington, DC: Government Printing Office, 1948), 10.

The application of "military expertise to the nation's urban needs," she contends, marked a "major turning point in American history."[5] Anthropologist Joseph Masco, meanwhile, argues that civil defense initiatives based on NSRB's *United States Civil Defense* (1950) were "less about the protection of citizens and cities than about the emotional training of the populace and the psychological conversion of U.S. citizens into Cold War Warriors." Through such initiatives, he suggests, Cold War government planners "engineered a new kind of militarized society" and "reorganized everyday life as permanent warfare."[6] The result of these programs, according to geographer Matthew Farish, was a "deep and subtle militarization of everyday life."[7]

From the perspective of the genealogy of emergency government, NSRB's central role in planning measures to reduce domestic vulnerability in the early Cold War is both puzzling and intriguing. Why was this office—created, as we saw in chapter 2, to plan for mobilization in a future wartime emergency—issuing reports on such topics as industrial dispersal and civil defense? What did "national security resources" have to do with domestic vulnerability? In investigating these questions, this chapter uncovers a little-known episode of US governmental development. Although military expertise on vulnerability plays a role, it is not a story of the "militarization" of civilian life. Instead, our account follows a path that leads from the management of economic depression (during the New Deal) and industrial mobilization (during World War II) to a new problem of emergency government in the early Cold War: preparedness for an atomic attack on the United States. From this perspective, the much-studied and publicly visible policies of civil defense and urban dispersal appear in a different light. Federal authorities undertook these initiatives reluctantly, spurred by public demands to reduce the vulnerability of the civilian population. Meanwhile, out of public view, a small group of officials, planners, and experts, working both inside and outside government, pursued a different set of measures that they considered to be more strategically important. These measures focused on ensuring the continuous operation of governmental activities, services, and industrial processes seen as essential to national survival and recovery in a future war. In these initial and often fragmentary attempts to ensure the continuous operation of vital systems, federal officials formulated a novel problem, which remains central to emergency government today: finding political, legal, and technical means to prepare for domestic catastrophe.

The New Problem of Preparedness

Chapter 3 traced how economists working in industrial mobilization planning and air intelligence constituted the vulnerability of vital systems as an object of specialized knowledge. The chapter focused on the period from the end of World War II through the early Cold War, documenting how experts gathered data on vital systems, analyzed their interdependencies, and assessed their vulnerability. The present chapter examines how, over the same period, federal officials defined the reduction of system vulnerability as a task of government administration. We show that this problem of system vulnerability corresponded to a novel political imperative: government was expected to maintain ongoing preparedness for a future event that threatened to catastrophically disrupt the systems underpinning collective existence.

The governmental norm of preparedness was not entirely new to the early Cold War. Military readiness to fight anticipated wars had long been discussed as a matter of "preparedness." Insofar as it related to civilian administration, preparedness thus referred to military-industrial mobilization: procuring equipment and critical materials, increasing production capacity, and marshaling manpower. Discussions of preparedness, thus understood, had intensified in the run-up to the two world wars, when large-scale military expansion was required. After World War II, "mobilization preparedness" referred to measures to ensure that a similar industrial expansion could be successfully organized in a future war that, military strategists thought, might break out at any moment. Initially, the new problem of preparedness—related to system vulnerability—was still tied to industrial mobilization. The systems in question were military-industrial production systems, which, planners assumed, would be targeted in a future war. Many of the experts and officials who worked on preparedness after World War II were veterans of wartime mobilization planning. They drew on concepts and practices that had been invented for wartime emergency government—such as resource evaluation, "balance sheet" analyses of supplies and requirements, and policies such as stockpiling and production controls—to address the problem of preparedness for an enemy attack.

At the same time, addressing this new form of vulnerability raised novel challenges. First, it required knowledge about an unprecedented future catastrophe. Beyond the atomic bombings in Japan, which had been probed for lessons and examples, no prior experience offered insight into the situations the government would confront in a future war involving an attack on the United States. Addressing vulnerability to atomic attack also raised legal and political challenges, which echoed the issues that Charles Merriam and other Progressive reformers (discussed in chapter 2) had faced in setting up emergency government in the 1930s. Passive defense measures like urban-industrial dispersal or emergency preparedness blurred the line between times of normalcy and times of emergency; between a theater of war—the sphere of military authority—and the domestic space of civilian government; between governmental control and private property; between the jurisdiction of the federal government and that of the states; and between the executive and other branches of the federal government. With these issues in mind, Hanson Baldwin, the military editor for the *New York Times*, warned in 1947 that the implication of "home-front passive defense" measures was "totalitarian." Implementing such measures would require "the endowment of government" with enormous "powers of compulsion." Invoking sociologist Harold Lasswell's well-known term, Baldwin argued that from these powers "could grow the 'garrison state,'" which he called "the great specter of the atomic age."[8] The first secretary of defense, James Forrestal, similarly argued that "home-front passive defense" measures presented a fundamental challenge to American political institutions. "A democratic people, who have always imposed the minimum of restraints on individual freedom and entrusted our security to geographical position, industrial strength, and free institutions," he observed, had "come face to face with problems which existing machinery [of government] seems ill-adjusted to handle."[9]

In what follows, we examine how officials and experts sought to address these technical and political challenges in the late 1940s and early 1950s. During this period, no agency had clear overall authority for implementing preparedness or vulnerability reduction measures. "Passive defense" was pursued in an array of government offices and through a tangle of overlapping initiatives. Our account focuses on the activities of the National Security Resources Board, which was created in 1947 by the National Security Act, the landmark legislation that shaped the postwar defense establishment. Alongside the Board's primary responsibility for mobilization preparedness, it was also charged with advising the president about the "strategic relocation" of "industries, services, Government and economic activities" whose continuous operation was essential to "the Nation's security." This reference to "strategic relocation" pointed to a specific passive defense policy that was widely discussed and debated after World War II: the dispersal of industry and government facilities outside dense urban centers. But particularly in the early 1950s, NSRB's work on nonmilitary defense expanded to include a range of programs related to government preparedness, civilian protection and relief, and postattack industrial rehabilitation.

Across these programs, which at first glance seem disparate and unconnected, NSRB mobilized a common set of concepts and practices. First, the Board employed the knowledge forms and technical practices of resource planning to constitute diverse activities—whether government administration, urban services, or industrial production—as vital systems, whose critical nodes could be identified. Second, it developed novel techniques for generating anticipatory knowledge, such as damage models and scenario-based exercises, to assess the vulnerability of vital systems, identify gaps in readiness, and develop plans to address such gaps. And third, NSRB drew on governmental devices that had initially been invented by Progressive reformers to address the emergencies of economic depression and war to accommodate these nonmilitary defense measures with existing forms of American government and economic organization. Among these governmental devices were the expert evaluation of emergency needs, a distributed organization of emergency government, and methods for selectively intervening in what Charles Merriam had called the "strategic points in a working system."

The chapter begins by describing how experts and officials conceptualized American vulnerability to enemy attack in the immediate aftermath of

World War II. These discussions laid the conceptual and practical ground-work for NSRB's charge, under the National Security Act, to advise the president on the "strategic relocation" of industries, essential services, and government activities. The first section examines a series of demobilization studies in the Office of the Provost Marshal General, which was asked to develop recommendations for postwar resource evaluation and civil defense. The next section turns to the activities of a loose network of civilian experts, including atomic scientists like Ralph Lapp as well as economists who had worked on mobilization planning and air targeting during World War II. After World War II, these civilian experts became leading advocates for reducing American vulnerability to enemy attack through measures to ensure the continuous operation of vital industrial facilities, services, and government activities in a future emergency.

The remaining sections of the chapter then trace the succession of projects through which NSRB worked to institute this framework of nonmilitary defense in government planning and policy. In these now-forgotten projects, NSRB staff defined a distinctive schema of emergency government, oriented to identifying and reducing the vulnerability of American vital systems and preparing government to perform the functions that would be required in a future war involving an attack on the United States.

Nonmilitary Defense: "The Whole Economy of the Nation at War"

As in many areas of national security strategy, postwar deliberations on American vulnerability were framed by the US Strategic Bombing Survey, the massive study of the effects of the American air war in Germany and Japan. In their summary of the study's findings, the Survey's directors noted that in the face of Allied air attack, German and Japanese industrial production had been surprisingly "resilient" due to the existence of unused capacity and stocks of critical materials, the active use of substitutions, and the reconstruction of damaged facilities. The Survey's report on the bombings of Hiroshima and Nagasaki concluded that "despite its awesome power," the atomic bomb had "limits of which wise planning will take prompt advantage." In anticipation of a future war, it was "essential to prepare to minimize the destructiveness" of a future attack, and to "so organize the economic and administrative life of the nation that no single or small group of successful attacks can paralyze the national organism." The report highlighted several measures "to cut down potential losses of lives and property," such as the accumulation of "reserve stocks of critical materials" and "a reshaping and partial dispersal of the national centers of activity" to ensure that "production of essential manufactured goods" was not "confined to a few or to geographically centralized plants." It also called for measures to reduce the vulnerability of civilian populations in cities, such as "timely decentralization of industrial and medical facilities, construction or blueprinting of shelters, and preparation for life-saving evacuation."[10]

The US Army's Office of the Provost Marshal General (OPMG) conducted the first postwar examination of such measures, beginning its research while the Strategic Bombing Survey was still under way. As we have seen, during World War II OPMG's Internal Security Division oversaw measures to ensure "the procurement of all military supplies" and "adequate provision for the mobilization of material and industrial organizations essential to war time needs" based on resource evaluations by the Resources Protection Board. Amid the broader demobilization that began in the last months of World War II, Undersecretary of War Robert Patterson ordered OPMG to examine whether and how these internal security responsibilities should be continued after the war's conclusion.[11] This question became particularly urgent when, in early 1945, the chair of the War Production Board, J. A. Krug, ordered that the Resources Protection Board be shut down.[12] In response, OPMG officers produced a series of reports and memorandums that described wartime resource evaluation and suggested how it could contribute to the reduction of domestic vulnerability after the war. In considering the postwar fate of resource evaluation, these reports also addressed a broader question: What measures should be taken to reduce American vulnerability and prepare for a future war?

FROM INTERNAL SECURITY TO NONMILITARY DEFENSE

Across the military, demobilization studies were based on a common planning assumption: in a future war, American urban and industrial centers would be "the initial objective" of a massive enemy attack, which would come suddenly and "without formal declaration of war."[13] Considering this scenario, OPMG's demobilization studies emphasized the strategic significance of American vulnerability to an attack concentrated on vital systems. As a staff memorandum put it, an enemy in a future war was "unlikely to risk planes, men, and materiel to destroy a small fraction of an industry." But an enemy might risk a surprise attack if it were possible to "annihilate or even halve an industry essential to the maintenance of some part of the US war machine."[14] Colonel Alton C. Miller—who served as chair of OPMG's Demobilization Planning Board—argued that this strategic vulnerability would grow more acute with the development of long-range bombers that could deliver ever-more powerful weapons anywhere in the United States. "The trend of modern weapons," he wrote, "indicates that the whole nation must become a 'line of defense' the moment that an attack is threatened and that all defensive forces must be deployed as effectively as possible before the first lethal blow is delivered." Peacetime preparedness measures would have to identify and protect the "points of dangerous vulnerability" in American production systems.[15]

Given these strategic considerations, OPMG officers argued, the "comprehensive type of industrial reconnaissance" that had been carried out by resource evaluation experts during World War II was "essential to the future security of the United States."[16] These "methods of reconnaissance" could provide an "ever contemporary picture of national industry as well as a plan for industrial security whenever it becomes needed."[17] To preserve the data collected by the Resources Protection Board, in August 1945 OPMG ordered that its files be transferred to the military.[18] OPMG also proposed a new peacetime resource evaluation agency, to be located in the "upper echelons" of the military establishment, that would keep these files current. This agency would maintain an inventory of American production facilities and assess the vulnerability of production chains based on factors such as the location of vital facilities in target areas, the concentration of vital production in a small number of facilities, or the dependence of vital production on vulnerable facilities and essential services.

OPMG demobilization studies also outlined how the wartime practice of vulnerability assessment could be adapted to the new strategic problems of the postwar period. During World War II, resource evaluation had a residual function. Once mobilization planners made decisions about building or expanding specific facilities, the size of stockpiles, or the location of storage facilities, resource evaluation specialists examined the resulting structure of the military-industrial economy to identify areas of dangerous concentration. On this basis, they prioritized security measures, such as posting armed guards to stop saboteurs and preparing plant-level accident management plans. But in the postwar context, OPMG officers argued, such measures would be of limited value. Armed guards and plant-level accident management plans could not assure production continuity in the face of a massive air attack. Thus, they proposed, resource evaluation could also be used to "guide federal agencies" in their decisions on mobilization policy measures such as plant expansion, the stockpiling and allocation of critical materials, and the allocation of manpower to priority activities.[19] In short, resource evaluation could be used to *design* a more resilient industrial production system by incorporating the vulnerability reduction measures that would be the most "potent safe-guards against sudden attack" into government-wide mobilization plans.[20] Plant expansion policies, for example, could be used to encourage construction in dispersed locations. Vital military production could be secured by distributing industrial production across many smaller facilities. Critical materials could be stockpiled near sites of priority use. Such measures would increase what wartime air-targeting specialists had referred to as the "cushion" in the American industrial production system, thus making the system better able to withstand or adaptively respond to an enemy attack.

OPMG's demobilization studies also proposed that the method of resource evaluation—which during World War II had focused on industrial production—should now be employed to assess the vulnerability of other vital systems. Most importantly, it should be used to assess what OPMG referred to as "essential services," such as railroads, inland waterways, electric power and gas, communications, water systems, highways, and petroleum pipelines. In a future "attack aimed at the United states," argued Colonel Miller, such systems would be "primary targets," since their destruction would affect both critical military production and the life of the civilian population. Services such as heat, electricity, and gas, he reasoned, were "necessary to the health and essential needs of civilian war workers." Greater study and recognition of the "military significance of the vital humanitarian functions performed by these facilities" would result in the "multiplication of facility ratings"—that is to say, in the number of essential services that were deemed critical vulnerabilities.[21] Miller's report also underscored the significance of such services for large cities. "Water systems," it noted, "are immediately indispensable to the maintenance of life in any metropolitan area." Rehearsing an example that instructors in the Air Corps Tactical School had used (see chapter 1), the report observed that "cutting off the water supply of New York City . . . would force its evacuation in short order." Indeed, the report concluded, the United States' "continued functioning as an organized society" depended on the ongoing function of essential services and vital facilities.[22]

"DEFENSE AGAINST ENEMY ACTION
DIRECTED AT CIVILIANS"

While its demobilization studies were underway, OPMG was given another, broader assignment to develop plans for postwar nonmilitary defense. As the war wound down, military leadership was convinced of the strategic significance of nonmilitary defense but divided about whether the military should take responsibility for it.[23] Thus, given a lack of initiative from the Truman administration, and without any statutory mandate, military leadership instructed OPMG to examine the wartime activities of the already-shuttered Office of Civilian Defense, as well as "comparable agencies" in both allied and enemy countries, and to make recommendations concerning "civilian participation in defense against enemy action directed at civilians, their installations and communities."[24]

At the outset of their investigation, OPMG's planners determined that American civil defense during World War II did not provide a relevant model for addressing the problems that the United States would likely face in the aftermath of a future enemy attack.[25] The wartime Office of Civilian Defense had focused on organizing disaster relief at the local level. Such a narrow conception of civil defense seemed inapt for addressing a large-scale atomic attack, which would quickly overwhelm local governments.[26] In search of relevant experience, Colonel Miller, who organized the OPMG study, looked to the civil defense study team of the Strategic Bombing Survey, which returned from Nagasaki in early 1946. He recruited the director of the study team, Lieutenant Colonel Barnett W. Beers, to lead the OPMG Civilian Defense Branch and to coordinate their own study.[27]

Beers's report, issued in 1946, rejected the widespread perception that nonmilitary defense would be ineffective against atomic attack. "Many members of the press, as well as some strategical experts," it argued, "initially inferred that the explosion of the two atomic bombs in Japan destroyed in that nation the last faint spark of will to resist"; their conclusion was that "there is no defense against the atomic bomb." The report allowed that this was true insofar as "active" defenses such as bomber detection and interception were concerned. But the report held that "defense against any bomb"—that is to say, a completely secure system of active defense—and "protection against the effects of its explosion are two entirely different things." Defining a vast new area of government activity, the report concluded that "it is a grave and fundamental responsibility of our government to develop means of protection and to create plans to put those means of protection into effect without delay."[28]

The report's proposals for nonmilitary defense began from the strategic assumptions that underpinned all postwar national defense planning. The next war, it posited, "will be a total war which may begin at any moment"; the initial enemy attack on the United States would target "the civilian populations and installations first, or simultaneously with its attack against the military." This attack, moreover, would meet with "considerable initial success, inflicting enormous casualties, mostly among civilians, and corresponding destruction of property, affecting materially our ability to produce for war." The country's capacity to "withstand that attack" would "determine the outcome of the war and the future existence of the nation." Everything would therefore turn on the "thoroughness and efficacy of plans prepared by the national government for their organization to resist and survive the attack."[29]

In light of these considerations, the report argued for a broad strategic conception of nonmilitary defense in an atomic age.[30] "The discussion thus far," it observed, "has concerned itself mainly with the protection of civilians and their property." Postwar nonmilitary defense, by contrast, would have to be "an integral and essential element of over-all national defense," encompassing "the mobilization of the entire population for the preservation of civilian life and property . . . and the rapid restoration of normal conditions in any area which has been attacked." In sum, an effective civilian defense program would address "the protection of industrial plants, water supplies, transportation and communications systems, and, in fact, the whole economy of a nation at war."[31] Such a program would also include plans to set up "a national unified organization under one command with complete directing and coordinating authority and responsibility in all matters pertaining to civil defense and protection."[32] In this conception, nonmilitary defense was an integral part of national security in an air-atomic age, involving both the management of resources and the operations of emergency government in a future war.

The Technopolitics of Dispersal

At the same time that OPMG was conducting classified studies of nonmilitary defense, a group of civilian experts was also working on the problem of how to protect industry and the population against an atomic attack. Many of these experts had been involved in the war in various capacities:

physicists who had participated in the Manhattan Project; mathematicians and engineers who had contributed to weapons development and operations research; and economists who had worked in the fields of mobilization planning and air targeting. After the war, they were clustered in various centers of the emerging military-industrial-research complex, such as universities, think tanks, and government advisory boards. From 1945 through the late 1950s, these experts played a prominent role in debates about US national security, writing in public forums such as the *Bulletin of the Atomic Scientists*, testifying in Congress, and serving on the new advisory committees that were being created within the federal government, such as the Research and Development Board in the Department of Defense, and the National Security Council. In these various capacities they were vocal advocates for nonmilitary defense.

In the immediate aftermath of World War II, civilian experts' advocacy for nonmilitary defense focused on the dispersion of industry and population outside dense urban concentrations. The most ambitious proposals for dispersal sought to transform the structure of metropolitan areas.[33] But calls for radical dispersal encountered resistance from critics who claimed that such a program would be unacceptably costly and, if made compulsory, would undermine American democratic institutions.[34] An alternative approach—which focused not on urban form but on the vulnerable nodes of vital systems—was formulated by a group of economists with backgrounds in wartime air intelligence and mobilization planning. They suggested that the tools of wartime resource evaluation offered technical means to identify the most vital government activities, essential services, and industrial production facilities. Preparedness measures such as stockpiling, dispersal, or plant-level security could then be used to reduce the vulnerability of these critical nodes in vital systems. Such "selective" dispersal, they held, could also address the political question of how to organize peacetime nonmilitary defense in a democratic society: by dispersing only the most vital facilities and by employing economic incentives—such as rapid amortization and defense loans—rather than government compulsion.

"TOTAL" VERSUS SELECTIVE DISPERSAL

The first civilian experts to assume a prominent public role in postwar discussions of national security were physicists who had participated in developing the atomic bomb. In September 1945, a group of these physicists formed the Atomic Scientists of Chicago, and would soon launch the *Bulletin of the Atomic Scientists*, an important public forum for debates around postwar national security strategy—particularly regarding the role of atomic weapons.[35] The stated purpose of the Atomic Scientists of Chicago was to "explore, clarify and formulate the opinion and responsibilities of scientists in regard to the problems brought about by the release of nuclear energy." In their official, collectively authored statements, these scientists argued that given their role in developing the bomb, they bore a special responsibility to stem the proliferation of atomic weapons.[36]

Against the claims of airpower advocates such as Air Force General Curtis LeMay—who argued that only overwhelming offensive force could deter enemy attack[37]—the atomic scientists contended that military advantage alone could not guarantee national security. No nuclear "secret" could prevent a future enemy from acquiring atomic weapons, and in the next war the United States would likely face an adversary that possessed thousands of atomic bombs. As physicists David L. Hill, Eugene Rabinowitch, and John A. Simpson wrote in 1945, "If both sides in a conflict have enough atomic bombs to wipe out each other's cities, they are in approximately equal position, even if the one has three times more bombs than the other." Indeed, as "a peaceful, democratic, highly industrialized country" that was unlikely to be the first to attack, and whose population was "concentrated in comparatively few metropolitan centers," the United States would be "at a great disadvantage" in a future atomic war.[38] As an alternative to the accumulation of ever-larger nuclear stockpiles, the atomic scientists initially placed their hope in political solutions to the threat of a global nuclear arms race, such as international control of nuclear weapons.[39] But as the Cold War intensified, and hope for such political solutions faded, the atomic scientists emerged as influential exponents of nonmilitary defense.[40]

According to Eugene Rabinowitch, a biophysicist who worked on the Manhattan Project and was editor of the *Bulletin of the Atomic Scientists*, the memorandums written by atomic scientists in 1945 contained "perhaps the earliest discussions" of urban dispersal, and the scientists' articles and other publications were "the earliest occasion on which the public was acquainted with this problem."[41] In *The Atomic Bomb* (1946), a jointly authored pamphlet, the Atomic Scientists of Chicago argued for a "large scale effort to disperse the population, industry, government and transportation systems of all our cities and industrial concentrations [and] to spread them out uniformly over the habitable area of the country" as a means to "greatly minimize the effects of an attack upon us."[42] Hill, Rabinowitch, and Simpson described such a program as "the most efficient defensive measure that a single nation can adopt."[43]

A 1946 article in the *Bulletin of the Atomic Scientists* by economists Jacob Marschak and Lawrence Klein and physicist Edward Teller elaborated this proposal for radical urban and industrial restructuring. From the perspective of vulnerability reduction, they argued, the "ideal" situation would be "to have our population dispersed evenly over the 3 million square miles" of habitable area in the United States. Each dwelling unit would occupy 1/13 square mile, with "a distance of about ¼ mile between any two neighbors." Marschak, Klein, and Teller's vision of dispersal was a counterpart to James T. Lowe's musings about a targeting strategy in a future atomic war (see chapter 3). Lowe had described an approach to target selection in which a country would be divided into equal "little squares" that could be evaluated and selected for "obliteration." In the "ideal" dispersal advocated by Marschak, Klein, and Teller, every "little square" would have the same value, thus robbing a future adversary of any economic principle of target selection. Acknowledging that such "complete dispersal" was impracticable, the authors proposed the resettlement of the national population into clusters of "a thousand to ten thousand houses each," separated by a distance that would "just exceed the diameter of the destruction area of the bomb." In support of this proposal for dispersed satellite cities, they contended that "the difference in the degree of social and cultural upheaval implied in the cluster scheme as compared to complete dispersal is obvious." Though it would be necessary to "abandon many of the habits acquired in the course of a century" of dense urban living, the costs implied by such radical steps were "perhaps not a serious price to pay for safety."[44]

In the late 1940s, the dispersion of the population into small clusters or satellite cities along the lines proposed by Teller, Marschak, and Klein continued to be discussed among national security experts. But another line of thinking about vulnerability reduction—which focused on vital systems rather than on urban form—would prove to be more significant for the eventual direction of American emergency government. In 1946, the Social Science Research Council (SSRC) formed the Committee on the Social Aspects of Atomic Energy, chaired by statistician Winfield Riefler.[45] Riefler had worked on both mobilization planning and air targeting during World War II and led the Economic Warfare Division in London which housed the Enemy Objectives Unit. Riefler and other social scientists on the SSRC Committee offered an alternative to a radical program of urban dispersal. Drawing on the knowledge practices of mobilization planning and air targeting, they proposed the selective dispersal of vital functions as well as a range of other vulnerability reduction measures.

Riefler described the rationale for this approach, pointing to the daunting challenges involved in preparing for a nuclear attack. "The scientists who developed the bomb have been as one," he wrote, "in their warning that there is no defense against its destructive power except in widespread decentralization or deep underground shelter." Taking such drastic measures would "require . . . a change in our way of life so complete as to exceed the capacity of the imagination to envisage them." But the problem of atomic vulnerability, Riefler argued, was "much too complicated to be judged in light of such sweeping generalizations." Efforts to reduce vulnerability need not lead either to "widespread dispersal or to 'cave man' living"; there were "many alternatives, involving different costs, and differing degrees of disruption." The key question, according to Riefler, was how to ensure the continuous functioning of vital industrial and governmental systems. Here, Riefler referred to the practices of resource evaluation that had been developed in wartime mobilization planning and air intelligence. In the "vast and impersonal system of accounting imposed by the requirements of total war," he explained, "no resource, human, material, psychological or political, escaped evaluation." Mobilization planning decisions were made by evaluating the extent to which a particular resource "contributed to war potential." Meanwhile, in air targeting each potential target was evaluated to determine whether the "damage to be inflicted justified the expenditure of effort involved in attack." The same tools of economic assessment, Riefler suggested, could be used to reduce domestic vulnerability. Given the United States' "pattern of urban centralized production, and centralized administration of governmental functions, both civil and military, such targets may exist, so few in number that their surprise elimination would yield the prospect of a *decisive* military advantage."[46] The task was to identify these targets and to reduce their vulnerability to attack.

For Riefler and other economists on the SSRC Committee, this selective approach to dispersal provided a model for a broader program of nonmilitary defense, which would be based on a detailed economic analysis of sites of industrial vulnerability. In a February 1946 meeting, Riefler suggested that a study of "economic vulnerability to atomic warfare" was "probably a more suitable approach from which to study this range of problems than that of dispersal of populations."[47] Instead of considering models for "optimum dispersal," he explained, it was possible to "treat the points of critical

vulnerability" through a strategy that included industrial dispersal alongside a number of other measures, such as "duplicating industrial facilities,"
addressing "transportation bottlenecks," and "building up stockpiles at the
point of consumption."[48] A proposal along these lines was presented by
another participant in the SSRC meeting, economist Lewis Dembitz, who
had worked on air targeting under Riefler in the Economic Warfare Division in London during World War II. Dembitz described an "economic"
approach to vulnerability reduction, employing concepts such as depth and
cushion that drew on wartime target analysis (see chapter 1). Pointing to
the example of steel—one of the "controlled materials" in World War II
mobilization—Dembitz proposed that an economic analysis of vulnerability
reduction would consider the range of possible measures to "assure a certain
continuing supply of steel products during any prospective war in order to
assure surviving the war." It would be possible, Dembitz posited, to build
small dispersed plants, to construct physical protections for large plants,
to accumulate stockpiles of semifinished steel near every steel-consuming
industry, or to make plans for substitutions that would bolster the "cushion" in the system of production. Assuming that all these measures were
"physically feasible," it was an "economic problem to determine which is
preferable, considering the extent to which each would call upon abundant
or expansible resources, and thus involve a minimum in real cost to our
economy."[49] Domestic vulnerability reduction—like strategic air targeting—
could be economized.

"THE PROBLEM OF REDUCING VULNERABILITY
TO ATOMIC BOMBS"

Based on its initial discussions, the SSRC group commissioned a study of
American economic vulnerability to atomic attack and possible measures to
address it. The Committee assigned this work to Ansley Coale, a graduate
student in economics at Princeton University who would go on to become
a leading American demographer. Coale presented his results the following
year in an article in the *American Economic Review*, as well as in a book, *The
Problem of Reducing Vulnerability to Atomic Bombs*. His analysis extended
the tools of mobilization planning and air intelligence to the novel problems
of mitigating industrial vulnerability and ensuring the continuous functioning of vital government activity and essential services. Beyond that, Coale
presented a comprehensive outline of a governmental approach to nonmilitary defense that addressed both vulnerability reduction measures and the

political problems that such measures might raise. While Coale's study was not conducted for official government purposes, it served as a key point of reference in later plans for nonmilitary defense.[50]

Coale began by spelling out now-familiar strategic assumptions. A war that broke out in the next several years, he posited, would be a lengthy conflict like World War II, in which the United States would "plan to translate its superiority in industrial capacity and technology into a superiority in effective military strength."[51] The strategic aim in mobilizing for such a war would be to design a "war structure strong enough to win." Coale described this war structure as an "organic complex within the nation," composed of "combat personnel and their accumulated equipment; the industrial structure to produce the further materials needed for the armed forces; the labor force, materials, services, and productive equipment to support the war-producing economy; and the administrative nucleus to keep the organism in proper function." The task of vulnerability reduction was to ensure that this "organic complex" would be "impervious to the destructive attacks upon it" and "able to perform in the conditions that would exist" after an attack on the United States.[52]

Like Riefler, Coale considered and rejected calls for radical urban dispersal. Since a "primary source of vulnerability is the existence of large industrial clusters and of high population densities," he reasoned, it might be possible to "draw up a design for a new configuration of our industrial economy" that was "devoid of large concentrations of plant or people." The problems with such a proposal began with its economic costs: an "optimal" pattern of dispersal would necessarily abandon "economies of scale or other special advantages of concentration."[53] But the "real difficulties of such a program," Coale argued, would be political. The compulsory dispersal of concentrated urban centers would "require a revolutionary interference by the government in individual choices, a revolution for which popular support can hardly be envisaged." Coale doubted whether the US government "has, or even should have, the needed power of enforcement."[54]

As an alternative to such radical urban restructuring, Coale considered a program of selective vulnerability reduction, exploring how it would be instituted through concrete knowledge practices and governmental mechanisms. Selective measures focused on the most vital and vulnerable nodes of the military-industrial production system, Coale argued, would require "knowledge of the structure of the future war economy," understood as "an integrated series of industrial processes from the refinement of raw materials through the manufacture of minor parts, the sub-assembly of major components, and the final fabrication and assembly of the completed product" that together constituted a "self-sufficient producing mechanism." This knowledge, he explained, could be built out of the practices of data collection and evaluation that had been developed as part of wartime mobilization planning. The records of the War Production Board, for example, contained "detailed information on the capacity and realized production rates of plants," thus providing "a starting point for a continuing inventory of the active and reserve capacity for production of most war materials." Existing information on "the flow of product from the earliest stages of production to final war equipment"—the product of wartime industrial surveys—would enable planners to estimate "how much of each type of production is needed for a given bundle of finished munitions."[55] Such studies would examine a large number of factors that, in their complex interrelationship, determined the effects of bombing a particular plant on overall production: What material inputs does a particular industrial process rely on? What are its products used for? What possibilities for substitution are available in the event of damage? How long would it take to build or convert plants to replace damaged ones? How long would it take to repair damaged plants? Was a given industrial process reliant on large amounts of electricity, natural gas, or other essential services that might themselves be disrupted?

Coale suggested that such a systematic analysis of industrial production could be coupled with a new kind of anticipatory knowledge: plans for domestic vulnerability reduction would be tested by simulating likely patterns of enemy attack. Such simulations would resemble a wartime feasibility test in that they would examine whether, in a future emergency, resources would be sufficient to meet requirements. But in place of a surge of production that might result in bottlenecks or scarcities of critical materials—as in World War II and the Korean War—the future contingency to plan for would be an atomic attack on the United States. "Each time a set of plans is completed," Coale proposed, "it could be tested crudely by assuming that the plans have been put into effect, and that five atomic bombs have landed on each of fifteen largest cities." Through such tests, critical vulnerabilities would be found where a "substantial proportion [of national production] is conducted in a small number of plants, or in plants found in a few clusters each of small radius compared to the destructive radius of atomic bombs."[56]

How would this knowledge be translated into vulnerability reduction measures? Coale suggested that existing tools of mobilization policy could be used to reduce the vulnerability of the military-industrial production apparatus. In cases where the government was providing financing to new facilities, whether through defense loans or the rapid amortization program, this financing could be used to encourage "design of that expansion for minimum vulnerability." The costs of such locational decisions—namely, the "diseconomies imposed by the requirements of safe location"—would be assumed by the federal government. Other aspects of industrial mobilization plans could also be "carefully drawn to include protection as an essential." For example, in the area of transportation, measures to eliminate bottlenecks and create redundant lines "would be an integral part of the design of a war economy capable of withstanding attack." But if transportation should be interrupted by attack, the "war economy must be prepared to continue." As a remedy, Coale proposed stockpiling critical inputs near sites of production to reduce "the probability that interruption of transportation would be disastrous," and to allow the "economy to continue uninterrupted while damaged sources of supply are being restored."[57]

Coale extended this method of vulnerability analysis from industrial production to government administration and to basic infrastructures for sustaining the civilian population, which he also understood as complexes of vulnerable, vital systems. The vulnerability of the civilian population, Coale argued, could be assessed in relation to its importance as "manpower"—that is, as an input to industrial production. Do the workers employed in a given process reside in an area that might be targeted for attack? Are "alternative supplies of labor" available in the event of attacks that produced very high casualty rates? Such an assessment would also evaluate the population's vulnerability to "devastation as such," not because civilian survival was vital to the outcome of a war, but "simply because life and the necessary supports of living are in themselves worth saving." The atomic bomb, he observed, "has a greater advantage over chemical explosives and incendiaries in destroying people than in destroying property," and it would be "small comfort to survivors . . . to know that careful planning had enabled large-scale production of atomic bombs and airplanes to continue right up until victory was assured, if meanwhile the ability of the nation to produce the necessities of civilian life had disappeared." The method of analyzing interdependency and vulnerability that economists had used to evaluate industrial systems could also be used to carry out a "complex study of the structure of civil life, particularly urban life." Such "complex study" might take stock of "the goods and functions which are essential"—such as food and water, medical supplies and equipment, and utilities such as power and transportation—as well as "the least quantity of these items that will allow a tolerable level of living." Such an analysis would also have to consider the vulnerability of "establishments which would be called upon to provide relief for bombed cities: hospitals, fire-fighting organizations, and the like."[58]

Turning to the vulnerability of government activities, Coale warned that nonmilitary defense must also "guard against disruption of organization itself." In part, this meant that plans for emergency government must be put in place in advance. Officials would have to ensure that "preparations and plans to ensure that blueprints for expansion and conversion were ready, that data and statistics were complete and up to date, and that competent personnel were available for the needed administrative functions." But such "careful provisions might prove to be fruitless" in the event of an atomic attack on Washington, DC. "If a hostile nation possessed even two or three atomic bombs," Coale wrote, "accurate delivery of these bombs on the capital would have at least the effect of disrupting unified war activity until the nation could improvise emergency substitutes." The government's capacity for "unified action" could be "destroyed by crippling the central direction of war activity or the system of internal communication, or by weakening morale to the point that unity of action became impossible."[59] Coale outlined a series of anticipatory measures to ensure such "unity of action." Executive departments could be transferred to dispersed and secure locations. Emergency operations facilities with secure communications systems could be established. Files and records could be prepositioned in emergency locations. Personnel and facilities could be protected. Lines of succession could be established for key decision makers. Government, too, could be understood as a vital but vulnerable system, whose continuous operation must be assured.

The Continuous Operation of Vital Systems

The studies of nonmilitary defense by the Office of the Provost Marshal General (OPMG) and by Ansley Coale were not tied to any specific government policies or regulations. Indeed, at the time of their writing, no organization in the federal government had responsibility for nonmilitary defense. During World War II, some nonmilitary defense functions had been carried out by civilian agencies in the executive branch—primarily the Office of Civilian Defense—and by offices in the military, such as the OPMG Internal Security Division. But by late 1945, these wartime agencies had been disbanded. A central question raised by the postwar studies, then, was where nonmilitary defense functions might be housed.

The OPMG demobilization study proposed that nonmilitary defense was "properly chargeable to the military establishment." But military leaders were reluctant to take on this substantial new responsibility and wary of the likely political reaction to an expanded military role in domestic security. In early 1947, a military study of civil defense directed by General Harold Bull offered a different proposal: a high-level civilian planning agency should lead nonmilitary defense efforts.[60] The Bull report speculated that a National Security Resources Board—then under consideration in broader discussions of the postwar national security establishment (see chapter 2)—could be charged with "making the necessary high-level policy" for nonmilitary defense alongside its role of mobilization preparedness planning.[61]

The clauses in the 1947 National Security Act (NSA) that created the National Security Resources Board and defined its functions took up the Bull report's proposal—at least in part. Though the Act did not address some elements of nonmilitary defense, such as the protection of critical facilities and civil defense, it did charge the Board with advising the president on the "strategic location of industries, services, Government and economic activities, the continuous operation of which is essential to the Nation's security." The language in the NSA largely repeated President Roosevelt's charge to the Provost Marshal General in 1942—to undertake internal security measures that would ensure the "continuity of production and delivery of materials required by the armed forces."[62] But where wartime internal security had focused on ensuring the continuity of production by bolstering the physical security of facilities, the NSA referred to strategic relocation or the dispersal of vital facilities as the key mechanism to reduce industrial vulnerability. And where wartime internal security focused primarily on industrial facilities, NSRB's charge was expanded to include "industries, services, Government and economic activities" whose continuous operation was "essential to the Nation's security." NSRB's responsibility, in short, was to develop plans to ensure the continuous operation of the nation's vital systems.

The Board's early nonmilitary defense projects—on the dispersal of industry and government—hewed closely to its statutory mandate to plan for strategic relocation.[63] In these initial projects, NSRB staff formulated a distinctive approach that they would apply in other nonmilitary defense programs. They identified critical nodes in vital systems and, on this basis, developed measures to reduce vulnerability and prepare for a future attack.

In its work on industrial dispersal, NSRB focused on the publication and circulation of a pamphlet, *National Security Factors in Industrial Location* (1948). Because the Board had no means to compel or incentivize industrial dispersal at this time, the pamphlet provided factory owners with tools to assess the vulnerability of their existing facilities and to identify secure locations for new facilities. Drawing on the strategic assumptions of wartime resource evaluation, the pamphlet described what areas would likely be targeted in a future atomic war. "The strategic planner of a potential enemy in appraising the area in which your facilities are located," it explained, would evaluate "the actual use our military forces will make of the items produced and the amounts of such items as compared with the actual requirements, and the possibilities for employing substitutes." Thus, in a future war, an enemy might target "an individual plant, producing a large percentage of a highly critical item, which if destroyed, would have immediate far reaching adverse effects on the production of that and other related items." A future adversary might also target essential services, including "public utilities such as power plants and water systems which serve a general industrial area." In either case, the "best target for atomic or other modern weapons" was an area with a "high concentration of industry or population," in which several vital facilities were located.[64]

Following the arguments made by the atomic scientists, the pamphlet maintained that the best defense against atomic bombing was dispersion outside urban centers. The attacks on Hiroshima and Nagasaki had destroyed "almost everything" within a half-mile radius and caused moderate damage up to 1.5 miles from the "zero point" of a detonation. With the future development of atomic weapons technology, the area of "heavy damage" might extend up to three miles from the bomb's detonation point. Based on these estimates of damage radii, the pamphlet explained, factory owners could conduct an aiming point analysis to assess whether their facilities were in a secure area. On an area map, the owner was to draw a circle with a three-mile radius around an existing industrial facility, and then identify "everything within this circle" that might be "of interest to the strategic planner of a potential enemy" (see figure 4.3). A concentration of facilities within this radius would indicate that the plant was vulnerable to a future attack. Similar considerations could guide the selection of new sites. Since it would not be "economically feasible" for an enemy to attack industrial concentrations of less than five square miles or urban concentrations of fewer than fifty thousand people, such sites would be suitable for the construction of new facilities.

NSRB used similar techniques in a project on ensuring the postattack continuity of the activities of the federal government. In this early project, the Board understood the continuity of government as a matter of the security of "the Nation's capital in time of war." Could vital functions be carried out in Washington, DC, following an atomic attack? A 1948 report

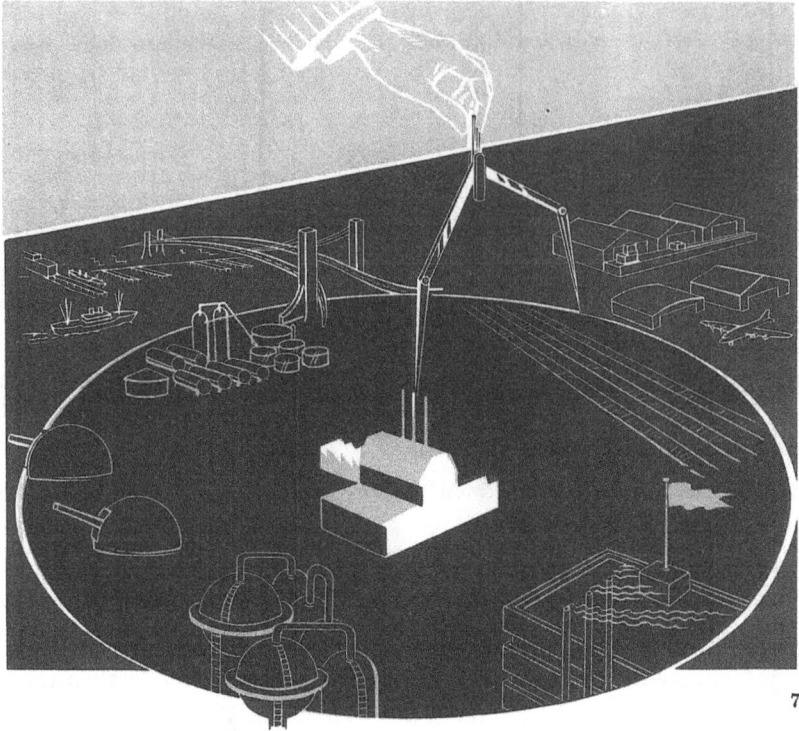

FIGURE 4.3. Industrial vulnerability assessment. The National Security Resources Board sought to reduce American vulnerability through the selective dispersal of vital industrial facilities. The board instructed the owners of facilities to study whether current or prospective facility sites were within the damage radii of detonations centered at likely targets of enemy attack. *Source*: National Security Resources Board, *National Security Factors in Industrial Location* (Washington, DC: Government Printing Office, 1948), 7.

explained that the Board's approach to this problem followed the procedure that "the Government has recommended that industry should follow in planning the construction of new plants": identifying critical functions, assessing the vulnerability of these functions, and relocating essential activities to secure locations.[65] The Board's analysis of the vulnerability of the nation's capital drew on a 1948 aiming point study conducted by the Atomic Energy Commission (AEC), which identified three likely targets in a future war: the Pentagon, the Capitol, and an area around the White House in which key government departments, including the Departments of State, Treasury, and Navy, were located.[66] The report then drew on techniques of bomb damage assessment to model the probable effects of an atomic detonation on buildings in which these government organizations were housed. Focusing on the AEC's own building as a case study, it anticipated that in the event of an atomic attack, "windows, partitions and probably walls would become missiles." Eighty percent of the workers in the building at the time of the detonation would be killed immediately by the shock wave, while an additional unknown number would die from radiation exposure. The detonation would knock out basic services. The "usually dependable telephone, telegraph, electric service, and transportation would cease to exist." Fire services would be incapacitated, as "water pressure would fall to nothing," and equipment would be "crushed by the collapse of the firehouses." As a result, the many fires started by the initial blast "would speedily merge," and firestorms would engulf the area.[67] Given this assessment of likely bomb effects, the Board concluded that vital government activities would be crippled by an atomic attack on the capital, and that it was therefore necessary to develop a long-term plan for their permanent relocation outside the vulnerable center of Washington, DC. NSRB assigned this task to the Federal Works Agency, which in turn commissioned urban planner Tracy Augur to formulate a master plan for Washington, DC, in which key federal facilities would be relocated to new satellite settlements.[68]

These early NSRB projects on industrial location and government continuity addressed long-term planning and avoided contentious political issues. The Board's studies of government continuity, thus, were conducted out of public view and were shared with a small circle of planners and policymakers. They did not touch on sensitive issues—such as the relationship between federal and state or local governments, or between civilian and military authorities in a future national emergency—that would become flashpoints in government continuity planning during the 1950s.[69] Meanwhile, the Board's discussions of industrial dispersal emphasized the noncompulsory nature of its recommendations, in an attempt to avoid controversy. Thus, in his foreword to *National Security Factors in Industrial Location*, NSRB chair Arthur Hill meekly assured factory owners that the NSRB provided no "panaceas" but rather offered "some suggestions" for their "perusal." The pamphlet itself explained that, since the United States was a "democratic Nation dedicated to the principles of free enterprise," the government could "neither dictate nor finance" any "large-scale change in the industrial pattern."[70]

But the Board was soon drawn into work that addressed more immediate and contentious problems of national security policy in the early Cold War. Events such as the Berlin Blockade (which began in June 1948), the first Soviet atomic test (August 1949), and the outbreak of the Korean War

(July 1950) convinced federal officials that, as NSRB vice chairman Edward Dickinson put it, "the glorious haze of the immediate postwar period was over," and the United States "faced a real threat from the Soviet dominated areas." The Board was thus increasingly consumed with "short-term, day-to-day work" to address the "critical world situation."[71] On the one hand, particularly as American involvement in the Korean War expanded, NSRB had to take operational responsibility for many aspects of the ongoing mobilization program. On the other hand, the Board was pulled into increasingly heated debates around nonmilitary defense, either because nonmilitary defense measures that were part of its initial charge moved to the center of political attention—as with industrial dispersal—or because it was handed tasks that were not part of its original statutory mandate, as in the case of civilian protection and relief. In these areas, NSRB faced complex cross-pressures. Local governments, the public, and some members of Congress demanded immediate federal action to reduce US vulnerability. But specific proposals for nonmilitary defense provoked accusations that the federal government was impinging on the prerogatives of local governments and private enterprise and taking steps toward the establishment of a "garrison state." Thus, particularly in the early 1950s, NSRB's projects on nonmilitary defense became one site of contest over the fraught relationship between the demands of emergency government and what were understood to be the American traditions of limited government, decentralized sovereignty, and free enterprise. Correspondingly, NSRB had to address not only the technical problem of reducing vulnerability and preparing for emergency response, but also the political challenge of adjusting US governmental institutions to the new problem of preparedness.[72]

Economizing Vulnerability Reduction

When NSRB circulated its 1948 pamphlet on industrial dispersal, *National Security Factors in Industrial Location*, the Board had no means to compel or incentivize strategic relocation, since wartime statutory authority to control or finance production had lapsed. The Board did little work on industrial dispersal over the next two years.[73] But with the ramp-up of military production for the Korean War in the second half of 1950, and the appropriation of vast sums of money for rearmament, the federal government once again had powerful tools at its disposal to restructure industrial production. The Truman administration instructed the agencies involved in mobilization to use the war production program to encourage a dispersed pattern of industrial development. In a September 1950 memorandum, the administration emphasized that rapid tax amortization, loans for military producers, and priorities power "should be of particular significance in inducing firms to select locations for new plants with primary regard for national security rather than economic considerations."[74]

NSRB did not take immediate action on this instruction, likely due to the overwhelming pressures it faced in mobilizing industrial production for the Korean War. Thus, in December 1950, NSRB reported to a congressional committee that it planned to consider "dispersion and strategic vulnerability as among other factors" in granting access to rapid amortization and defense loans, suggesting that it had not made much progress in the prior months.[75] But following the transfer of operational management of the mobilization program to the Office of Defense Mobilization (ODM) in late 1950, NSRB staff worked more intensively on industrial dispersal. The Board's approach to dispersal drew on existing elements from the domains of air targeting and mobilization planning. These elements included facilities ratings, damage assessment, and the regulatory devices of mobilization planning. In assembling these elements and linking them to channels of financing in a comprehensive dispersal program, NSRB created the first mechanism in the US government for identifying and mitigating the vulnerability of vital systems to a future catastrophe.

INDUSTRIAL DISPERSAL AND THE POLITICS OF NONMILITARY DEFENSE

NSRB's work on dispersal in the first half of 1951 was spurred, in part, by a contentious congressional debate that pitted advocates of an aggressive, government-led dispersal program against critics who claimed that such a program would violate American traditions of free enterprise and limited government. NSRB responded by formulating a proposal for selective dispersal, based on expert assessments that identified both likely enemy targets and the most vital facilities whose continuous operations would have to be assured in the event of a future attack. The Board understood selective dispersal as both a technical intervention to reduce vulnerability and a techno-political solution to the tensions between emergency preparedness and American economic and political institutions.

The congressional controversy around industrial dispersal began with an investigation by the House Subcommittee on Government Operations. This Subcommittee was responsible for overseeing the issuance of certificates of necessity, which gave private enterprises access to various federal loans and tax benefits that supported war production. The Subcommittee found that, notwithstanding Truman's instructions, the Office of Defense Mobilization and the Defense Production Administration—which had taken over the mobilization control program in late 1950—were not taking security factors into account in their decisions on certificates of necessity. "No effort is made by the Government," the Subcommittee charged, "to control the location of new facilities." The selection of the site was left "entirely to the applicant." Plants that had received support through the mobilization program were concentrated in the densest metropolitan regions of the country. The existing policy thus resulted "in further concentration of industry in areas favored by economic factors" and neglected "considerations of military security or the avoidance of knock-out blows in the event of a sudden enemy attack."[76] In an attempt to force the administration to take action on industrial dispersal, a congressional joint committee proposed an amendment to the Defense Production Act, requiring that new plants be located in areas that were "relatively less vulnerable to enemy attack by reason of geographic location, or the absence of heavy concentrations of population or vital defense industry."[77] The amendment's sponsor, Democrat Albert Rains of Alabama, argued that "if you believe there is danger of an atomic attack it is utterly foolish to take the taxpayers' money—and have the Government build defense plants in dangerous congested areas."[78]

Rains's proposed amendment to the Defense Production Act sparked an angry reaction from legislators who opposed industrial dispersal. Representatives of industrially developed areas of the country, concerned that economic activity would be relocated away from their districts, charged that the policy threatened the American free enterprise system. "This country has become great," argued Herman P. Eberharter, a Democrat from Pennsylvania, "because the financiers and industrialists have taken advantage of the natural resources of the country and placed industrial plants where they should belong and where it is most advantageous to have them, and where they have produced the greatest good, both financially and militarily." Republican James T. Patterson of Connecticut charged that a government-directed dispersal program would "nullify the laws of nature and of economics." Critics further argued that the proposed industrial dispersion policy would grant the federal government unaccountable powers to control the economy. California Republican Gordon Leo McDonough warned of a "grave danger that once this authority is put on the statute books of this Nation, Washington bureaucrats could . . . further plan the economy of the country by moving industries now located in other areas to a location of their choice for some specific reason." The amendment was an "opening wedge" for "authority to nationalize our industry and socialize the Government."[79] Another California Republican, Donald L. Jackson, proclaimed that precedents for such "arbitrary action are not difficult to find, nor historically remote. Every dictatorship in history from the Nile Valley to the planned socialism of England have depended upon an absolute power in the hands of the rulers to direct employment [and] the location and production of industry."[80] These critics defeated the Rains amendment, leaving unresolved the question of whether and how a dispersal program would be implemented.

Amid this political fray, NSRB issued a "Proposed Statement of Industrial Dispersion Policy" that it circulated to federal agencies for comment. In describing the policy, the statement sought to assuage critics, emphasizing how it differed from the Rains amendment. Existing facilities, it explained, would not be forced to relocate except in rare cases involving the most vital and vulnerable production and service facilities. Meanwhile, new facilities would be dispersed within metropolitan areas rather than between regions or states, so that no part of the country would be "denuded" at the expense of another. Dispersal would thus be primarily a matter of local planning.[81] The federal role would be limited to providing incentives and formulating guidelines for determining the "relative strategic vulnerability of the major industrial areas of the United States" and evaluating "existing production programs with respect to vulnerability."[82]

Advocates of nonmilitary defense outside government backed NSRB's proposal for selective dispersal. Likely at the Board's prodding, Ralph Lapp wrote an article in the *Bulletin of the Atomic Scientists* that underscored the urgency of a dispersal program.[83] "A logical person might assume," Lapp wrote, that in the context of a massive military-industrial mobilization, the federal government would act to "insure industrial invulnerability by locating new plants in non-target areas." But in "the present defense effort," industrial dispersion had only been "talked about and not put into effect." A "high order of insanity," Lapp lamented, "must invest a society that assiduously builds bigger bombs and blithely goes about making the best targets for such weapons."[84] A year earlier Lapp had written that "realistic defense measures" would "spell the end of the metropolis as we know it."[85] Now he argued for selective dispersal based on "a cold-blooded analysis of what is critical to our national defense."[86]

Lapp invited his readers to imagine the decisions government officials would have to make in the aftermath of a devastating atomic attack. In this extreme situation, political factors would be "out the window and sheer survival is the problem." Department stores, laundries, and restaurants would be "written off as non-essential." Priority would be given to "those defense plants which can produce the war goods most urgently needed by the armed forces." What was required in advance of attack, thus, was to identify the most vital facilities whose continued operation would have to be ensured in the event of an emergency. "A critical survey of this type," Lapp urged, "should be conducted now and those installations selected as most critical to the national defense should be given top priority in relocation." Such a selective program of dispersion, he explained, would involve "the potential sore spots or bottlenecks of industry and while percentagewise only a small part of defense industry may thus be made invulnerable through relocation, what is thus dispersed will be of crucial importance once an attack takes place."[87] Echoing the arguments of Progressive reformers, Lapp suggested that emergency preparedness did not require arbitrary or exceptional powers that would undermine American governmental and economic traditions. Rather, government could be equipped with the expertise required to assess critical vulnerabilities, so that limited interventions could be precisely targeted.

RESOURCE EVALUATION AND VULNERABILITY REDUCTION

The Truman administration adopted NSRB's proposal for selective dispersal as official policy in August 1951. In his message promulgating the new policy, Truman warned that "dense agglomerations of industrial plants were inviting targets for the enemy," and that plants "separated in space would better survive an atomic attack." Referring to NSRB's 1948 pamphlet on dispersal, he argued that the urgency of industrial vulnerability "had only increased given developments" in the few years since its release. With the growing recognition that an enemy air attack could "penetrate any defense," and the outbreak of war in Korea, which underscored the "semi-peace conditions under which we are living," what previously seemed to be merely "desirable objectives" had become "a subject of major concern, and one vital to our national security." Gesturing to ongoing criticism of government-directed industrial dispersal (Rep. Jackson charged that the promulgation of the policy so soon after the defeat of the Rains amendment suggested that the administration held Congress in "utter contempt"[88]), Truman emphasized that the administration's approach was a "common sense" measure to serve "the national security in the atomic age" and was "consistent with the

American system of competitive free enterprise."[89] A new NSRB pamphlet issued along with Truman's statement, entitled *Is Your Plant a Target?*, reiterated the urgency and strategic necessity of industrial dispersal while emphasizing the limited and selective intervention it implied. The administration's program, the pamphlet explained, would focus on new construction rather than on moving existing industry and would not build up one region of the country "at the expense of another." Arguing that "American industry has become great through local and individual initiative," the pamphlet again emphasized that the policy was "in accord with the initiative of private enterprise" and would largely be organized by local government.[90]

In the administration's dispersal program, two key functions remained with the federal government. First, federal officials would identify the industrial facilities that were most vital to military production. Second, they would develop standards for identifying target zones and for defining safe "dispersal areas" to which vital plants located in such zones could be moved.[91] These functions implied a series of practical challenges, which Ramsay Potts of NSRB outlined in a memorandum to staff working on the dispersal program. It was accepted, Potts wrote, that the tools of mobilization planning, such as "defense contracts, defense loans, certificates of necessity," should be used to encourage private businesses to locate in dispersed locations. But it was not clear how such a strategy could be instituted. What organizations would "pass on the industrial security factor"? What "standards of reference" would they use? "It is highly important," he emphasized, to work out "in detail exactly how the technique would work, who would administer it, who would make the decisions, etc."[92]

An initial set of dispersal criteria was formulated by Oscar Sutermeister, a regional planner who had served in naval intelligence during World War II and worked as a consultant for NSRB on various nonmilitary defense programs.[93] Sutermeister proposed that the dispersion policy would apply to federal investments greater than $1 million in facilities that represented a significant "concentration" of production (accounting for more than 15 percent of domestic production of a war-essential product). Based on current assumptions about the likely size of Soviet atomic bombs, these facilities would have to be located thirteen miles from "target areas," which Sutermeister defined as highly industrialized or densely populated centers.[94] But it was still necessary to devise technical procedures for identifying the vital facilities that would be required to meet dispersal standards and to define dispersal areas for specific localities.[95]

Identifying critical facilities. NSRB addressed the first of these challenges—identifying which critical facilities would have to be relocated in a program of "selective" dispersal—by drawing on the practices of wartime resource evaluation. As we have seen (chapter 3), during World War II military officers used resource evaluation to plan facility-level security measures, such as the hardening of facilities or the positioning of armed guards, to reduce the threat of sabotage or industrial accidents. In the late 1940s, an Industrial Security Committee within the National Security Council took over planning for such plant-level security measures.[96] A 1948 report described the rationale for this Committee's work in terms of system vulnerability. "There exists," it warned, "a huge and vulnerable industrial target for saboteurs." The interdependence of the "American industrial machine" meant that an entire industrial sector could be "halted by the action of a single enemy agent."[97]

To promote policies that could address this vulnerability, the Industrial Security Committee pushed to set up planning bodies that could identify the most vulnerable facilities and recommend site-specific protective measures. By 1951, the Committee was overseeing a Facilities Protection Board—which was charged with developing policy on site-specific security measures—and an Industry Evaluation Board, which carried out resource evaluation to identify the most vital points in industrial production systems.[98] The charge of the Industry Evaluation Board was directly modeled on the wartime Resources Protection Board: it was to use "economic and technical criteria" to assess "the relative importance to war production and essential civilian wartime economy of key industrial facilities, as well as materials and products upon which a war mobilization, war production, and wartime civilian economy depend."[99] Kurt E. Rosinger, who had worked on wartime resource evaluation, was appointed to organize and run the Industry Evaluation Board (IEB).[100] In testimony to a congressional committee, Rosinger explained that IEB's job was to "make use of screening methods, which were as complete and scientific as possible, breaking military end-products into their constituents and subconstituents, down to the raw materials." IEB analysts would then identify "a relatively few truly critical points within the entire industrial economy for special security supervision."[101]

For mobilization planners working on dispersal, IEB's system of ratings offered ready-to-hand tools and a body of data for assessing "existing production programs"—in other words, the ongoing industrial mobilization—"with respect to vulnerability."[102] Robinson Newcomb, another economist who had worked in wartime mobilization, described the rationale for using IEB's criticality ratings in a September 1951 memorandum.[103] IEB maintained a "small list of types of plants with which it is extremely concerned," Newcomb explained. A plant that was of "relatively little significance from the standpoint of national security" could be "located with relatively less reference to the security factor and a maximum reference to production requirements." By contrast, a plant that was "absolutely vital to the national defense" could be located with "much more reference to security factors." Newcomb emphasized that national security demands could be accommodated with economic efficiency if dispersal policy focused only on the most vital and vulnerable facilities. "The dispersion policy as we envisage it," he wrote, "is one which seeks to secure compromise between conflicting requirements which will result in the erection of plants so as to give a maximum of production and of safety."[104]

Dispersal standards. The second federal responsibility under the Truman dispersal program was the establishment of national standards for dispersal and for the identification of target areas from which critical plants would have to be moved. NSRB contracted technical work on such standards to the Stanford Research Institute (SRI), one of the new national security–oriented think tanks of the early Cold War. In this work SRI developed criteria for defining "critical target areas" and the minimum distance vital plants should be located from them.[105] These criteria served as the basis for a method that local planners could use in identifying target areas and dispersal zones in a given city. First, on a map of the city, local officials were to indicate the number of workers at "defense supporting plants" and the population by census tract. Using transparent circular cutouts whose diameters corresponded to four miles on the planning map, local officials were then to draw lines around areas that encompassed either sixteen thousand industrial workers or a residential population of more than two hundred thousand and connect the centers of these circles (panels 1 and 2 in figure 4.4). The resulting shapes—the boundary of the "densely populated" and "highly industrialized" sections of the city—defined a "target zone" (described in panel 3). In a final step, local officials would draw a curvilinear form whose points were at least ten miles away from any point in the target zone (panel 4). Plants that were rated as critical by the Industry Evaluation Board could receive government support only if they were located outside the area defined by this form.

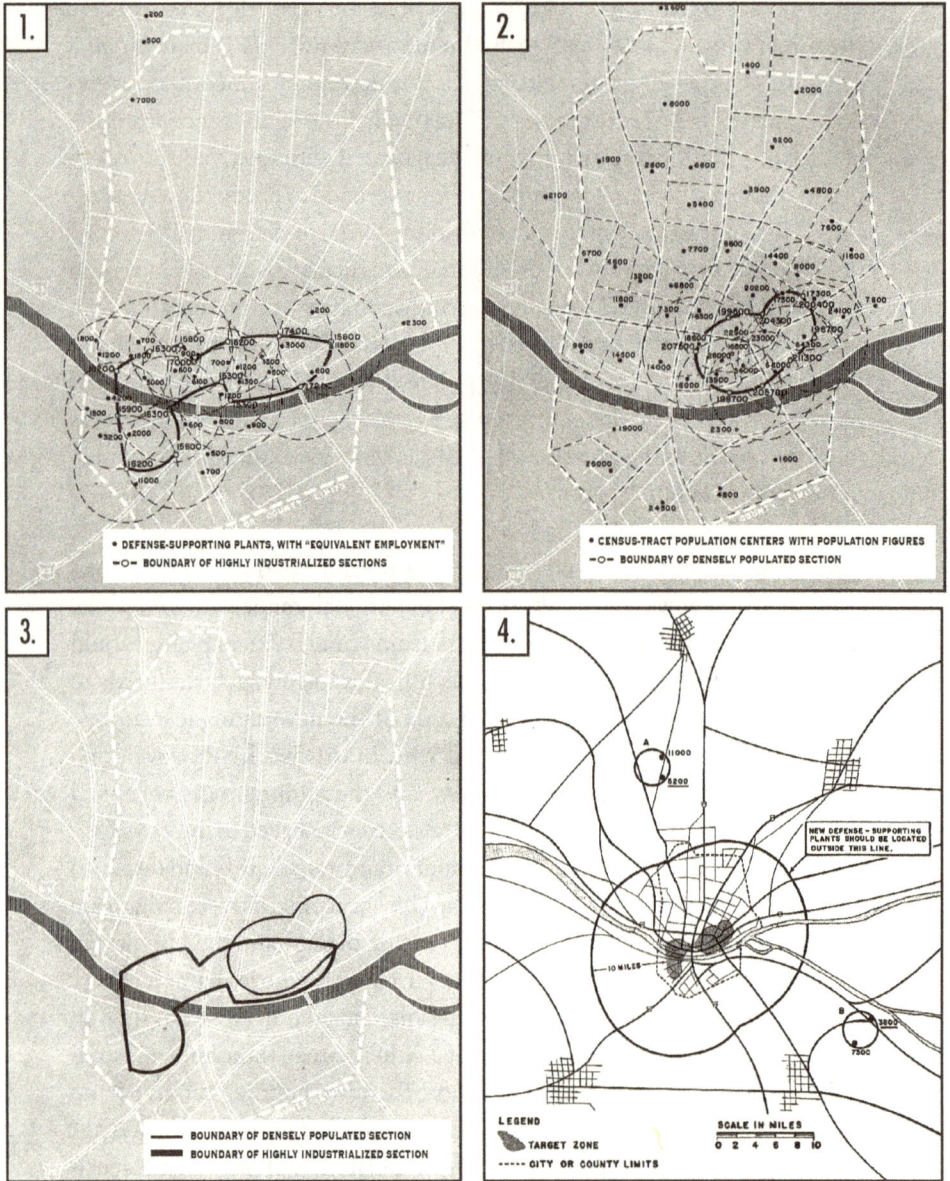

1.
• DEFENSE-SUPPORTING PLANTS, WITH "EQUIVALENT EMPLOYMENT"
—○— BOUNDARY OF HIGHLY INDUSTRIALIZED SECTIONS

2.
• CENSUS-TRACT POPULATION CENTERS WITH POPULATION FIGURES
—○— BOUNDARY OF DENSELY POPULATED SECTION

3.
—— BOUNDARY OF DENSELY POPULATED SECTION
—— BOUNDARY OF HIGHLY INDUSTRIALIZED SECTION

4.
NEW DEFENSE — SUPPORTING
PLANTS SHOULD BE LOCATED
OUTSIDE THIS LINE.

LEGEND
▨ TARGET ZONE
----- CITY OR COUNTY LIMITS

SCALE IN MILES
0 2 4 6 8 10

FIGURE 4.4. Method for determining target areas. The Stanford Research Institute developed a methodology that city planners could use to determine safe "dispersal" areas. Planners were to identify concentrations of population and industry and then define a dispersal line beyond which critical facilities should be located. *Source*: US Department of Commerce, *Industrial Dispersion Guidebook for Communities*, Domestic Commerce Series No. 31 (Washington DC: Government Printing Office, 1956).

Civilian Mobilization: Preparedness for Emergency Government

The prior two sections described the template NSRB developed for assessing and reducing vulnerability in its projects on strategic relocation—the nonmilitary defense policy that was specified in its statutory mandate. The remaining sections of this chapter show how the Board extended this template to a range of other nonmilitary defense measures. The first of these measures related to the protection of the civilian population in a future war. During and after World War II, the provision of civilian protection and relief was generally associated with "civil defense" and with local organization for emergency response. By contrast, NSRB approached civilian protection and relief as a matter of national resource planning for what it called "civilian mobilization." What were the requirements for managing the effects of an attack? What resources were available? How could a balance between supplies and requirements be ensured?

POSTWAR CIVIL DEFENSE

After World War II, planning for the protection and relief of civilians in a future war had a complex and shifting relationship to the overall project of nonmilitary defense. Postwar studies by the military analyzed the protection of civilians and the organization of postattack relief as part of the broad strategic problem of how to mobilize for total war in the atomic age. Thus, the Office of the Provost Marshal General study, *Defense against Enemy Action Directed at Civilians* (1946), argued that the protection and relief of civilians was an element in the broader project of ensuring the continuous operation of the "entire economy of the nation at war." Rejecting the model of the wartime Office of Civilian Defense—which focused on local organization for disaster relief—this study proposed to establish a powerful federal civil defense agency that would serve as a "pre-planned and well-qualified fountainhead of security leadership and direction." During a future war emergency, this office would "direct the implementation and execution" of preparedness plans through "a national unified organization under one command with complete directing and coordinating authority."[106] The subsequent Bull Board report, completed in early 1947, also approached civil defense as part of a broader strategic conception of nonmilitary defense. It thus understood civil defense to include the full range of measures to "minimize the effects of enemy action directed against communities, including industrial plants, facilities, and other installations," and to "maintain or restore those facilities essential to civil life, and to preserve the maximum civilian support of the war effort." The Bull Board report argued for a distributed structure of nonmilitary defense that would involve "all levels of government," while emphasizing a strong federal role.[107]

Following the passage of the National Security Act, which did not address responsibility for civilian protection and relief, military leadership was concerned that it would be saddled with responsibility for civil defense in a future war. In an attempt to spur the Truman administration to take action, in early 1948 Secretary of Defense James Forrestal created an Office of Civil Defense Planning and ordered it to conduct yet another study of civil defense. Forrestal instructed this office to develop "detailed plans" for an "integrated national program of civil defense," and to provide states and localities with "guidance and assistance in civil defense matters."[108] The report that resulted from this study—referred to as the Hopley report after its chair, telephone executive Russell Hopley—abandoned the broad strategic vision of civil defense in an air-atomic age that had been articulated in prior military studies. Instead, it focused on short-term preparation for a war based on the current capabilities of potential adversaries, foremost the Soviet Union, which did not yet possess either atomic weapons or long-range bombers.[109] After encountering resistance to a strong federal role in its meetings with state officials, the Hopley Committee concluded that "the federal government should not attempt in any way to direct local activities in peacetime."[110] It recommended that operational responsibility rest with "states and communities," while the federal government's role would be limited to a "small but capable staff that would furnish leadership and guidance in organizing and training the people for civil defense tasks."[111] The bulk of the Hopley report described a pattern of local civil defense organization—to be set up immediately—that would focus on planning for "wartime disaster relief" and on enlisting civilians in civil defense activities.[112]

The Hopley report garnered public attention to (and support for) civil defense. But its recommendations were resisted by the Truman administration, which felt that the military, by sponsoring planning for an operational civil defense organization, was intruding on the prerogatives of civilian leadership.[113] Moreover, at this point Truman did not consider an air attack on the US "to constitute a 'clear and present danger' demanding immediate precautionary action."[114] He doubted, in any case, that a future war would be anything like World War II—rendering pointless a civil defense organization like the one proposed in the Hopley report. Thus, in March 1949, when

Truman announced that Louis A. Johnson would succeed Forrestal as secretary of defense, he rejected the Hopley report's recommendation to set up
an operational civil defense organization. Noting that civil defense planning
was "related to, and a part of, overall mobilization planning of the Nation in
peacetime," Truman instead directed the NSRB to "assume such leadership
in civil defense planning and to develop a program which will be adequate
for the Nation's needs."[115] In response, NSRB developed an approach that
treated civilian protection and relief as one element in a broader effort to
ensure the continuity of vital industrial and civilian activities in a future war.

CIVILIAN MOBILIZATION AS EMERGENCY
RESOURCE MANAGEMENT

NSRB housed its work on civil defense in an Office of Civilian Mobilization
that was set up in 1949 with a small staff.[116] As the name of this office suggests, the Board approached civilian protection and relief as one aspect of
mobilization planning: conducting surveys of national resources, estimating
requirements in a future emergency, and recommending measures such as
stockpiling and facility expansion to ensure that supplies would be available
to meet these requirements. Paul Larsen, who was appointed head of the
Civilian Mobilization Office in early 1950, thus explained, "Before realistic
mobilization planning can be accomplished it is necessary to make at least
a rough inventory of the Nation's resources available in wartime and then
compare this inventory against the anticipated needs." These anticipated
needs for civilian protection and relief would have to be included in the
broader "balance-sheet operation" of mobilization planning.[117] Indeed,
civilian mobilization was "as important in the overall national picture as is
military defense." Unless such a program was "properly planned and properly operating," there was no way that the military could "properly plan
their . . . activities in the event of war." Industrial production for military
goods would have to continue "irrespective of what disasters occur within
the confines of our country."[118]

Given this orientation to civilian protection and relief as a problem
of emergency resource planning, NSRB criticized the Hopley report for
its failure to consider the substantive problems that civil defense would
address. What was the strategic aim of civil defense? What functions should
it include? And what practical measures were needed to ensure that these
functions could actually be carried out? The Hopley report had offered no
"practical guidance" for the solution of those substantive problems it did

identify, such as the provision of emergency medical care to casualties of an attack. And it excluded key questions that NSRB considered to be within the purview of nonmilitary defense, such as reducing economic vulnerability and ensuring the continuity of essential services. Finally, the report had failed to undertake "a preliminary study of available resources and antici-pated needs." An effort to address these problems, argued NSRB, should have "preceded the planning of an organization."[119]

Based on this assessment, in late 1949 the Board wrote to state gover-nors, informing them that the federal government was not in a position to provide guidance on an operational civil defense program. Instead, it would conduct planning studies to "determine first, what must be done to accomplish the objectives; second, what resources of manpower, materials, and equipment are needed and what resources available; and then, finally, what organization is best suited to place the plans in operation." To guide this planning process, NSRB circulated bulletins that instructed state and local governments to determine "existing resources of major importance to civil defense plans."[120] These local studies would focus on systems that were vital to sustaining civilian populations: water supply; means of communica-tion; streets and highways; hospital and first-aid facilities; firefighting equip-ment; and manpower resources. Within these categories, local government officials were to survey resources in current use and in reserve and estimate postattack requirements. Based on the findings of these surveys, federal and local planners could identify measures—such as stockpiling, resource conservation, and the use of substitutes—to aid in "repairing or restoring facilities disrupted or damaged during enemy attack." In sum, NSRB applied industrial mobilization planning policies to the infrastructures of civilian life and emergency response.[121]

The Civilian Mobilization Office's work on health resources, which was conducted by a Health Resources Division, illustrates NSRB's approach.[122] Following the template of mobilization planning developed in World War II, the Health Resources Division conducted surveys of resources in three areas: manpower, supplies, and facilities. Thus, for example, the survey of health supplies covered instruments, drugs, chemicals, biologicals, antibiotics, glassware, and textiles, examining product inventories as well as "manu-facturing potentials" and "available materials" for producing these items. Health Resources staff drew on this appraisal of existing resources to esti-mate the volume of critical supplies that would be available in a future war emergency. On this basis, they sought to develop a "balanced" program in which available supplies would match requirements. In industrial mobilization, balance sheet planning began with estimates of requirements based on military "tables of organization," which defined the equipment, materiel, and other resources required to provision a military unit (a tank division, for instance). But as the director of NSRB's Health Resources Division, Norvin C. Kiefer, pointed out, existing military tables of organization were of little use in resource planning for an atomic attack on the United States. No nation had "ever suffered a sudden localized disaster of comparable magnitude" to the one that strategic planners assumed the United States would face in a future atomic war. The only two examples of atomic detonations in cities—the bombings of Hiroshima and Nagasaki—offered few useful lessons. "First-aid medical and other health services broke down so badly," Kiefer explained, that these prior atomic blasts did not offer a guide to "the numbers of health personnel which would be necessary for adequate services." Health resource planning was further complicated by "unpredictable factors, such as the degree of success of penetration of our defense by enemy planes, or the accuracy of enemy bombing."[123]

To address this challenge of planning for the unprecedented and uncertain future of an atomic attack, NSRB staff employed two practices of anticipatory knowledge: bomb damage assessment based on aiming point analyses and governmental preparedness exercises. As described in the prior chapter, aiming point analysis was developed in the late 1940s by air intelligence specialists and experts in atomic weapons effects. To instruct local officials in the use of these techniques, NSRB circulated a number of manuals, such as *Civil Defense Urban Analysis* and *Health Services and Special Weapons Defense*.[124] The latter manual, which was prepared by the Health Resources Division, described how aiming point analysis could be used to conduct supply-requirements studies of health resources. The manual explained that local civil defense planners should begin by determining the number of casualties they would have to manage in the aftermath of a future attack. To do so, they would prepare a series of maps showing urban population densities at different times of the day and would place transparent sheets with concentric circles over the maps (see figure 3.5, p. 221). Then, drawing on estimates of the "casualty percentages based on the various degrees of destruction probable for each encircled area" they could gauge requirements for emergency services in different parts of the city. For example, within a half mile of the detonation, planners should base their estimates on a 90 percent fatality rate, with 10 percent injured; between one half and one mile, they should assume a 50 percent fatality rate, with 35 percent injured.[125]

The manual illustrated this technique through a "hypothetical bombing example"—a daytime attack with a single atomic weapon on an "average city." The hypothetical attack resulted in sixty thousand casualties, forty thousand of whom were alive after one day. The manual then used military estimates of health resource requirements per casualty—the equivalent of the military table of organization—to determine total resource requirements.[126] For example, drawing on estimates for blood requirements per casualty, planners could project that forty thousand units of blood would be needed in the first week after the hypothetical attack. An equal amount would be required in each of the following weeks. These estimates of post-attack requirements, explained the manual, pointed to the "serious problems to be solved by civil defense blood-supply officials in planning against multiple attacks on congested cities." Since "large reserves of whole blood could not be stored," only 20 percent of requirements on the first day could be "furnished as whole blood." The remainder would have to be supplied through "plasma, serum albumin, or blood plasma." Following a "mobilization of donors" and shipment of blood to stricken areas—a complex logistical task that would itself require careful planning—the percentage of casualties that could be treated with whole blood would rapidly increase.[127] The manual recommended that local planners carry out similar studies of other resources that would be essential for postattack survival. These resources included medical facilities (eighty maternity beds per one hundred thousand people under disaster conditions, for example), a standard package of medical supplies required per thousand casualties (including medicines, bandages, instruments, etc.), and a range of other facilities necessary to care for the civilian population, such as emergency shelter (forty square feet of free floor area per person), emergency latrines (one per twenty-five persons), potable water (a minimum of one gallon per person per day), and nutrition (2,000–2,200 calories per person per day, but 3,500 per person per day for civil defense workers with long shifts).

The second technique of anticipatory planning that NSRB developed in its work on civilian mobilization was the "test exercise" to assess governmental readiness.[128] Drawing on hypothetical attack patterns and aiming point studies, the Board provided local officials with situation assumptions that specified the number of deaths and other casualties, damage to physical facilities, and remaining resources.[129] On this basis, local officials could identify both postattack actions that would have to be taken and gaps in resources and capabilities that would have to be addressed in future preparedness planning. NSRB piloted this technique in a handful of US cities. One test exercise, conducted in Chicago in September 1950, was based on an attack scenario involving three atomic detonations—two air bursts and a ground burst— at critical points in the city. NSRB presented local officials responsible for particular sectors (such as electricity and water supply, sanitation, health services, etc.) with information about the postattack situation in their area. These officials were instructed to prepare a presentation about the steps that they would take in response.[130]

In the area of power supply, for example, the exercise scenario specified the effects of the attack on the city's electricity generation, transmission, and distribution system. It anticipated total destruction of local substations and above-ground distribution systems located in the areas nearest to the ground detonations (areas A and B of figure 4.5). While no electric power generation facilities were located within these bomb radii, the scenario noted that power generation would be severely curtailed by the north air burst, which "put the Northwest Generating Station out of service indefinitely." In their response to the exercise assumptions, officials responsible for Chicago's electricity supply stated that their mission was to restore electricity to "essential services such as: water supply, hospitals, transportation, communications and other services that are determined to be essential to the welfare of the community." The officials then detailed the resources that were available for response efforts, including specialized personnel in each part of the city (field foremen, specialized workers who could conduct repairs on lines, poles, and facilities, etc.) and specialized equipment (fifty-six compartment-type trucks, six cable-pulling trucks, four water-pumping trucks, etc.). Their response to the scenario also described the steps they would take both immediately prior to the attack (alerting the superintendent of power supply to suspected subversive activity, dispatching personnel to headquarters and operational locations) and in its aftermath (assessing damage, dispatching repair trucks to disconnect live wires, activating mutual aid agreements with neighboring utilities, connecting emergency lines to substations to supply hospitals, transferring step-down transformers to restore power service, etc.).[131] Afterward, the exercise planners recorded what they considered to be the most valuable outcomes of the exercise: the compilation of an inventory of city resources, experience for key personnel in emergency planning, and lessons for government officials in other jurisdictions, many of whom participated in the exercise as observers.[132]

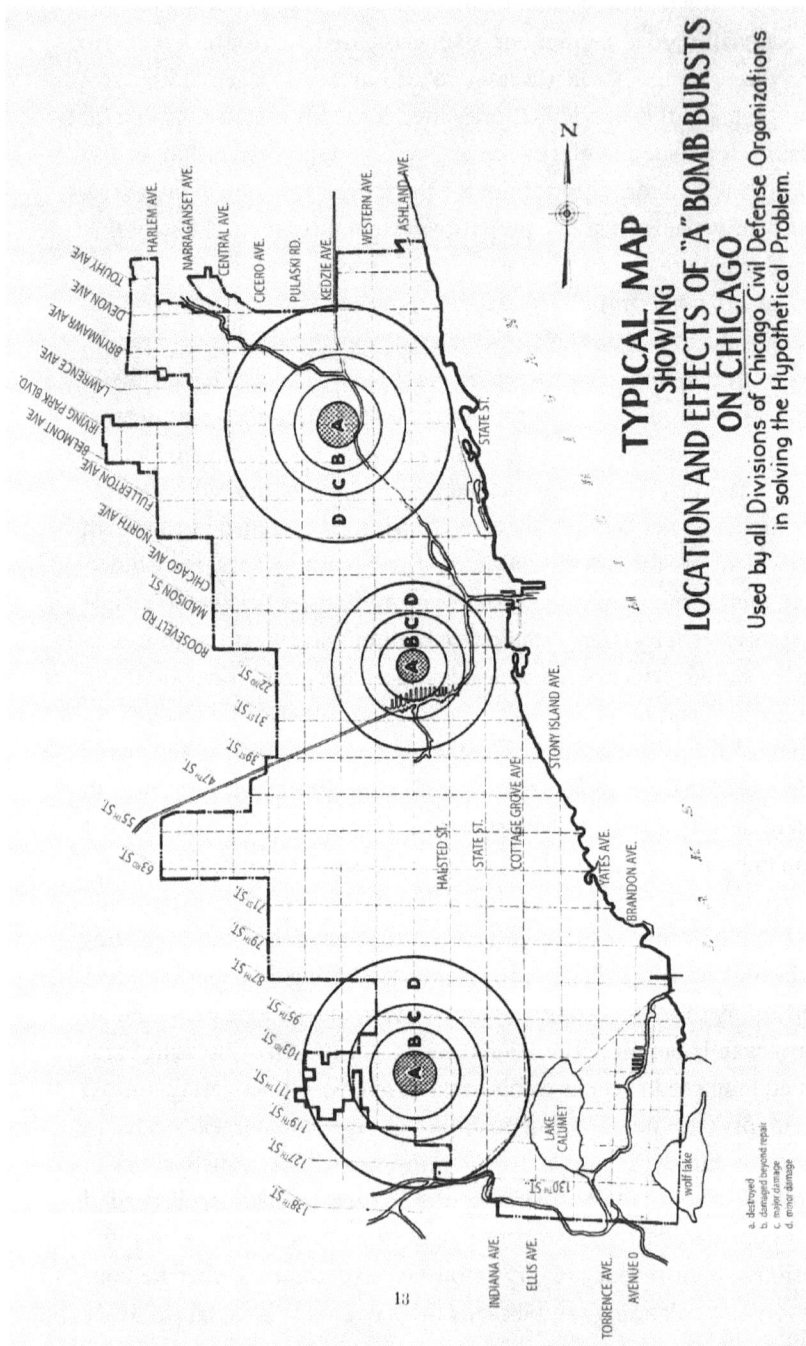

TYPICAL MAP
SHOWING
LOCATION AND EFFECTS OF "A" BOMB BURSTS
ON CHICAGO

Used by all Divisions of Chicago Civil Defense Organizations in solving the Hypothetical Problem.

a. destroyed
b. damaged beyond repair
c. major damage
d. minor damage

FIGURE 4.5. Location and effects of A-bomb bursts on Chicago. The National Security Resources Board ran test exercises in cities as part of its work on civilian mobilization. The scenarios for these exercises envisioned a certain number of atomic detonations at different points in a city. Municipal officials were then asked to outline the steps that they would take to deal with the effects of an enemy attack. *Source:* Chicago Civil Defense Committee, *Chicago Alerts: A City Plans Its Civil Defense against Atomic Attack* (Chicago: Chicago Civil Defense Corps, 1950).

FROM CIVILIAN MOBILIZATION TO THE
CIVIL DEFENSE BLUE BOOK

The schema of preparedness for governing a future emergency situation that NSRB developed in its work on civilian mobilization would guide later work on nonmilitary defense and preparedness planning, both by NSRB and its successor agencies (see chapters 5 and 6). But in the early 1950s, external demands pushed federal planning for civilian protection and relief in a different direction. From the outset of its work on civilian mobilization, NSRB had faced urgent calls from local governments and members of Congress to take action on civilian protection. Advocates of immediate action were alarmed by Truman's decision to reject the Hopley report's recommendations, by his order to NSRB to conduct further study rather than setting up an operational civil defense organization, and by the apparently meager resources NSRB initially devoted to the Office of Civilian Mobilization.[133] Following Truman's announcement of the Soviet atomic bomb test in September 1949, prominent public figures such as the financier and policy advisor Bernard Baruch and the young congressman John F. Kennedy angrily denounced federal inaction. Kennedy famously warned of an "atomic Pearl Harbor" and declared himself "shocked" to learn of the small staff in NSRB's Office of Civilian Mobilization.[134] Partly in response to these pressures, NSRB added staff and named a prominent civilian expert, Paul Larsen, to serve as head of the Civilian Mobilization Office.[135] Under Larsen, the Board tried to satisfy demands for immediate action through planning activities and exercises—such as the above-mentioned "alert" exercises in US cities—while avoiding contentious political questions about which federal agency should take responsibility for civil defense, and how functions should be distributed between federal and local governments.

NSRB ultimately failed to navigate these competing pressures. In congressional hearings, Board officials faced harsh criticism, particularly from municipal leaders who demanded immediate steps to address what they perceived to be an imminent threat. This criticism came to a head with the North Korean invasion of South Korea in June 1950, which raised fears of a general war. Bowing to public demands, Larsen promised that his office would quickly complete a federal plan for civil defense. NSRB staff hastily assembled *United States Civil Defense*, widely known as the civil defense Blue Book. The Blue Book, according to historian Nehemiah Jordan, was designed to "satisfy local demands for planning directives," but not to address what advocates of nonmilitary defense saw as "the strategic needs of an air-nuclear age."[136] In place of the concern with national-scale emergency resource management and governmental preparedness that had guided NSRB's work on civilian mobilization, it focused on the local organization of civil defense. The Federal Civil Defense Act, based on the measures outlined in the Blue Book, passed in late 1950 with little congressional debate.[137]

Advocates of a stronger federal role in nonmilitary defense—including, prominently, some of the atomic scientists who had called for vulnerability reduction measures from the outset of the postwar period—publicly criticized the Blue Book. They charged that it neglected the central strategic problem of national security in an atomic age: ensuring the continued operation of the nation's vital systems in a future war. Thus, in an August–September 1950 issue of the *Bulletin of the Atomic Scientists*, the journal's editor, Eugene Rabinowitch, lamented the "immense gap" between the "well-understood . . . extent of the national catastrophe which an atomic attack on the US may produce . . . and the parochial organization which is being planned to deal with it." The "grim reality," he wrote, was that in the event "of a massive atomic attack the main problem will be, not how to save the greatest number of lives, but how to prevent the heartbeat of the nation from stopping. . . . Not the death of millions, but the disorganization of industry and transportation, will be the main threat to our survival." Consequently, priority should be given to measures aimed at "maintaining the country as a living organism . . . to keep the country as a whole a going concern." What was required, above all, was not a schema for local organization but a "nation-wide plan of maintaining . . . war-essential production and transportation after an atomic onslaught on our industrial cities, harbors, and railroad centers." This "master plan"—which Rabinowitch described as "analogous to the industrial mobilization plan for war"—should be prepared by "economic, military, scientific, and political experts of the highest caliber," who could formulate a "blueprint for national survival."[138]

In his contribution to the same issue of the *Bulletin of the Atomic Scientists*, physicist Ralph Lapp offered a similar assessment. "The rapid cascade of recent events," he wrote, "looms as the overture to war"; nonmilitary defense had become "a matter of life-and-death seriousness." Thus far, however, the federal government had only "toyed with the problem by writing a few reports on civil defense administration," accomplishing "nothing of any substance." The Blue Book was merely "a new edition of the Hopley report" that "by its very bulk" was meant to show that "official Washington is hard at work on civil defense." But this approach, with its emphasis on the local organization of postattack relief, was meaningless in an air-atomic age. "You cannot provide effective defense if you wait for the attack to occur and then rush around after the fact trying to reduce the casualties," Lapp wrote. "Yet this is precisely the philosophy of the few people who have emerged from Washington to advise local authorities on civil defense planning."[139]

Upon the passage of the Federal Civil Defense Act and the establishment of the Federal Civil Defense Administration (FCDA), advocates of nonmilitary defense both inside and outside government sought out other channels through which they could pursue what they saw as the key strategic measures of nonmilitary defense in an atomic age. For example, the military's Research and Development Board, on which Lapp and other civilian experts served, commissioned studies of nonmilitary defense by expert advisory bodies. One such study, Project East River (completed in 1952), shaped the Eisenhower administration's approach to nonmilitary defense. Another, Project Charles (1951), included an appendix on nonmilitary defense written by economists James Tobin and Carl Kaysen, veterans, respectively, of wartime mobilization planning and air targeting.[140] Kaysen and Tobin argued that the model of civil defense outlined in the Blue Book and enacted by FCDA was "oriented more toward disaster relief after bombs fall than toward measures that will lessen the weight of the blow" and advocated a program of selective dispersal to mitigate industrial vulnerability.[141]

The actions of NSRB officials following the completion of the Blue Book and the passage of the Federal Civil Defense Act suggest that they also saw these steps as inadequate.[142] The Board cosponsored Project East River, and many of its staff members worked on the project and shaped its influential report. Meanwhile, in its own work programs, NSRB focused increasingly on nonmilitary defense, particularly after it handed off its responsibilities for civil defense and for operational management of the mobilization control program—along with a significant percentage of its staff—to FCDA and the Office of Defense Mobilization.[143] By mid-1951, NSRB had consolidated its nonmilitary defense activities in a Special Security Programs Office led by Ramsay Potts, a former air commander, who served as deputy to NSRB chair Stuart Symington. During the next two years, a small group of staff in the Special Security Programs Office worked on projects in three interrelated areas of nonmilitary defense. They continued NSRB's work on industrial dispersal and initiated ambitious new projects on postattack industrial rehabilitation and on governmental preparedness.[144] Through such efforts, NSRB reconstructed the diagram of government that it had developed in its earlier work on civilian mobilization in a program of federal preparedness planning for a future national emergency.

Postattack Industrial Rehabilitation

In mid-1951, having completed its proposal for an industrial dispersal policy, NSRB's Special Security Programs Office turned to another phase in non-military defense planning: maintaining continuity of production following a large-scale atomic attack. The office addressed this problem in a program on Post-Attack Industrial Rehabilitation (PAIR), which was initially overseen by the head of the Special Security Programs Office, Ramsay Potts.[145] Potts explained that his approach to postattack rehabilitation was shaped by his experience with strategic bombing during the war and by his participation in the interrogation of high-ranking Nazi officials, notably Albert Speer, who directed German industrial mobilization during World War II.[146] "The most effective way to have post-attack rehabilitation," he recounted, citing the surprising resilience of the German wartime economy, "is to have pre-attack programs such that you have a large amount of cushion in your economy." The United States must, in other words, plan for an industrial production system that could absorb the shock of an enemy attack and continue to function.

Following this general orientation, NSRB's initial discussions of postattack industrial rehabilitation emphasized measures to bolster the resilience of the industrial production system, such as stockpiling critical materials and dispersing key industrial facilities outside dense urban zones. But the aim of ensuring "a large amount of cushion" in the industrial economy soon opened up a much broader field of problems and possible responses. William H. Stead, an NSRB consultant who led the PAIR project for several months in early 1952, recalled that the "initial approach"—focused only on industrial facilities—"was shifted to encompass planning of nonmilitary defense for all major segments of the economy," including "transportation, banking, and credit, electric power, communications, housing and community facilities, income maintenance and labor supply."[147] In this expanded conception, planning for postattack rehabilitation encompassed all necessary means of ensuring the recuperability of the US military-industrial system following an attack, including the mechanisms of "community support" on which industrial production depended.

As it began work on the PAIR program in late 1951, NSRB commissioned two studies by outside consultants to elaborate its approach. One, conducted by the RAND Corporation, was led by economist Shaw Livermore, who had worked in the War Production Board during World War II and who would later play a central role in translating the tools of mobilization planning into emergency preparedness during the 1950s (see chapters 5 and 6). Livermore's study outlined strategies for the rehabilitation of the steel sector as a test case of planning for postattack rehabilitation. It proposed a range of vulnerability reduction measures, including plant-level security, planning for postattack reconstruction and operational continuity, and the accumulation of off-site stockpiles or on-site inventories of vital inputs. A second study commissioned by the Board, conducted by George T. Hayes, William J. Platt, and William Hosken of Stanford Research Institute (SRI), had a broader charge: elaborating the "basic principles to be used in formulating a national policy and program for post-attack rehabilitation of industry."[148]

In 1952, SRI completed a report based on its study that laid out a broad strategic and conceptual framework for postattack rehabilitation. The SRI study was closely coordinated with NSRB—its authors essentially served as staff on the PAIR project committee. Consequently, its report suggests how Board staff reconceived postattack rehabilitation to encompass the broad strategic problem of nonmilitary defense.[149] Building on NSRB's prior work in areas such as industrial dispersal, the continuity of government, and civilian mobilization, the SRI report outlined a program to ensure the postattack continuity of vital government activities, industrial production, and essential services. Two broad aims guided the program and would go on to structure US emergency government throughout the Cold War: ensuring the resilience of vital systems following a domestic catastrophe and maintaining ongoing governmental preparedness.

VULNERABILITY AND RESILIENCE

The SRI report began by diagnosing the sources of American vulnerability, analyzing the "industrial system" as a "complex" of interdependent processes. "The product of one plant," the report explained, "is an ingredient in the next production process." Changes "such as might be caused by atomic bomb damage" in a given location "have repercussions throughout the system." This initial observation suggested a distinction between two kinds of loss. "Primary loss" referred to "the reduction in war potential caused by damage to facilities and personnel in the immediate effects zone"—that is, the direct impact of blast, fire, and radiation as these atomic weapons effects unfolded in space. But the "high degree of industrial interdependence" in US military production meant that "many plants throughout the country, not in the immediate vicinity of the bomb damage will find their flow of materials and products affected." Drawing on concepts from the New Deal

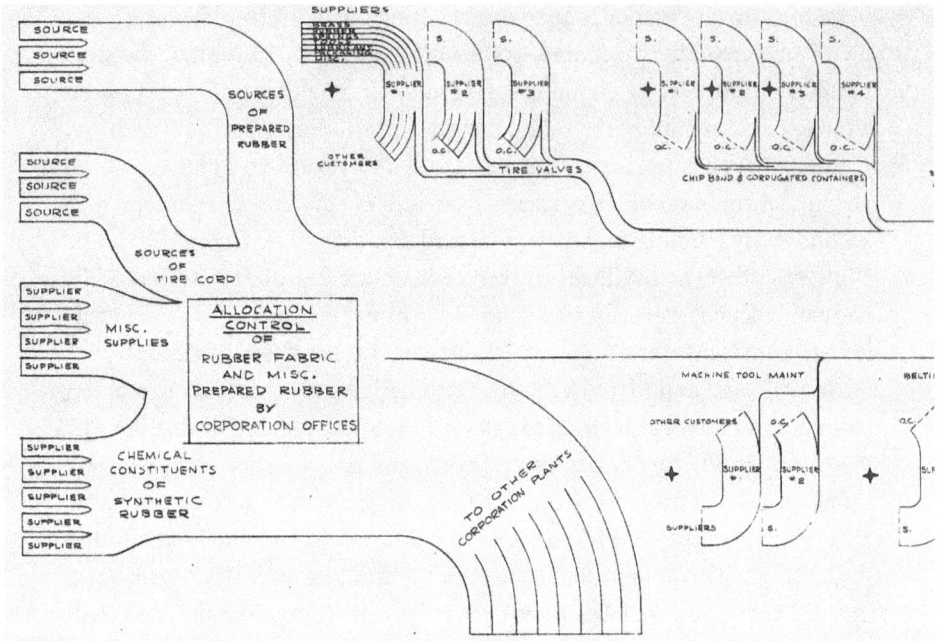

FIGURE 4.6. A typical rubber products plant: Supplier-customer relations. After World War II, mobilization planners analyzed the flow of materials through various industrial sectors, as New Deal planners had in the late 1930s. But they used these analyses for a different purpose: identifying the effects that the destruction of a particular facility would have on total production. *Source*: George T. Hayes, William E. Hosken, and William J. Platt, *Report on Problems of Post-Attack Rehabilitation of Industry*, January 4, 1952, report prepared for the National Security Resources Board, Project Number 511, Stanford Research Institute.

science of flows (described in chapter 2), the SRI report observed that as primary losses were "amplified in passing through the economy," they generated "secondary loss," defined as the "reduction in war production in the dependent economic system outside of the effects zone."[150] For example, if supplier plants were "completely destroyed" or "incapacitated," then "deliveries to [an] undamaged plant cease." Similarly, a plant that was "close to the damaged area but untouched" might encounter "a shortage of utilities, interference with transportation and communications, and actual loss of personnel." It was possible for secondary loss due to such factors to "exceed primary loss by many times"—a bomb damage multiplier effect.[151]

To illustrate, the SRI report presented a flow chart that extended across a seven-page fold-out insert, which represented supplier and customer relations for a particular rubber products plant that manufactured tires, tubes, battery cases, and aircraft components (see figure 4.6). The flow chart mapped inputs, including industrial products (tire cord, machine tools, belting, lubricants) and basic services (power, heat, and electrical supplies). It

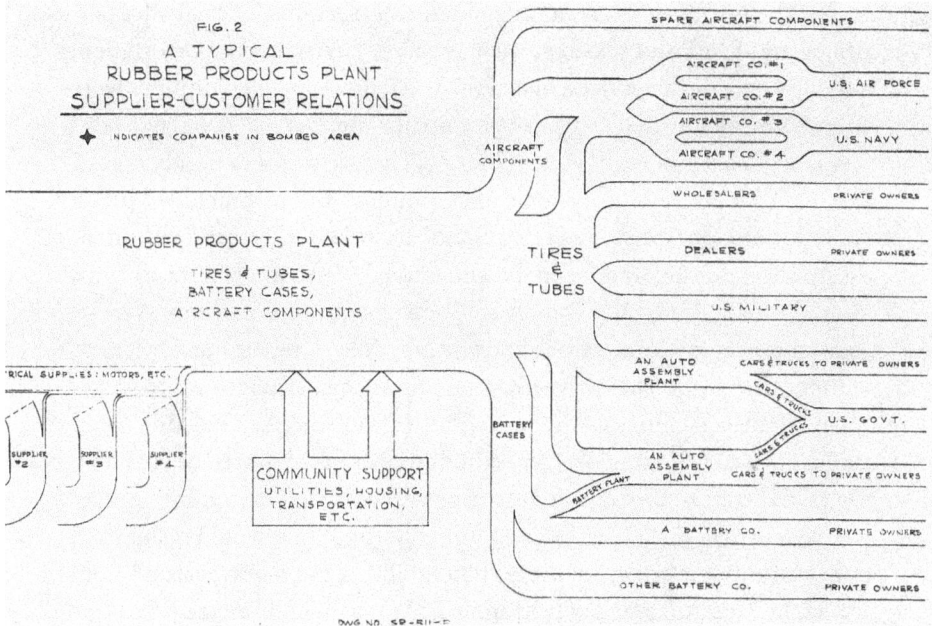

FIG. 2
A TYPICAL
RUBBER PRODUCTS PLANT
SUPPLIER-CUSTOMER RELATIONS

also indicated the plant's civilian and military customers, such as aircraft and automobile companies that supplied the Air Force and Navy. Similar visualizations had been used by New Deal economists to characterize the national economy as a complex of substantive flows, and by wartime mobilization specialists to trace "flows of materials" through systems of military-industrial production. The SRI analysis introduced a new element to this technique by identifying which of this plant's suppliers were located within target areas (identified by asterisks on figure 4.6) and were therefore likely to be destroyed in an atomic attack.

The chart demonstrated that although only a small number of facilities were located in target areas, these facilities were vital to rubber production and therefore were key vulnerabilities for the entire production system. For example, the only suppliers of belting and tire wrap were located within target areas. The case of rubber highlighted the problem of what wartime resource evaluation specialists had called "concentration." "There is a strong tendency," the SRI report observed, "for more than half of total output to be concentrated in the top four companies" within a given industrial sector. The case of rubber products also illustrated how vulnerabilities might be located in systems of "community support," a capacious term that referred to the entire complex of infrastructures and essential services that sustained social and economic life. Industrial production, the report explained, depended on cities, whose function as an "integrated unit" required "the effective per-formance" of "utilities systems" such as water, power, and transportation. Although outside the initial scope of the PAIR program, these utilities were "of such great importance to the operation of a community" that it had been necessary to examine the "vulnerability of these systems to bomb attack." Critical vulnerabilities in systems of community support included urban water networks (whose "vulnerable points" included local reservoirs, pump-ing stations, purification facilities, and mains), communications systems (central offices, switching equipment, power supply), electricity systems (central power stations were "almost impossible to repair quickly"), and transportation (marshaling yards, truck terminals, airports, railroad sta-tions). Medical facilities were another critical vulnerability. Existing medi-cal facilities in US cities, the report found, were inadequate to deal with casualties "comparable in extent to those experienced in London, Berlin, or Tokyo in World War II," to say nothing of those that would result from a single atomic air burst. All these vulnerabilities, the report warned, con-firmed the "country's lack of preparedness" for an atomic attack.[152]

To address these acute vulnerabilities, the SRI report argued, it was nec-essary to design a more "resilient" industrial production system. As we have seen, the concept of resilience had previously appeared in the report of the Strategic Bombing Survey to describe how the German economic system adaptively responded to Allied bombardment through measures such as the use of substitutions for industrial inputs and the rapid repair of production facilities. The Survey had analyzed resilience from the perspective of an economics of target selection: more resilient systems of military-industrial production were less attractive targets for strategic bombing. SRI's report addressed the problem from a different perspective: How could government selectively intervene to *bolster* the resilience of American military-industrial production systems?

The report observed that the "strength of the American industrial system is its resilience," which was a product of the "resourcefulness and enterprise of independent persons and organizations," and the "flexibility and adap-tiveness inherent in our competitive economy." In some cases, therefore, postattack industrial rehabilitation measures would only have to "encour-age the basic ingenuity of industry and the resilience afforded by alternates and substitutions." But in other cases, it was a "proper function of [govern-ment] rehabilitation planning" to undertake "preparedness and post-attack measures" to bolster the resilience of the industrial production system.[153] For example, given the "multiplicity of suppliers and users and the diver-sity of alternates that exist with respect to processes plants, and materials," planners could identify alternative supplier relations that might be set up in the event that key plants were damaged. The reserve capacity that was "inherent in a standard of living as high as that of the United States" could be tapped through restrictions on nonessential production to "tak[e] up the slack to maintain output." The "inventory and transportation practices of industry" might also produce a "significant degree of preparedness against secondary loss," as firm managers could draw on stocks, substitutions, and alternative sources of supply. Similar measures could address the vulnerability of essential services and systems of community support. Water utilities could stockpile parts, such as valves and pipes, or build alternative control centers. Electricity providers could maintain "facilities with excess capacity and interconnecting power grid systems." The effects of an attack on health infrastructure, meanwhile, could be mitigated by using "independent power, light, heat, gas, and water supplies," or by setting up facilities in buildings that were suitable for "war-time conversion to medical use, both inside and outside city centers." In sum, the toolkit of mobilization planning—the accumulation of stockpiles, the planning of substitutes, the use of priorities power and rationing to direct scarce materials to vital production—could be employed for a new purpose: to reduce the vulnerability and bolster the resilience of the nation's vital systems.[154]

MAINTAINING A "STATE OF PREPAREDNESS"

The task of designing a more resilient industrial production system pointed to another aspect of postattack rehabilitation: maintaining a "state of preparedness" by anticipating the problems that would arise and functions that would have to be carried out in the wake of a future attack. What the SRI report referred to as "pre-attack preparedness" partly entailed resources planning measures, such as marshaling raw inputs and equipment, accumulating stocks and inventories of critical materials, and establishing emergency facilities. But the central object of preattack preparedness was government itself. It would be necessary to maintain ongoing readiness to establish an emergency organization that could manage the "new situation" created by an attack. Since an attack could come at any moment, the report argued, "preparedness plans had to be operative 24 hours a day, 365 days a year." Given changes that would "continuously occur in war materiel requirements and in criticality of plants," as well as in the overall war situation, the government organization responsible for postattack response would have to be "readily adaptable to continuously changing conditions." Among these changing conditions might be damage to the governmental apparatus. "Any organization which is set up," the report explained, had to prepare for "the possibility of loss in the attack of many key people, facilities, and much materiel." Provisions therefore had to be made for alternative work sites, as well as for "alternate commanding authority and a regular chain of command which functions regardless of which key personnel are lost."[155]

What governmental agency might take responsibility for such measures? And how would an ongoing program of "pre-attack planning and preparation" be maintained? The SRI report asserted that preparedness planning would have to be taken up "at all levels of management and government," especially at the level of the federal government, which had "the greatest knowledge of the over-all situation with regard to war plans needs and material outputs." It anticipated particular difficulty with enlisting local governments and private firms in maintaining a state of preparedness. "One of the toughest problems confronting the federal government in setting up this program," it explained, "is to convince American companies that pre-attack bomb damage rehabilitation programs are vital and require immediate action." To address this difficulty, the report suggested the use of movies, slides, and diagrams. By demonstrating "what a damaged plant looks like, how it can be repaired, the methods used to repair and the length of time required," such techniques would convey the urgency of taking action to reduce vulnerability. Similarly, test cases could demonstrate "what would happen" in the event of an enemy attack: "We can point to the man from Plant A and say 'If the bomb were to fall at this point, you would lose 15 percent of your people and 30 percent of your plant. . . .' It should help materially to arouse his interest and to obtain his cooperation in setting up a rehabilitation program for his own organization."[156]

The SRI report did not anticipate similar problems with the federal government, which it described as "the single agency responsible to the people for their welfare" and in the best position "to evaluate the danger" and "control the funds and materials essential to rapid rehabilitation." Referring to the executive branch offices then managing the Korean War mobilization, the report posited that "the organization already exists for adequate federal government participation," and that "the rehabilitation problems that will arise as the result of an attack will not be unlike" those that had confronted emergency agencies in World War II and the Korean War. It was merely a matter of reorienting existing activities to "include the new series of problems," of "alerting organizations to the nature of rehabilitation problems and their powers to meet them," and of "coordinating their planning to create a state of preparedness for post-attack situations."[157] But as it turned out, this matter of incorporating a "new series of problems" into the work of the existing emergency agencies was far from simple. Officials working on postattack rehabilitation soon concluded that federal agencies were perilously unprepared to perform essential functions after a future attack on the United States.

Unprepared

As SRI was conducting its study in late 1951, NSRB kicked off interagency planning for postattack recovery of the industrial economy by convening what it called a Conference on Post-Attack Industrial Rehabilitation.[158] The significance of postattack rehabilitation in NSRB's overall work program is indicated by the Board's representation at this conference, which included its chair, Jack Gorrie, his deputy, Edward T. Dickinson, and more than twenty staff members who had worked on various aspects of nonmilitary defense.[159] Also in attendance were officials from several other federal agencies. Alongside the Federal Civil Defense Administration, the group was dominated by offices with ongoing emergency responsibilities in the Korean War: the Office of Defense Mobilization, the Defense Production Administration, and emergency mobilization offices responsible for planning civilian requirements.[160] These were precisely the offices that, according to the SRI report, could readily incorporate postattack planning into their ongoing work.

In his opening remarks at the conference, Gorrie informed the assembled officials that NSRB was broadening the scope of its nonmilitary defense efforts. The Board's work on mitigating industrial vulnerability, Gorrie explained, had previously addressed the dispersion of industrial facilities outside urban centers (a program that, he claimed, was "working very well"). But industrial dispersion was "only one part of this greater problem of how we meet the situation" presented by a future atomic war: "what we are prepared to do or what we can do in the event of an attack on the country."[161]

Following these preliminaries, Gorrie turned the meeting over to NSRB consultant William Stead, who began his presentation by describing the vulnerability of the American industrial system to an enemy attack. "We have," Stead told his audience, "very large proportions of key components of our industries concentrated in a few places which makes us very vulnerable." Gesturing to an aiming point map of Chicago—likely borrowed from civilian mobilization test exercises of the prior year—Stead invited the assembled officials to "take a moment to consider what can happen" in a future attack. The primary effects would be extensive: 220 plants destroyed; another 2,000 damaged. But the most significant impact would come from secondary effects that arose from system interdependencies. "Think of what would happen," Stead implored, "if this hit was on the two or three major power stations and power centers or the stock yards or the centers of transportation. You can see the effect all around through the area." Stead illustrated his point by referring to the ongoing management of production for the Korean War, as many of the assembled officials were involved in some capacity. The shortages that resulted from restrictions on the use of critical materials for the mobilization control program "fan out into every single industry and into every part of our industrial structure." If instead of these regulatory restrictions, "you had the elimination of substantial parts of some of our key products, you can visualize what would happen, and it is a very real and vital risk."[162]

Stead expressed confidence that the agencies represented at the meeting were already developing plans for carrying out their postattack responsibilities, telling the conferees that NSRB's goal at this stage was simply to gather information. "We are vitally interested," he said, "in what you are doing, what the stage of development is." NSRB staff distributed questionnaires to be filled out by officials at each agency represented at the meeting. What were its responsibilities in a postattack situation? What was its program for meeting these responsibilities? What was this program's current status?[163]

Several weeks later, NSRB staff in a new Division of Post-Attack Rehabilitation completed a report on the agencies' responses. The survey revealed that a few agencies, such as the Federal Civil Defense Administration and the Atomic Energy Commission, had "clear-cut authority to do certain jobs now or to develop plans for their post-attack functions." But most federal agencies and offices either had no identifiable authority to perform their emergency functions or did not "believe they have important responsibilities in a post-attack rehabilitation project." Moreover, the survey indicated, they had no capacity to "project [their] activities into that kind of situation, and to visualize [their] responsibilities under those conditions," and had given little thought to estimating and reporting damage. The restoration of "the machinery of government" had "received almost no attention." Consequently, even those agencies that had considered their postattack roles could not be expected to have "the background and planning which an adequate job of restoring production would require." Indeed, NSRB staff concluded that it was "difficult to point to an agency which now has a document which could be said to contain: (1) an analysis of that agency's functions in a pre-attack period; (2) a statement of the problems it would have to deal with in the event of an attack; and (3) its proposals for dealing with them." This situation was "particularly alarming" because each of the federal agencies queried had "obvious post-attack functions which undoubtedly would be exercised tomorrow if an attack occurred tonight."[164]

It is not clear whether these results actually surprised NSRB staff—their purpose in conducting the survey may have been precisely to document federal agencies' lack of readiness as a means to marshal support for coordinated nonmilitary defense planning.[165] Whichever is the case, Board staff members argued that the survey results underlined the urgency of addressing current "gaps in planning." In order to address such gaps, NSRB linked its work on postattack rehabilitation—which focused on the continuity of industrial production—to another of its program areas: the continuity of government operations.

GOVERNMENT IN AN EMERGENCY

As we have seen, NSRB's work on government continuity had initially focused on the security of the nation's capital, which resulted in a long-term plan for the relocation of critical activities outside Washington, DC. According to Ramsay Potts, who ran the Board's continuity of government program, NSRB staff analyzed Washington's vulnerability by looking at the critical nodes of this "nerve center for the operation of government." In the capital, he explained, "almost all of the key operations that we will depend upon to carry on and conduct a war" were concentrated "in a very small space."[166] The Truman administration had advanced legislation to fund the relocation program, but its proposal stalled in congressional committee. Given this impasse in long-term planning, in late 1950 the National Security Council asked NSRB to develop "stop-gap" measures to ensure governmental continuity in the event that an attack "occurred prior to the completion of a dispersed office development."[167] This program of "alert" planning for the federal government resembled (and was likely modeled on) the alert exercises that NSRB's Civilian Mobilization Office had conducted with localities. Overseen by NSRB's Special Security Programs Office, the project sought to create a "plan to operate the U.S. government under the stress of nuclear attack."[168]

Initially, alert planning focused on preparing for the operation of federal agencies in relocation sites: ensuring that "each key agency" had "an assembly point outside Washington for its own key officials" that would be equipped with necessary communication equipment, files, and other logistical support. Potts described the aim of alert planning as one of identifying "serious gaps in the planning being done at the top level of Government to insure continuous and uninterrupted civilian control in the event of a disaster."[169] To conduct such planning, NSRB developed a "scientific approach" to ensuring "the continuity of government functions in the event of an attack." This approach—which NSRB termed "essential functions analysis"[170]—involved surveying the "heads of executive departments and establishments" in the federal government. One survey form asked each federal agency to identify its essential functions and to assign priority ratings based on the function's "importance or contribution to the total national effort in the event of war."[171] On a second form, agencies were asked to indicate resource requirements for maintaining each activity in a relocation site. Was it necessary to have direct contact with the president or with other federal agencies? Could these relationships be maintained from a dispersal site? How many government workers would have to be accommodated, and what facilities (equipment, office space, vital records) would they need to carry on their work? This form also instructed agencies to make plans for the delegation of authority in case top officials were killed or incapacitated—to provide for the "assumption, in the event of attack, of both intermediate and top command by key officials located outside of the central area." Finally, NSRB advised agencies to draft a "directive on delegation" that could be "formalized on short notice if required."[172]

By early 1952, NSRB planners had abandoned plans to permanently relocate key federal agencies outside of Washington, DC, given congressional resistance to funding such a measure as well as demands on resources for the Korean War. In this context, NSRB's program of alert planning, which had initially been conceived as a temporary measure until a fully elaborated relocation plan was put in place, became the centerpiece of its continuity of government program. The Board's chair Jack Gorrie submitted the short-term plan to President Truman as "a readiness measure which can be placed in effect immediately in event of attack or imminent threat of attack upon the Nation's Capital."[173] This readiness planning effort converged with work in the PAIR project on resource planning for a future war emergency. An April 1952 executive order issued by President Truman on "Preparation by Federal Agencies of Civil Defense Emergency Plans" indicated that the administration now understood alert planning and postattack rehabilitation as part of the same broad problem of emergency preparedness. The order instructed federal departments and agencies to "prepare plans for maintaining the continuity of its essential functions at the seat of the Government and elsewhere." This included planning to provide "personnel, materials, facilities, and services" to fulfill civilian defense functions, taking into account the "essential requirements of the Department of Defense." NSRB was charged with establishing "standards and policies" to ensure "uniformity of planning for the continuity of essential functions."[174] With this executive order, NSRB's statutory mandate for nonmilitary defense—initially limited to the narrow task of planning for strategic relocation—had been expanded to address a novel and capacious problem: emergency preparedness planning for the entire US government.

DISTRIBUTED PLANNING FOR GOVERNMENTAL PREPAREDNESS

In the second half of 1952, NSRB organized its continuing work on postattack rehabilitation under a Central Task Force. The membership of the Task Force suggests that NSRB and other national security planning agencies saw it as an urgent priority. Edward T. Dickinson, the deputy director of NSRB, took over leadership of PAIR due to "the extreme importance of this program."[175] Officials from the Federal Civil Defense Administration and the Office of Defense Mobilization (ODM) served as co-vice chairs of the Central Task Force. ODM was represented by Shaw Livermore, who had initially joined the PAIR program as a consultant for RAND and had since become assistant to the director of the Office, Charlie Wilson, and a member of the Planning Board of the National Security Council.

The Central Task Force began by defining the "post-attack contingencies which should be provided for," grouping them in a series of "broad areas in which general problems can be anticipated" (see figure 4.7). These broad areas included direct inputs to productive processes (materials, equipment, and manpower); provision for the civilian population (the welfare of individuals, community facilities, and housing); essential services (electric power, communication, and transportation); and government operations, including measures for "reconstituting the machinery of government" after an attack. This delineation of problem areas raised an issue of organization: some of these substantive problems fell under the responsibility of multiple federal agencies while others were not the responsibility of any organization in the federal government. For example, measures related to postattack industrial production touched on both the internal security program of the National Security Council and NSRB's dispersal program, as well as on the

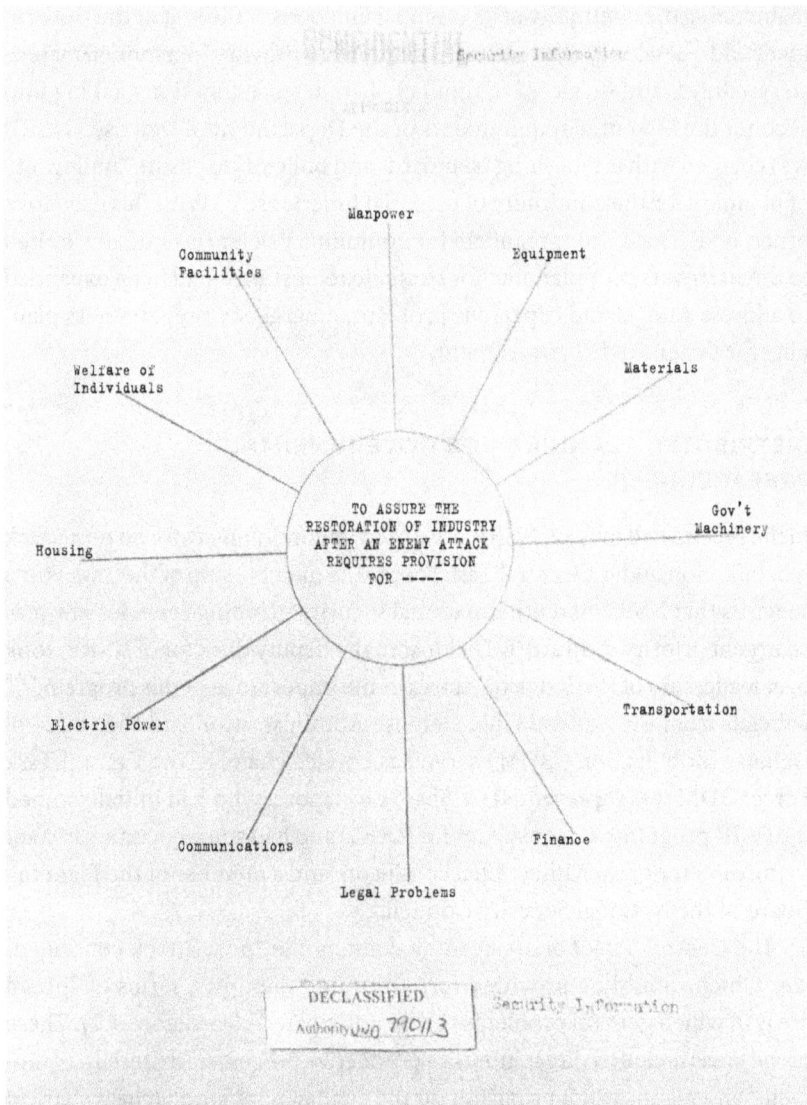

FIGURE 4.7. Postattack industrial rehabilitation problem areas. The National Security Resources Board's Post-Attack Industrial Rehabilitation project identified broad areas to be addressed in planning for governmental response to a future attack. These areas spanned the entirety of social and economic life, including critical infrastructures, housing and community facilities, industrial production, and the machinery of government. *Source*: Division of Post-Attack Rehabilitation, "Planning for Post-Attack Industrial Rehabilitation: Summary of Agency Reports," National Security Resources Board, Office of the Vice Chairman, Security: Classified Office File of Edward T. Dickenson, 1951–1953, Projects: Post-Attack Rehabilitation, entry 8, box 13, RG 304, Records of the Office of Civil and Defense Mobilization, National Archives, Washington, DC.

stockpiling and other mobilization planning programs overseen by emergency agencies and by the Department of Defense. The "training of mobile repair teams" and other measures designed to "minimize the effectiveness of an enemy attack" were not the responsibility of any agency or office. Meanwhile, FCDA and state and local governments involved in community planning all bore responsibility for "the rescue, evacuation, emergency feeding, clothing, and housing of individuals" and for "maintaining income, for social security programs and restoring the community facilities required to maintain the civilian population."[176]

The Central Task Force addressed this tangle of tasks by distributing responsibility for them across the existing organization of government. The planning process was organized through interagency committees whose charge was to "prepare recommendations for legislative or executive authorization and action." Four of the interagency committees dealt with specific resource areas: industrial operations; utilities and services such as communications, electric power, and water; manpower; and health and welfare, which was divided into sections on housing and medical care.[177] A fifth interagency Committee on Surveys, Communications, and Statistics for Damage Reporting was charged with developing techniques of preparedness planning that could generate anticipatory knowledge about the substantive requirements of postattack government and about current readiness to fulfill these requirements. One task group in this Committee—on damage assessment and vulnerability analysis—coordinated the work of government agencies with related programs run by FCDA, the Industry Evaluation Board, and the Air Force.[178] Another task group worked on alert planning to test governmental preparations for carrying out essential functions in the event of an attack. In mid-1952, NSRB vice chairman Dickenson recommended that all the agencies involved with planning for postattack rehabilitation conduct a "dry run" to identify "gaps in knowledge and requirements."[179] This test program was assigned to a special task group that included members from the Department of Defense, ODM, and FCDA.[180]

In sum, by the end of 1952, the scope of NSRB's project on Post-Attack Industrial Rehabilitation had expanded to include the "restoration not only of industrial plants" but also of the "closely allied segments of the national economy, such as transportation, communications, power and other utilities" that would be required to "mount and maintain the retaliatory attacks necessary to national survival."[181] The project encompassed a wide range of activities, dispersed across many federal agencies and offices. Some of these activities were part of existing nonmilitary defense programs such as civilian defense, internal security, and continuity of government, as well as related technical activities, such as resource evaluation and dispersal planning. Others fell under the purview of the Korean War emergency agencies, such as mobilization control programs, or other military preparedness programs, such as stockpiling. Still others—damage assessment, vulnerability analysis, and the proposed program of test exercises—were new to the federal government. The work of the Central Task Force was to organize these heterogeneous activities, which had in many cases been invented for other purposes, around a common problem: preparing to govern in a future emergency.

A New Diagram of Emergency Government

This chapter began with a discussion of how military officials and civilian experts understood nonmilitary defense in the immediate aftermath of World War II: in terms of the broad strategic aim of ensuring that vital military-industrial production could continue during a future war. In the PAIR project, security planners translated this capacious vision of nonmilitary defense into specific practices of planning and administration. Many elements of the postattack rehabilitation program were drawn from established government functions, including mobilization control programs, internal security, and the provision of protection and relief to the civilian population. But the broad aim of ensuring preparedness for the catastrophic event of an atomic attack did not fit readily into existing categories of government organization and activity. Shaw Livermore later described the novelty of the "pioneering work" on civilian planning that he and others had conducted in 1952. "People could never understand what we were doing," Livermore recounted. "They thought it was just old-fashioned, simple civil defense. Jump in a dugout, you know, or rush into a shelter."[182] What was new, and difficult, perhaps, for contemporaries to comprehend, was not specific policies and practices, many of which were familiar. Rather, the novelty lay in how these elements were linked together in a diagram of emergency government. During the New Deal and World War II, emergency government had meant economic management to address the downturn of depression or the production demands of total war. These experiences were formative for the officials and experts working on the PAIR project. In the PAIR project, however, the aim of emergency government shifted: to the problem of managing survival and recovery following a future domestic catastrophe.

This new diagram for emergency government was composed of two elements: emergency resource planning and governmental preparedness.

Emergency resource planning sought to ensure that vital systems would continue to function in the aftermath of a future catastrophe, and that critical resources—from industrial products to medicines, food, and housing—would be available to carry out essential relief and recovery functions. Such planning closely followed the procedures of industrial mobilization and resource evaluation: assessing the criticality of vital systems; anticipating emergency resource requirements; and taking measures to ensure that sufficient supplies would be available to meet these requirements. It also included new techniques to anticipate how an attack would affect the "balance" between supplies and requirements. The aim of governmental preparedness, meanwhile, was to ensure that federal and local agencies would be able to carry out emergency operations. Here, too, planners drew on familiar practices from emergency government in World War II and Korea: the delegation of emergency functions; the creation of emergency agencies; and the establishment of emergency powers. And here, too, planners combined these existing practices with new techniques, such as test exercises and planning for the continuity of government operations.

To say that planners and officials assembled this diagram of emergency government for the first time in the PAIR program is not, of course, to say that the problems of emergency resource planning and governmental preparedness were "solved." The point is not that at this moment, the federal government actually became prepared for a nuclear attack, or that the nation's vital systems were now resilient against disruption. Nor, even, can we say that the national security establishment accepted NSRB's approach to emergency preparedness as a framework for addressing the nation's vulnerability to atomic attack. Indeed, despite the importance that staff from NSRB and other agencies involved in PAIR assigned to their work, the project remained obscure, and progress in planning for specific resource areas was halting at best. Moreover, as Truman's term drew to a close at the end of 1952, the future of this project was uncertain. Perhaps in anticipation of its own dissolution with the change in administration, the PAIR task force assigned the work of its interagency groups to various federal agencies: Manpower to the Department of Labor; Industrial Continuity to the Office of Defense Mobilization; Housing and Community Facilities to the Department of Housing and Home Finance, and so on. NSRB envisioned that it would continue to lead work on damage assessment and reporting but noted that this role was temporary, "pending further study as to permanent assignment."[183] With the end of the Truman administration, the PAIR project was abandoned, at least in the form initially envisioned.[184]

But high-level planning bodies in the federal government soon accepted NSRB's approach to nonmilitary defense as a central element of national security in a nuclear age. In part this was due to the efforts of the officials working on postattack rehabilitation. They succeeded in influencing what historians have recognized to be some of the most formative statements of national security policy in the early Cold War. For example, NSRB staff who had worked on postattack rehabilitation shaped the findings of Project East River—completed in late 1952—which established nonmilitary defense as a key element of national security strategy for the next several years.[185] NSRB leadership also pushed Truman to release a statement on the administration's support for nonmilitary defense. At the end of 1952, Truman signed NSC 139 on Continental Defense, which, in a clear reference to NSRB's programs, stated that "military defense must be supported by well-organized programs of civilian defense, industrial security, and plans for rapid rehabilitation of vital facilities."[186]

As Dwight D. Eisenhower took office in 1953, he was presented with a number of reports that identified continental defense as "the Achilles heel of American national security" and that emphasized the central importance of nonmilitary measures in broader strategies for defense against Soviet attack.[187] An influential May 1953 memorandum for the National Security Council, authored by Paul Nitze and Carlton Savage of the State Department's policy planning staff, summarized these reports and pushed for immediate action on continental defense. Their recommendations included both active defense measures, such as an early warning system, and passive defense measures, such as "civil defense, reduction of urban vulnerability, post-attack rehabilitation, and continuity of Government."[188]

The final two chapters of this book show how, beginning in the second part of 1953, federal officials instituted this diagram of nonmilitary defense. The focus of our narrative shifts to the Office of Defense Mobilization, with which the NSRB was merged in 1953. Chapter 5 shows how NSRB's techniques of essential functions analysis, emergency action steps, and test exercises were brought together in a program of "administrative readiness"—the first systematic approach to governmental preparedness for a future emergency. Chapter 6 then traces how mobilization planners employed the tools of resource planning and the nascent practices of vulnerability and damage assessment to address a novel problem: ensuring the survival of the national population following a nuclear attack on the United States.

PART III

Cold War Planning for National Survival

5
Enacting Catastrophe

In an essay written in 1949—the year the Soviet Union detonated its first atomic bomb—political scientist Clinton Rossiter considered the implications of the atomic age for the organization of American government. Following his prior work on "crisis government," Rossiter asked what the threat of a catastrophic nuclear attack meant for executive power in a liberal democracy. He envisioned a dire postattack situation in which "an abrupt halt, even total collapse of all legislative and judicial activity—state, local, and national—and an all but total collapse of public administration are not a vague possibility but a reasonable certainty." The disruption of the country's "communications, transportation, industry, and entire pattern of living and working" would be "virtually complete." Indeed, the "whole order of normal activity would dissolve in chaos" along with the "agencies and procedures of normal government."[1]

Rossiter rejected the possibility that the United States might prepare for atomic war by dramatically changing its metropolitan, economic, or governmental structure. Referring to proposals for dispersal outside of dense urban centers—widely discussed at the time—he ruled out "any radical plans of decentralization and subterranean concealment, whether of population, industry, or government." The United States would confront its "first atomic war" with its "population, industry, and government in much their present condition and distribution." The question, then, was how to ensure that existing governmental institutions could manage a postattack situation. For Rossiter, the central issue was that the "complex system of government of the democratic constitutional state"—with its widely dispersed sovereignty and checks on executive power—was "essentially designed to function under normal, peaceful, conditions." But such institutions often proved "unequal to the exigencies of a great national crisis." This problem, he argued, had attracted far too little attention: "No one seems to have outlined the overall pattern that the American government would assume in the event of atomic war or indicated the workable adjustments that we might undertake now to prepare our constitutional system for this dreadful contingency."[2]

Here, Rossiter looked to "past experiences with emergency government," in which the United States had depended on "the executive for extraordinary action in the use of emergency powers." In particular, he pointed to the precedent of the World War II mobilization, and to the Office for Emergency Management (OEM), which housed emergency offices such as the War Production Board. During World War II, OEM served as a device through which President Roosevelt wielded emergency powers and coordinated planning across the federal government (see chapter 2). The administrative reformers who designed OEM assumed that the office would persist after the war, providing "advice for the President in his efforts to meet the nation's major and minor crises" and developing "plans to deal with future emergencies."[3]

Although by 1949 OEM had been demobilized and was "in a state of suspended animation," Rossiter argued that it offered a model for the type of government agency that was now needed to confront the specter of atomic attack. The president "could well use a permanent trouble shooter, and a revived OEM with one of his executive assistants at its head would fill the bill admirably." Rossiter anticipated that preparedness planning in such an office would not require a large staff or significant expenditures. "Some one could well be set to work at $10,000 a year," he wrote, "thinking up the answers to such questions as: Where shall the President go in the event that we are assaulted by atomic bombs? Who shall go with him? How shall he communicate with his military and civil defense commanders after he reaches his place of safety? What shall be his relations to the committees and persons who will be speaking for Congress? How shall he keep in touch with the state authorities and also keep them under control?"[4]

More systematic planning for emergency government in a future war did in fact begin over the next several years. Perhaps not surprisingly, this work took place within federal offices charged with managing war mobilization—the central task of US emergency government both during and in the immediate aftermath of World War II. As described in the prior chapter, in 1951 and 1952 the National Security Resources Board (NSRB) developed non-military defense programs in areas such as the continuity of government and the protection of vital systems. This work on nonmilitary defense culminated with the Board's project on Post-Attack Industrial Rehabilitation, which laid out a schema for governmental preparedness for a future atomic war.

Given that NSRB's statutory mandate was limited to advising the president, it did not have the power to implement its proposals for nonmilitary defense. But in 1953, President Eisenhower lodged NSRB's nonmilitary defense functions in a new office that had expansive authority to undertake peacetime planning and preparedness. With the conclusion of the Korean War, as the task of managing an operational mobilization program wound down, a government study concluded that the complex of mobilization agencies "should be consolidated, simplified, and reduced substantially in size."[5] Pointing to a shift in priorities "from the requirements of the immediate defense build-up to the development of an adequate mobilization base" to prepare for a future war, in early 1953 Eisenhower ordered a merger of NSRB with the Office of Defense Mobilization (ODM), which had been created as an emergency agency to manage the Korean War mobilization.[6] The president appointed Arthur Flemming, who had served on the War Manpower Commission in World War II and had been director of ODM's Manpower Policy Committee, to lead the merged agency. The new office retained ODM's name and high profile (its director continued to sit on the National Security Council), as well as many of the statutory authorities that derived from legislation such as the Defense Production Act. But "the primary importance" of Eisenhower's order, as government historian Harry Yoshpe observed in 1953, was to "permit the uninterrupted operation, under ODM, of the NSRB non-military defense program."[7]

During the 1950s, the reconstituted ODM was the center of emergency preparedness planning within the federal government. Through a transformed practice of mobilization planning, ODM officials sought to implement the nonmilitary defense measures that had been proposed by NSRB. This new approach to mobilization planning would no longer focus on military-industrial production during a lengthy conflict. Rather, mobilization would now mean ensuring the continuity of government and the survival of the population in the wake of a devastating nuclear attack. In this light, mobilization planners would have to prepare for entirely new resource management problems—concerning the provision of food, medicine, and shelter to the civilian population rather than the allocation of steel, aluminum, and copper to maximize military production. They would also have to develop new techniques for anticipating the unprecedented challenges that would arise in the aftermath of a future catastrophe.

The final two chapters of this book examine how mobilization planners assembled a new apparatus of emergency government in the mid- to late 1950s. The present chapter focuses on the process through which ODM developed mobilization plans for the scenario of a massive nuclear attack on the United States. In this planning process, ODM addressed the questions that Rossiter had raised in 1949: What form should the US government take after an atomic attack? How would it manage a catastrophic postattack situation? How could peacetime government maintain a condition of ongoing readiness for such an eventuality? Far from being undertaken by "one person earning $10,000 a year," as Rossiter suggested, the planning process proved to be a massive undertaking that required coordination with agencies across the federal government. In collaboration with the Federal Civil Defense Administration, ODM set up test exercises that enlisted thousands of federal and local government employees in an iterative process of formulating and revising plans. At the same time, mobilization planners also devised new techniques for managing resources in a future emergency. As we show in chapter 6, they combined existing tools of resource planning—such as balance sheet analyses—with novel methods of computer-based simulation to anticipate and address postattack resource needs.

The plan that resulted from this process—Mobilization Plan D-Minus— looked utterly unlike prior war mobilization plans. It was not a compendium of battalions to be formed and equipped, tanks and airplanes to be built, or munitions to be produced. Rather, it was a plan for survival and recovery in a future nuclear war. Plan D-Minus outlined the emergency organization of the federal government, the essential functions this organization would have to carry out, and the "action steps"—laid out in seemingly endless tables occupying nearly one hundred pages in the final draft—to be taken by the president and by emergency agencies to address the unprecedented problems that would be faced following an enemy attack. Although still referred to as a "mobilization plan," D-Minus is best understood as the first US national plan for emergency preparedness. In it, we find a schema of emergency preparedness that has persisted up to the present day.

The first sections of the present chapter examine the initial stages of work on postattack mobilization planning following the merger of ODM and NSRB. In the period 1954–1955, ODM planners worked to implement a new program of "administrative readiness," which involved identifying the emergency actions that agencies across the federal government would be expected to take in the aftermath of attack and then iteratively testing these actions through a series of scenario-based exercises. The chapter then describes how, in implementing this program of administrative readiness, planners confronted the problem that Rossiter identified in 1949: How would the authority be established for executive action to manage resources for relief and reconstruction in a future war emergency? And how would such executive authority be accommodated with American constitutional norms?

To address these questions, ODM staff initially proposed that, following an attack, the president would establish a hierarchical "chain of command" in which state and local governments would follow orders from the national executive. This approach was tested in the 1955 Operation Alert exercise, during which Eisenhower declared a state of "limited martial law" to ensure the orderly provision of aid to overwhelmed localities. The declaration provoked a strong backlash from Congress and from legal experts, who argued that it violated constitutional norms and undermined the system of distributed responsibility for emergency government. In response, ODM officials devised a techno-administrative "solution" to the challenge that emergencies posed to constitutional democracy. Their proposed solution followed the pattern of Progressive reform we have traced through the New Deal and World War II: a program of preparedness that would enable the government to manage a postattack situation without recourse to "exceptional" measures that would undermine constitutional order.

Mobilizing for a Different Kind of War

At the time President Eisenhower came into office, US national security strategists were becoming increasingly alarmed at the prospect of a sudden massive atomic attack on the continental United States. Reports from committees of civilian experts such as Project Charles and Project East River, completed in 1951 and 1952, advocated a program of "continental defense" to address this threat. This program would include both "active" measures, such as early warning systems and bomber interception, and "passive" (i.e. nonmilitary) defense measures to ensure the continuity of government, reduce urban vulnerability, and protect vital industry and essential facilities. By early 1953, as the new administration demobilized and rearranged the Korean War–era emergency agencies, officials in the national security establishment acted to institute such continental defense measures. On the one hand, the National Security Council revised its high-level war assumptions and incorporated nonmilitary defense into its policy guidance for mobilization preparedness. On the other hand, the reconstituted ODM began to overhaul its organization and planning practice—a first step in the mutation of mobilization planning into emergency preparedness.

CONTINENTAL DEFENSE

In May 1953, Paul Nitze and Carlton Savage argued for the urgent need for improved continental defense to meet the growing Soviet atomic threat in a memo to the National Security Planning Board (discussed at the end of chapter 4). Despite "an increasing realization of the danger to the United States from Soviet atomic potentialities and of a consequent urgency for continental defense measures," they warned, there had so far been "inadequate preparation to meet the danger," and the Soviet nuclear stockpile had "grown faster than our capacity to protect the homeland from attack." The "survival of our Republic and the entire free world" depended on putting in place measures to protect the continental United States. Nitze and Savage recommended the formation of an ad hoc committee to "develop for NSC consideration a balanced continental defense program with specific cost estimates for its implementation."[8]

In June 1953, the National Security Planning Board acted on this recommendation, establishing a Committee on Continental Defense, chaired by retired general Harold Bull. The ODM representative to the Bull Committee was William Y. Elliott, a mobilization specialist who had been vice president of the War Production Board in charge of civilian requirements during World War II.[9] The Committee's report, NSC 159, completed in July, argued for the implementation of the program of nonmilitary defense that had been proposed by NSRB. Observing that a few bombs could "destroy the ability of the government to perform those operations essential to waging war, maintaining order, and preventing anarchy," the report advocated "drastic steps to provide as rapidly as possible for the continuity of government."[10] The report also addressed the continuity of industrial production, emphasizing three distinct but related policy areas: first, industrial evaluation, that is, the identification and rating of facilities and products by the Commerce Department, in accordance with their relative importance to national security; second, industrial vulnerability reduction, which included a program of industrial dispersion; and third, a program of postattack industrial rehabilitation "directed toward advance preparation for actions which must be taken following attack to restore essential industry and maintain the wartime production effort, including manpower, resources and finance."[11] The newly reorganized ODM, the report argued, should assume responsibility for supervising these nonmilitary defense programs.[12]

At the time that NSC made its recommendations, ODM's core mobilization planning effort was still based on the assumption that a future war would be similar to World War II: American industrial facilities would not be damaged, the military-industrial complex would be able to freely use all national resources, and mobilization requirements would be defined by the needs of military-industrial production for a multiyear conflict. But this assumption was increasingly out of step with national security strategists' understanding of what a future war would look like. As NSC 159 argued, "The presently planned mobilization base for fighting a war is jeopardized by the vulnerability of industrial and population concentrations." Consequently, "all facets of mobilization base policy must be revised to provide against a new and greater strategic danger."[13]

In September 1953—less than a month after the first Soviet H-bomb test—the National Security Council incorporated the recommendations of NSC 159 in a "statement of policy" instructing federal agencies to orient their preparedness planning to a scenario involving a large-scale attack on the United States. Soon thereafter, the Council circulated a set of "interim defense mobilization planning assumptions" based on this scenario.[14] A future "global war," according to these assumptions, would involve "nuclear attack on and massive destruction to selected major urban areas of the United States wherein our principal Government centers and a large portion of our productive capacity and population are located."[15] ODM director Arthur Flemming urged the council to officially adopt these new planning assumptions in order to provide guidance to his agency on formulating its plans and programs.[16] NSC responded to Flemming's request in February 1954, releasing a continental defense strategy that focused on "the protection of our vitals and . . . the survival of our population and our Government in the event of attack."[17] The vulnerability of the nation's vital systems had been established as a central concern of national security policy.

THE MOBILIZATION PLANNING ADVISORY COMMITTEE

The release of NSC's new war planning assumptions provided an impetus for ODM to overhaul its approach to mobilization. As Flemming put it, a "new and overwhelming program problem" had been introduced by the threat of an enemy nuclear attack.[18] ODM would now have to ask: What kind of mobilization would be possible under the unprecedented circumstances of an atomic war? What distinctive resource problems might arise? Who would be responsible for managing them? And what new legal and administrative issues would this contingency present? Such planning, as ODM official Merrill J. Collett put it, required the "analysis of conditions which as yet can only be imagined, but clearly calling for discard of preconceived notions built up out of the World War II experience."[19]

Flemming worked with a group of veteran mobilization specialists on what Shaw Livermore called "a major turn-around in the internal functioning of ODM." Livermore noted that at this time, it was "obvious that some of the most capable generalist thinkers on the problems were by then outside government service" and could "render invaluable service as informal . . . advisors . . . of what was a wholly new area of governmental planning."[20] Flemming recruited several of these thinkers, among them Livermore and Edwin George—who had led work on the War Production Board's system of "vertical controls" during World War II—to a new Mobilization Program Advisory Committee (MPAC).[21] Livermore was a particularly influential member. As a former assistant to ODM chair Charlie Wilson and a member of the NSC Planning Board, he had participated in devising a new schema of governmental preparedness through his work on NSRB's project on post-attack rehabilitation. On MPAC Livermore worked to implement this schema as a new framework of American emergency government.

MPAC undertook a detailed analysis of what George described as "the substantive problems that will arise in a postatomic attack situation and that still await solution." The official title of the group's final report on this analysis was *Planning for Control and Direction of Total Wartime Resources by Non-Military Authority*. MPAC members referred to the report as "the Bedsheet," since it consisted of a list of problem areas that they had worked out collaboratively "on a vast expanse of white paper."[22] In the Bedsheet, MPAC continued to conceptualize mobilization as a problem of "resource management," following the World War II model of industrial mobilization for total war. The kind of war for which resource managers would now have to plan, however, was completely different. In particular, MPAC emphasized "several new concepts that should guide ODM in the future." Most importantly, the Committee advised ODM to shift its emphasis from industrial mobilization during a lengthy conflict to a "future war of annihilation." The "'tone' of ODM's thinking," MPAC urged, "should be oriented toward survival needs in a thermo-nuclear war rather than toward the development of a program having overtones of World War II planning."[23] George characterized this reorientation as "a fundamental shift from a mobilization base to a subsistence and recuperation base."[24]

Through their work on the Bedsheet, MPAC members identified several novel problem areas that ODM would have to address in planning for postattack resource management. For example, a section on "the structure of organization for war" addressed emergency government: Where would civilian war management be located after an attack on the center of government? What communications system would be used? Who would staff the wartime agencies? More generally, what kind of wartime governmental organization would be necessary to manage national resources? Another Bedsheet section, on "human resources," addressed the "the whole problem of safety and survival of the population." This category included the provision of "assistance and succor to victims of attack"—medical care, food, water, and shelter—as well as "the whole question of manpower, the manning of industrial plants." Yet another area, "essential supporting services," concerned the postattack restoration of systems, such as electric power and transportation, that were necessary for the survival of the population and for economic recovery.[25]

MPAC also argued that ODM should adopt new planning tools to envision the governmental capacities that would be necessary for managing the aftermath of a nuclear attack. In contrast to traditional mobilization planning, which was based on deterministic assumptions about the future, these tools were geared to an uncertain future event. As George put it, "entirely new and grotesquely different functions will leap into being with no human experience behind them, and we must imagine the realities for which structural provision must be made."[26] The Committee proposed two techniques for imagining such a future. First, it suggested that the "old military device" of war gaming could be applied to governmental preparedness for a future emergency. As we have seen, NSRB had experimented with such war games, or exercises, in its work on civilian mobilization and, in the context of the PAIR program, proposed that they be used in federal nonmilitary defense planning. Now, MPAC suggested that ODM could use the "civilian war-game" to iteratively test and revise mobilization plans based on the assumption of a large-scale attack on the United States. Second, MPAC recommended that ODM implement the system of bomb damage assessment based on computer simulation that had been proposed by the Interindustry Research Program in the Air Force. At MPAC's urging, in early 1954 ODM hired statistician H. Burke Horton, who had led the Air Force program. As we will see, Horton would direct a new National Damage Assessment Center that developed techniques for simulating a future nuclear catastrophe.[27]

The Bedsheet provided a template for federal officials during the 1950s as they recast mobilization planning into what we now recognize as emergency preparedness. As Livermore later observed, the schema for emergency government it described could be seen in the "evolving internal organization of ODM and its successor agencies," such as the Office of Emergency Planning and the Office of Emergency Preparedness, and it underpinned the 1964 *National Plan for Emergency Preparedness.*[28] More immediately, the recommendations in the Bedsheet guided Flemming and other ODM officials as they took steps to implement the "major turn-around in the internal functioning of ODM" that he and members of MPAC had proposed in 1953.[29]

Administrative Readiness

Beginning in the middle of 1954, ODM began to formulate a mobilization plan for a future war involving a massive attack on the United States. This program evolved into what would be called Mobilization Plan D-Minus.[30] D-Minus planning was based on the assumption of a sudden and devastating atomic attack in which there would be no time to mobilize the military-industrial production system, and in which survival and recovery rather than industrial production would be the central problem. As Flemming put it in 1957, when the first version of this plan was finally completed and circulated to agencies across the federal government, "The primary purpose of Plan D-Minus is to enable the nation to retaliate against and to survive and recover from atomic attacks on the United States."[31]

ODM planners recognized that adjusting to new war-planning assumptions would require an overhaul of ODM's internal operations and organization. Nonmilitary defense considerations would have to be integrated with ODM's existing work on emergency resource management. This new orientation posed major challenges: planning for dealing with the consequences of a future attack, according to a staff memo, was "totally new and not accepted" by officials working on traditional military-industrial mobilization. But with the adoption of new war-planning assumptions, ODM officials concluded that the "time may now be at hand when some major adjustments can be made and the entire organization of ODM reflect the infusion of non-military defense considerations into every mobilization program."[32] They proposed that functions such as industrial evaluation, the assignment of essential functions, and planning for emergency action steps and postattack rehabilitation be housed in a new "Plans and Readiness" office responsible for nonmilitary defense planning.

In 1954–1955, a small group of ODM officials worked to incorporate D-Minus assumptions into mobilization planning. Along with Flemming, this group included veteran mobilization planner Vincent Rock, who served as director of programs and reports (and was also in charge of coordinating ODM meetings with MPAC); General Willard S. Paul, who led the Plans and Readiness Office; Merrill J. Collett of the Continuity of Government Division; and Innis Harris, who was in charge of ODM's test exercise program. Following the approach established in wartime mobilization planning, officials organized their work according to a series of resource areas in which critical postattack emergency actions would be needed, such as health, food, fuels and power, housing, and communications. They then oversaw a series of interrelated projects to prepare for emergency government action in these resource areas: first, to define the structure of a future wartime government; second, to delineate the essential functions that would have to be carried out during an emergency; and third, to compile a checklist of the actions that would have to be taken by the executive branch at the outset of the emergency.

Each of these projects had been initiated separately under NSRB. ODM integrated these various projects in a program of "administrative readiness," developing them "in harmony with each other" and with the rest of its mobilization planning activities.[33]

THE WARTIME STRUCTURE OF THE EXECUTIVE BRANCH

The first area of ODM's work on administrative readiness concerned the wartime structure of the executive branch.[34] For ODM staff, the wartime resource management offices—the War Production Board and the Korean War–era ODM—offered models for a central resource planning agency that would coordinate mobilization across the array of government agencies with emergency responsibilities, and for the wartime establishment of emergency offices organized around resource areas such as transportation, food, and fuels. In the early Cold War, NSRB had developed plans for quickly setting up a similar organization in a future emergency. According to these plans, the Board would be renamed the Office of National Mobilization, and would serve as "the apex of coordination and integration of the various elements of resources mobilization into a comprehensive war program."[35] Indeed, as we have seen, ODM was initially established at the outset of the Korean War—based on these prior plans—as precisely such an emergency resource management agency.

In their work on nuclear preparedness, ODM planners envisioned a similar model of emergency government organization. In a June 1954 memorandum to Flemming, Willard Paul wrote that in a future wartime emergency, the ODM would shift from a planning role to become "predominantly coordinating and directing in nature." To reflect this shift in function, Paul suggested that the "emergency" incarnation of ODM be named the National Mobilization Authority. ODM planners later went through a range of alternative names for their office's future wartime incarnation, settling by the mid-1950s on the Office of War Resources. The task of this office, Paul explained, would be to exercise "the portion of the President's command function which involves the coordination and direction of agency programs involving industrial, agricultural, commercial and financial mobilization, rehabilitation and recovery from attack, and minimal services basic to [the] national will to fight."[36]

The envisioned Office of War Resources was at the center of ODM's plan for a "wartime organizational structure." The plan assigned emergency functions to existing agencies (such as the Department of State and the Department of Defense) and to a cluster of new agencies that would be established at the onset of an emergency. These latter agencies corresponded to the major tasks of emergency resource management: a War Manpower Administration, a War Communications Administration, a War Energy and Natural Resources Administration, a War Food Administration, a War Health Administration, and a War Transport Administration (see figure 5.1). In this wartime organization of government, the Office of War Resources would assume responsibility for—and command over—areas that were normally the prerogative of peacetime agencies such as the Departments of Agriculture; Health, Education, and Welfare; and Commerce.

ODM leadership recognized that wartime emergency resource management presented distinctive organizational challenges. Responsibility for implementing "military and foreign relations policies," Flemming explained in a July 1954 memorandum, was "concentrated within well defined, monolithic organizational structures." The Department of Defense and the Department of State were responsible for these areas during peacetime and would continue to be during a future emergency. By contrast, the organization responsible for emergency resource management was "complex, being found in a large cross section of the Federal departments and agencies."[37]

WARTIME STRUCTURE OF THE EXECUTIVE BRANCH

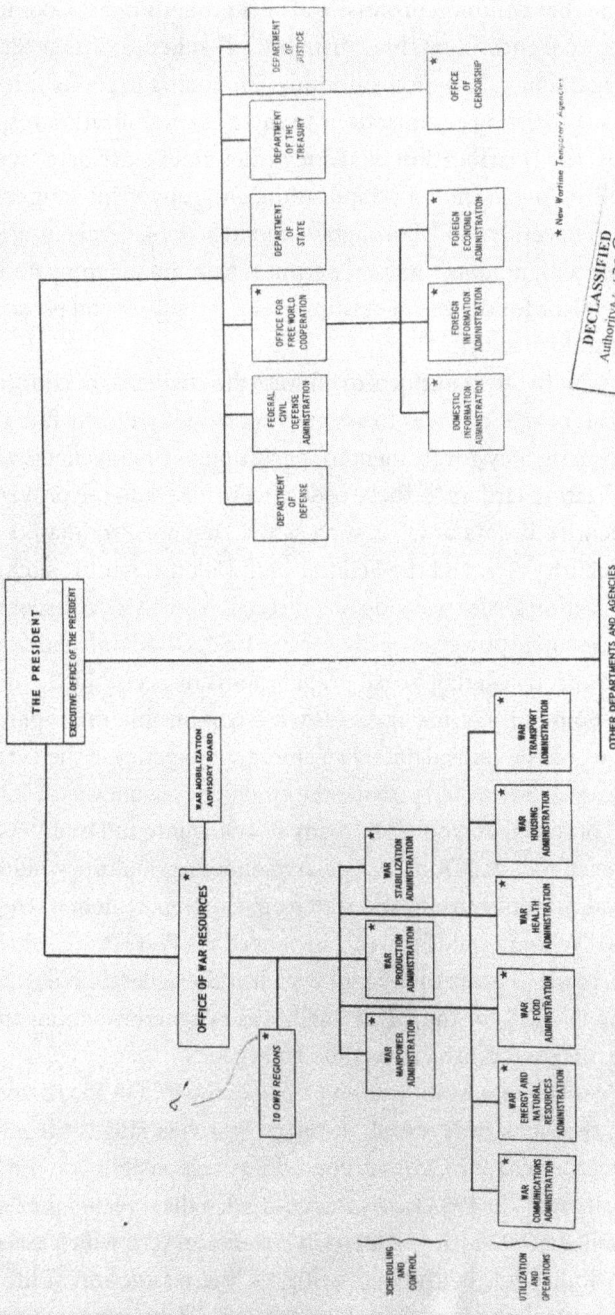

THE PRESIDENT
EXECUTIVE OFFICE OF THE PRESIDENT

WAR MOBILIZATION ADVISORY BOARD

* OFFICE OF WAR RESOURCES

* 10 OWR REGIONS

* WAR MANPOWER ADMINISTRATION
* WAR PRODUCTION ADMINISTRATION
* WAR STABILIZATION ADMINISTRATION

* WAR COMMUNICATIONS ADMINISTRATION
* WAR ENERGY AND NATURAL RESOURCES ADMINISTRATION
* WAR FOOD ADMINISTRATION
* WAR HEALTH ADMINISTRATION
* WAR HOUSING ADMINISTRATION
* WAR TRANSPORT ADMINISTRATION

SCHEDULING AND CONTROL

UTILIZATION AND OPERATIONS

OTHER DEPARTMENTS AND AGENCIES

DEPARTMENT OF DEFENSE
FEDERAL CIVIL DEFENSE ADMINISTRATION
* OFFICE FOR FREE WORLD COOPERATION
DEPARTMENT OF STATE
DEPARTMENT OF THE TREASURY
DEPARTMENT OF JUSTICE

* DOMESTIC INFORMATION ADMINISTRATION
* FOREIGN INFORMATION ADMINISTRATION
* FOREIGN ECONOMIC ADMINISTRATION

* OFFICE OF CENSORSHIP

* New Wartime Temporary Agencies

ODM - August 8, 1956

SECRET

FIGURE 5.1. Wartime structure of the executive branch. As part of its nuclear preparedness planning, the Office of Defense Mobilization envisioned the organization of a future emergency government. The lefthand side of the chart depicts an array of emergency offices, organized according to resource areas such as energy, production, transport, housing, health, food, and manpower, that would operate under an Office of War Resources. The righthand side depicts permanent agencies—such as the Department of Defense, the Department of State, and the Federal Civilian Defense Administration—that would have major emergency functions. *Source:* "Wartime Structure of the Executive Branch," August 8, 1956, Wartime Organization folder, entry 63a, box 209, Record Group 396, Records of the Office of Emergency Preparedness, National Archives, College Park, MD.

Emergency resource management in areas such as food, health provision, transportation, power, and communications would be "directed by administrators with neither common professional backgrounds nor . . . common administrative operandi." For this "distributed" scheme of emergency resource management to function smoothly, Flemming explained, it was necessary to alert "existing organizations to their full mobilization responsibilities." Thus, the "clarification of agency roles at an early date" would make it possible to "avoid friction, duplication, and gaps." This assignment of roles was also necessary to "thoroughly indoctrinate Federal employees in their wartime assignments" and to "permit maximum planning and the testing of these plans for the use of existing Federal facilities and personnel in the event of sudden attack."[38]

A starting point in ODM's efforts to address the challenge of coordinating across disparate agencies was to assign responsibility for the functions of a postattack emergency government to specific peacetime agencies, using Defense Mobilization Orders.[39] These orders were based on the provisions of various executive orders or laws, such as the Defense Production Act, the National Security Act, and the Federal Civil Defense Act. As such, the Defense Mobilization Order was a device through which ODM distributed and exercised executive power. Already in early 1954, ODM staff had drafted several such orders, delegating broad preparedness responsibilities to specific agencies. Some peacetime agencies were responsible to prepare for functions that would be carried out by an emergency agency in the event of a war. In such cases, they would provide the emergency agency with "nuclei organizations" of staff that would be "ready to swing into full mobilization action."[40] For example, staff from the Department of Agriculture would run a new War Food Administration, and staff from the Department of Health, Education, and Welfare would form the nucleus of the War Health Administration. In other cases, a peacetime agency, such as the Federal Civil Defense Administration (FCDA) or the ODM itself, was assigned functions that it would have to carry out during a wartime emergency.

A March 1954 Defense Mobilization Order to the FCDA illustrates the standard form that such orders took. It began by rehearsing the legal and administrative grounds for FCDA's preparedness responsibilities: the 1947 National Security Act, the 1953 Reorganization Plan that created the "new" ODM, and Public Law 920 (the Federal Civil Defense Act), which assigned FCDA responsibility for measures relating to "the protection of life and property against attack and for dealing with the civil defense emergency conditions arising out of attack." The order then laid out specific preparedness functions that FCDA would assume in a future wartime emergency: esti-mating requirements for civil defense purposes; coordinating with other departments to ensure that adequate capacity existed to produce "civil defense materiel"; meeting manpower requirements for emergency relief and postattack rehabilitation; assisting with planning for emergency relo-cation sites; assisting state and local governments in planning measures to protect vital production facilities; and participating in the development of a "unified system for assessment and reporting of attack damage."[41] ODM issued similar Defense Mobilization Orders to other federal agencies, thus distributing responsibilities for emergency preparedness across the federal government.

ESSENTIAL FUNCTIONS

These broad delegations of responsibility led to a further question: What specific functions would emergency agencies have to perform in the after-math of a future attack? ODM addressed this question through a second area of its administrative readiness program, the analysis of essential wartime functions. As Flemming explained, essential functions analysis sought to "define the wartime job" as it related to "direct military operations," the pro-vision of "emergency disaster relief," and the "bedrock" operations of civil-ian government.[42] Work on essential functions began in early 1954, when, as part of its continental defense strategy, the National Security Council instructed ODM to develop—with "the utmost urgency"—"emergency plans and preparations to insure the continuity of essential wartime functions of the Executive Branch."[43] ODM soon reported back to the council on its progress. The office had completed an emergency relocation plan for those agencies "having essential wartime functions and [that are] located at the seat of government." In the event of emergency, according to the plan, such agencies would evacuate to relocation sites, from twenty to three hundred miles from "the Washington target zone."[44]

But the bulk of the work on essential functions, ODM reported, remained to be done—it was still necessary to define the "wartime job." To do so, ODM set up a process of distributed planning that relied on "the Departments and agencies which constitute the working force of the Federal government." In line with its more general effort to reorient mobilization planning toward preparedness for survival and recovery, ODM advised the agencies conduct-ing self-assessments to avoid "reliance on or acceptance of World War II administrative plans, organizations, and procedures in view of the radically new weapons and conditions of war which will pertain." To assist the agencies in imagining such an unprecedented future, ODM included in its request a scenario of a future emergency. The initial air attack would be "massive" and would be "made with sufficient strength and types of weapons . . . to attempt to paralyze the United States as a war machine." The apparatus of governmental administration would be severely damaged. Government personnel living in principal target areas would be endangered. Communication and transportation between cities would be so disrupted that regional offices of the federal government would have to operate on their own. Washington, DC, would be a primary target, and the federal government would be reduced to "a bedrock level of operations." All activities "not absolutely essential to the bedrock continuity of this Government" would have to be terminated.[45]

To standardize this work across disparate agencies, ODM staff circulated criteria for evaluating "the essentiality of Department or agency functions to bedrock operations of this government"—defined as those functions required to assure "national survival under attack, the complete exploitation and mobilization of basic resources, and military retaliation." Some of these functions were essential for "the conduct of military operations" necessary "to secure military victory." Others had to do with the management of essential resources and the continuity of essential services: the allocation of manpower; the production, control, allocation, and distribution of food; and the management of transportation, communications, and energy. Still other essential functions concerned relief and recovery: the restoration of utilities and public works, as well as the provision of medical care, water, sanitation, food, and shelter.[46]

By early 1955, ODM had completed what National Security Advisor Robert Cutler described as "colossal work on this project."[47] Lists of essential functions had been compiled for twenty-six federal government agencies and their 175 constituent bureaus—the array of offices housed within the agencies. The project had identified essential functions for 1,400 "functional groupings."[48] These lists of essential functions were compiled in a thick binder, organized according to the agency responsible for each listed function.[49] They offered an exhaustive accounting of the problems that would arise in a future war emergency and the governmental functions that would be required to manage them. Essential functions analysis thus generated a new understanding of the organizational structure, normative requirements, and performance of civilian government from the perspective of its activities in a future national emergency.

In some cases, essential functions designated broad areas of responsibility. For example, the essential functions of the Department of Health, Education, and Welfare included responsibility for "policy formulation, executive direction, and overall operation and coordination of civil defense emergency actions"; as well as "direction and coordination of operations, operational intelligence, [and] planning for future requirements." In other cases, much more specific functions were identified for "constituent bureaus." For example, the Public Health Service would assist FCDA in determining health requirements, provide aid to states on mass evacuation, arrange training in monitoring for radiological fallout, and develop plans for environmental health; water, milk, and food sanitation; and waste and vector control.[50] The Social Security Administration would ensure that records of "each covered individual" would be maintained and that it could continue to distribute benefits in a war emergency. The Department of Commerce and its constituent bureaus identified 148 distinct functions. Some comprised vast areas of activity, such as the Business and Defense Services Administration's responsibility for developing and managing a defense materials system and for determining wartime civilian needs, including recommended "levels of essential civilian goods, estimates, material requirements."[51] Other functions of the Department of Commerce were highly specialized. For example, the Bureau of the Census was responsible for machine tabulation and for the operation of its UNIVAC computer, which would be used to support the emergency management functions of other agencies.[52]

EMERGENCY ACTION STEPS

A third component of ODM's administrative readiness program was the "emergency action steps" project, which addressed how emergency agencies would be set up, and the authority on which wartime actions would be taken. Concretely, the project involved the compilation of lengthy checklists of the specific actions that would be taken by the president and by executive branch agencies at the outset of a future emergency. These lists of action steps eventually formed the core of ODM's postattack mobilization plan, D-Minus.

Like the wartime organization and the essential functions programs, the emergency action steps project began in NSRB. This project was initiated to provide the president with "an instrument to aid him in assuring coordinated and rapid executive action in an emergency," as NSRB director Jack Gorrie explained in 1952.[53] Its aim was "to strengthen the unity of actions throughout the Executive Branch in the event of crisis."[54] The emergency action steps project moved to ODM in 1953. As with the determination of essential functions, the project was organized through a distributed planning process.[55] Each executive branch agency was instructed to compile a checklist of actions within its "field of responsibility" that would have to be initiated "within the first two weeks of a general emergency."[56] Agencies were to follow a standardized tabular format in compiling these checklists, indicating the action step to be taken, the "enactments"—such as statutes or executive orders—that provided the authority for a particular action, and the current state of planning. The goal of this process was to provide "a continuous indication of the existing degree of readiness for executive action in the event of war or of total mobilization," and to "reveal the areas and levels at which additional preparations were needed."[57]

A briefing guide that ODM circulated to the participants in a test exercise illustrates how the checklists of emergency action steps functioned in preparedness planning. After a short initial section that laid out the scenario of an enemy attack, the bulk of the briefing guide consisted of a long list of actions to be taken by government officials at the outset of a future war. According to the guide, the president would take the first actions, to set up the emergency organization of government and authorize emergency powers. These presidential action steps included strategic decisions concerning issues such as the use of atomic weapons, the provision of military assistance to civil authorities, and the "mobilization of the Nation's Resources." The president would also issue an executive order to "establish War Organizations and assign functions to new and existing agencies." Once these presidential actions had been taken, ODM would take steps to set up an emergency government. The director of ODM would issue Defense Mobilization Orders charging the newly created emergency agencies—such as the War Communications Administration, the War Transport Administration, the War Stabilization Administration, and the War Health Administration—with wartime functions.[58] The checklist of action steps also included measures that ODM would take to ensure the continuity of government, such as activating relocation plans and setting up a damage assessment center.

A final set of action steps outlined the measures that would be taken by emergency government agencies to carry out essential wartime functions. These steps were organized into "resource areas" that corresponded to the vital systems whose continuity would have to be assured in a future emergency: communications, food, health, housing, manpower, power and fuels, transportation, and so on. For example, in the area of power and fuels, emergency actions included items such as assessing damage to electric power facilities, channeling available power and fuel supplies to designated emergency uses, and ordering utilities to direct services to the most urgent military, industrial, and essential civilian needs. In some resource areas, emergency actions would be taken by peacetime agencies with significant wartime responsibilities. Thus, the guide indicated that the Department of Agriculture would estimate the effects of the attack on food requirements and supplies and set up and manage an emergency food distribution system, using the tools of wartime mobilization planning: priorities ratings, anti-hoarding orders, management of food stockpiles, and government requisitioning. In the area of transportation, the Department of Commerce and the Interstate Commerce Commission would "marshal all available equipment and materials without regard to ownership," authorize restoration of "rail lines of national importance, clearance of waterways as post-attack requirements indicate," and "arrange for temporary bridging of streams to restore most essential highway routes." The briefing guide also outlined action steps to be taken by emergency agencies. Thus, after ODM issued orders to create a War Health Administration and appointed a war health administrator, that office would work with mobilization and civil defense authorities to "govern the coordination, allocation, and use of all health facilities."[59] In the process of mobilizing for a future emergency, thousands of such action steps were assembled and assigned to peacetime agencies for planning purposes.

Imaginative Enactment: The Test Exercise

To test its plans for emergency government, ODM established a program of scenario-based exercises. Cold War–era test exercises, such as the large-scale Operation Alert exercises held annually beginning in the mid-1950s, are best known for their role in civil defense planning. Civil defense officials used them to enlist the public in nuclear preparedness, albeit with limited success.[60] A less well-known but arguably more significant aspect of these exercises was their central role in ODM's program of administrative readiness.[61]

Vincent Rock described the rationale for such "mobilization war games" as a means to "insure continuously improved mobilization plans." These games, he explained, would "test the Government's readiness to meet the situations envisaged in the agreed-upon mobilization war plans."[62] Similarly, the Mobilization Program Advisory Committee argued that a program of exercises would enable ODM to test out specific issues of concern and thereby find gaps or weaknesses in existing plans. This testing process could point to "inadequacies in pre-attack mobilization and stabilization planning, procedures, and organization structure," leading to "suggestions for the operation of future exercises" (see figure 5.2).[63] In this way, a "systematic war-gaming undertaking" would, Shaw Livermore suggested, provide "a continuous and effective way of enlisting and applying brain power to the many ramifications of an unknown problem."[64]

According to Innis Harris of ODM's Plans and Readiness Office, the program of mobilization exercises gained support from President Eisenhower, who was familiar with the war-gaming technique from his military career. After the release of the Bull Committee's report on continental defense and the first Soviet H-bomb test, ODM presented the president "with what was called a mobilization plan." As Harris recounted, the president responded: "No plan is any better than it works. Test it." For Eisenhower, exercises, or war games, were common-sense tools for testing out plans. But as an instrument of civilian planning for a future emergency, exercises were, as Harris put it, "an activity entirely new to the Federal Government."[65]

ODM initiated its program of test exercises in mid-1954. In its first exercise, ODM engaged in a "very limited" test focusing on agency relocation plans.[66] "A few high level officials," Harris reported, "assuming several hours['] warning of a hypothetical attack, left Washington with a checklist of possible actions." They assembled in a cave in a remote location, "with water dripping from the ceiling and oozing from the walls." The exercise lasted only a few hours, but "a great deal was learned," according to Harris, about the next steps in mobilization planning. ODM designed the exercise to focus attention on its current projects, specifically its work on "alerting procedures, transportation, adequacy of essential records and communications." After the exercise, Harris recalled, a "new push was given to the development of physical features for carrying on government" such as "relocation sites, an interagency communication system, and the further development of plans."[67] ODM officials concluded that the exercise had "succeeded," at least in the sense that it revealed gaps that could be addressed through further planning. As Flemming put it, the test "was invaluable" to ODM "in focusing a need for more intensive planning to make [participating agencies'] emergency operations feasible."[68]

ODM conducted its next test exercise, Operation Readiness, in November 1954. In contrast to the first exercise, which included only a handful of officials, Operation Readiness involved "a sizable portion of Federal employees involved in essential functions of Government." By placing these officials in their relocation sites and presenting them with a detailed scenario of an

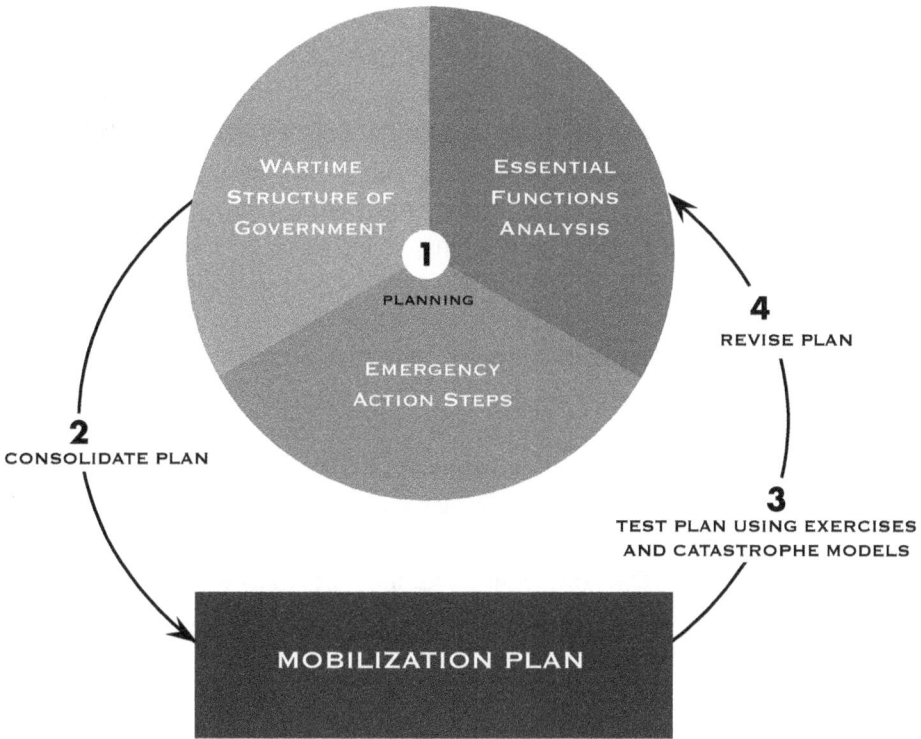

FIGURE 5.2. Administrative readiness. ODM's program of administrative readiness worked through an iterative process of formulating, testing, and revising its mobilization plan. This process gave mobilization officials a way to anticipate an uncertain and unprecedented event—by enacting a future catastrophe. *Credit*: Janice Yamanaka-Lew.

attack, the exercise aimed to imaginatively enact a future catastrophe, providing participants with an "experience" of the situation they might face in a future war. Described as a "continuity of government" test, the exercise addressed the challenges that would arise in relocating key government agencies outside Washington, DC. In advance of the exercise, ODM planners prepared a set of "substantive 'problems' and war-gaming activities" that would structure participants' experience.[69] The substantive problems and the attack scenario were laid out in a document entitled "Global War Plan," distributed to ODM staff in early November. In keeping with the new planning assumptions, the scenario enjoined participants to imagine that they faced the immediate aftermath of an atomic attack: "It is assumed that this attack took place today. With this background: What should we do? How ready are we to do it? How can we quickly fill the gaps?"[70]

Operation Readiness was held in the federal government's secret reloca-tion site, sixty miles west of Washington, DC, in Mount Weather, Virginia. At the beginning of the exercise, participants were briefed on "the magnitude of destruction to many cities and industrial complexes together with the heavy damage to communication and transportation services." The exer-cise focused on managing the immediate aftermath of a devastating enemy attack. Given "severe damage to Washington, DC and general disruption of the economic system throughout the country," the central challenge was not how to rapidly build up an industrial war machine, as in prior wars. "The restoration of central program direction"—that is, management of military-industrial production—would "require from three to six months" and was thus "beyond the limits" of the exercise. Instead, the scenario instructed participants to "address ourselves to the basic problems that the government would have to face during this immediate post-attack period."[71]

In 1955, ODM conducted a more elaborate test exercise, now fully inte-grated with FCDA's Operation Alert program, which had been initiated the prior year. As historian Guy Oakes documents, the exercise included the president and cabinet members, staff from most agencies of the federal gov-ernment, "scores of cities that had been marked for 'destruction,'" as well as business and labor organizations.[72] As a joint exercise between ODM and FCDA, Operation Alert 1955 linked the executive branch's emergency plan-ning to the overall governmental system, testing the relationship between federal agencies and state and local governments. As we will see, this exer-cise proved to be a significant moment in ODM's program of administrative readiness, exposing a gap in (and sparking a wide controversy about) the Office's work on a critical area of preparedness planning: how to establish central authority to manage resources for survival and recovery after a ther-monuclear attack.

Executive Authority in an Emergency

In their work on administrative readiness, ODM planners identified sev-eral problems that a nuclear attack would raise concerning relations of authority among disparate entities in the US governmental system. How, in the aftermath of attack, would the president establish centralized con-trol? How would decisions made by the envisioned Office of War Resources (OWR)—and the array of emergency offices working under it—be trans-mitted both to agencies in the federal government and to state and local governments?

A model for coordinating emergency actions among agencies within the federal government had been established during World War II: the machinery of delegation and coordination that was set up to manage wartime mobilization. In its projects on emergency organization and the assignment of essential functions, ODM adapted this existing machinery of delegation and coordination to address the novel challenge of nuclear preparedness. But mobilization planners warned that the World War II model would not be sufficient for a future war. As Flemming put it in June 1954, nuclear weapons presented a "new and overwhelming program problem." In areas that were attacked, he argued, state and local governments would be either overwhelmed or completely incapacitated; the challenge of meeting survival requirements would place "unprecedented strain" on "transportation and storage of medical and food supplies." The provision of relief in devastated localities would require centralized coordination and control, thus leading to a "re-examination of Federal-state relations."[73] The problem of nuclear preparedness also forced ODM to examine the role that the military might play in the federal provision of resources to stricken populations—and more generally, to consider civilian-military relations under war conditions. Indeed, the specter of nuclear attack led to a broader debate on the question Clinton Rossiter had raised in his 1949 reflection on American government in the atomic age: How could a postattack emergency be managed within the framework of the constitution?

THE CHAIN OF COMMAND

In early 1955, ODM turned to these problems of executive control, in particular to the issue of federal-state relations that Flemming identified. In a January 1955 memo to General Paul entitled "Federal Organization to Deal with the Bombing of Urban Target Centers," Merrill Collett of ODM's Continuity of Government Division described the coordination across governmental scales that would be required for postattack emergency response. The "basic question" that the government would face, wrote Collett, "is that of national existence." The substantive problem for ODM was "how to assure primary attention and total effort to meet the needs of the attack"—initially "in terms of keeping people alive," and then later to address the "problem of marshalling productive facilities, materials and manpower on a national scale to rehabilitate and to strike back."[74]

Collett moved methodically through the various resource areas around which ODM's preparedness planning was organized, outlining the coordination problems—both among federal agencies and between the federal government and local governments—that might arise in each area. For example, responsibility for emergency provision of food would rest primarily with mayors, and the provision of relief would be implemented through the "regular organization of city government." But many aspects of food supply would have to be coordinated with higher-level authorities. Stocks of food held by the federal Commodity Credit Corporation would be released "as local supplies prove inadequate"; field offices of the federal Department of Agriculture would coordinate food rationing; the War Food Administration, operating under the OWR, would manage the inspection for radiation contamination. Similarly, in the area of health resources, Collett anticipated that as local health systems were overwhelmed, "the immediate pull from the military will be felt." The transport of medical supplies, personnel, and other resources between regions of the country would "be subject to over-all direction from the national level." Power supply would also be immediately placed under regional control.[75]

How would this tangle of responsibilities be sorted out? And how would centralized authority for the allocation of resources be established? Collett proposed that the various organizations charged with emergency resource management in a particular region could be brought together in a "hard-hitting organization under single direction." Requests for resources to address the "human need problem" would be transmitted to a regional coordinator who would work with the OWR and the emergency agencies arrayed under it. Military assistance would be provided to civilian authorities if needed, but the military would not "enter this picture until Executive responsibility and authority for these areas has been clearly indicated so the Army may know from whom it will honor requests for assistance." Collett emphasized that this model would neither displace the existing structure of civilian government nor create a new one. "In times of crisis," he wrote, "new organizational arrangements cannot be superimposed or supersede existing governmental establishments at the local level or level at the point of action and get the job done."[76] Collett's proposal thus sought to build "the organization to meet an emergency on the firm foundation of what now exists," while establishing greater centralization and expert control of resource management decisions. "Recognizing the barriers to present action due to artificial boundaries, and in some cases inadequate and ineffective state organizations," Collett explained, his proposal would "[weld] together into a single regional War Resources office all elements which can be applied to the immediate attack problem and also the subsequent rehabilitation problem."[77]

General Paul transmitted the outlines of Collett's proposal to Director Flemming in a memorandum on "Governmental Chain of Command to Deal with Urban Problems in an Attack Emergency." Paul emphasized that the key question to be addressed was how the president would exercise his "full authority to pre-empt all resources of the country, as needed to meet the emergency." The structure of the "chain of command," he argued, "must be worked out in advance if Government (Federal, State and local) can hope to cope with the indescribable chaos of a thermonuclear attack." Following Collett's model of a "regional coordinator," Paul suggested that a single federal official would be responsible for marshaling and distributing all resources within a given area of the country. Crucially, Paul specified how the various governmental entities involved in survival and recovery would, as Collett had suggested, be "welded" together. The governor of each state would "be considered part of the President's emergency staff," carrying out orders transmitted by ODM regional officers.[78] In this vision, the president would exercise command authority over state and local officials, effectively eliminating the sovereignty of state governments established in the Constitution.

To develop procedures to implement this proposal for establishing a governmental chain of command after a nuclear attack, Flemming established an interagency Emergency Action Task Force in April 1955. The task force was initially charged to study "the problem of how to maintain the civilian chain of command from the President to the governors in the event of an attack."[79] Members of the task force represented the key agencies responsible for domestic preparedness in the federal government. Paul and Collett coordinated the task force's work. The group also included General Harold Huglin of ODM—the former director of the Air Force Management Directorate (see chapter 3)—Lieutenant General Frank Tharin of the Joint Chiefs of Staff, Hubert Gallagher of FCDA, and Colonel Barnett Beers, assistant for civil defense in the Office of the Secretary of Defense. Beers had led the civil defense study in the Strategic Bombing Survey at the end of World War II and had played a central role in discussions of postwar civil defense during the late 1940s (see chapter 4).

In addition to asking the Emergency Action Task Force to develop procedures for establishing a "governmental chain of command," Flemming also sent a memorandum to President Eisenhower seeking approval for ODM's approach. In a future war, Flemming wrote, the president must be in a position to "requisition all of the nation's resources—human and material." The governors of the states would "would respond to orders from the President relative to the utilization of our resources." In turn, by delegation from the president and the governors, local officials would be in a position "to utilize all of the resources available to them in an effort to make it possible for survivors in their area to recover from the initial shock."[80] In his response to Flemming, Eisenhower agreed that centralized executive authority would be required to manage an emergency response, given "the chaotic conditions that would [occur] in the event of widespread bombing attack on the United States." But he was uncertain about the authority on which such a structure would be based. Eisenhower instructed that the ODM proposal "be thoroughly studied by the Attorney General to determine whether any of the actions herein contemplated should be considered in excess of the authority granted to the President by the Constitution and by the Civil Defense Act and other laws."[81]

Upon review of the proposal, Attorney General Herbert Brownell declined to pass judgment on the general question of the limits of executive authority in a future emergency. This issue, he argued, could not be legally clarified in advance of the situation itself. "When it is recognized . . . that the full sweep of the President's powers has never been specifically defined and, in fact cannot be since their extent and limitations are so largely dependent upon particular conditions and circumstances," he wrote, "the difficulty of responding becomes evident." If the president did, indeed, have the authority to take the steps Flemming proposed, it did not rest in any statute but in the constitutional responsibility of the president—beyond any law—to ensure the survival of the nation. "Any program which comprehends the possible requisition of 'all of the nation's resources—human and material,'" Brownell wrote, "of necessity does not lend itself to definitive legal conclusions." Meanwhile, Brownell deemed Flemming's specific proposal that state governors would be considered "on the staff of the President" to be "legally indefensible." This arrangement collided "with the basic concept of the separate and independent sovereignty of the several States."[82]

Brownell's deputy, Assistant Attorney General Lee Rankin, later followed up with ODM on the question of the legal authority under which an Office of War Resources would manage the nation's resources in a postattack situation. Noting the wide range of planned OWR activities, including "control of the movement of goods," "control of basic materials," "control of telephone, telegraph and radio communication," and "control of civilian manpower," Rankin asked ODM to provide information concerning "the authority under which you believe the contemplated action could be taken, especially the statutory citations where relevant."[83] ODM general counsel Charles Kendall responded by arguing for the unbounded authority of the executive in an emergency situation, even in the absence of existing statutory authority. It "remains painfully evident . . . that there are areas of essential action in which statutory authority is insufficient or completely absent," he conceded. "However diligently we may search out bits and pieces of statutory authority for this or that contemplated action, we will end up relying upon authority other than statutory if an attack comes tomorrow." For Kendall, the president's constitutional obligation to ensure the survival of the nation justified the exceptional authority of the executive in an emergency: "That the President is authorized—in fact obligated—to take actions necessary to national survival even in the absence of legislative authority, seems clear. The Constitution does not contemplate surrender or present the alternatives of constitutional defeat or unconstitutional victory." The authority of the executive, Kendall claimed, was based on "the necessity for the action taken."[84] Nothwithstanding Kendall's arguments, in the absence of a definitive legal judgment from the attorney general, the status of ODM's proposal to establish a hierarchical command structure in a future war remained uncertain.

EMERGENCY AND THE SPECTER OF MARTIAL LAW

ODM's deliberations on the locus of authority in a future wartime emergency—particularly on the relation between federal and local government—unfolded in parallel to a broader debate among federal officials, legislators, and civilian experts about preparedness for survival and recovery after a thermonuclear attack. This broader discussion concerned questions about the organization of nonmilitary defense that had not been resolved in the aftermath of World War II. Should the provision of relief to the civilian population be the responsibility of the military establishment, or should it be assigned to a civilian agency? Should the federal government take a leading role in civilian preparedness? Or should responsibility for civilian relief fall primarily to states and localities?[85]

The Federal Civil Defense Act (1950) had offered a temporary resolution to these questions, assigning operational responsibility for civilian relief and recovery to city and state governments, and charging the Federal Civil Defense Administration with coordinating local activities and providing technical assistance. The Act also referred to potentially vast emergency powers that would be wielded by the FCDA administrator in a future wartime situation, such as the power to procure and distribute resources for civil defense purposes "without regard to the limitations of existing law." But the Act defined these powers only in general and vague terms and provided no specific means for their exercise.[86] Many advocates of nonmilitary defense were critical of this arrangement from the beginning and assumed that in an actual attack, these provisions for civil defense would prove insufficient. For instance, the 1952 report of Project East River argued that the Civil Defense Act's emphasis on local planning would be totally inadequate to a future war fought with thermonuclear weapons.

These criticisms intensified as the Soviet nuclear arsenal grew larger and more powerful.[87] The Continental Defense Planning Group—a military advisory committee under the Joint Chiefs of Staff—argued that a thermonuclear attack on a large city would create overwhelming challenges to civil defense capacities. "The dead from a single bomb," the group argued, "would number in the hundreds of thousands to millions. The burned and injured would be in equally great numbers. Every hospital, ambulance, doctor, nurse and all available air transportation within a range of five hundred or more miles would be needed to cope with the problem. Housing and food would be problems of unrealized magnitude." For the military planners who had been "living with this problem," reported a newspaper article that described this group's deliberations, "the failure of Congress and the people to understand the danger, and the need to finance and to organize on a large scale to minimize it," was "incomprehensible." Since the military was the only agency "in the United States capable of handling the problem," according to the planning group, such a disaster could only be managed by "a dictator in the area, backed by martial law." The armed forces did not "like to take on these responsibilities," which detracted from "their main business of war." But given the lack of governmental preparedness—and given that the military was called in "for every disaster, large or small"—it could not "escape being forced to act in the greatest of all disasters."[88]

The question of how national resources would be brought to bear at a local level to provide postattack relief to the population was addressed in hearings held by a subcommittee of the Senate Armed Services Committee, chaired by Estes Kefauver. The subcommittee's final report, issued in June 1955, as ODM was deliberating on the emergency "chain of command," echoed the Continental Defense Planning Group's view that unified action by the federal government to aid stricken localities would be necessary to manage a postattack situation. "An H-bomb disaster," the report argued, "would require a vast amount of direct Federal assistance to the cities and States." The Civil Defense Act had sought to "relieve the military" of civil defense. But testimony at the hearings had indicated that "the present state of unpreparedness of the civil defense effort would leave no alternative in case of an attack except for the President to declare martial law and use the Armed Forces—which historically has been the disaster organization for the nation."[89]

A report by the National Planning Association (NPA), submitted as evidence in the Kefauver hearings, also warned that it might be necessary to invoke martial law following a future attack given the current state of preparedness. "There is good reason to fear," the NPA study warned, that "if adequate preparations are not made in advance, an attack disaster would perforce bring military administration in its wake." If civilian government were to "buckle under the burden of attack, a cry would go up for the President and State governors to declare martial law." Such a declaration, the report contended, would be a clumsy and disastrous recourse "at a time when and in locations where resourceful and imaginative administration would be most needed." The way to avoid "such evils" was to "take steps now to determine how our civil institutions can best be prepared to withstand a blow, and then to proceed, promptly but calmly, to make the needed preparations."[90] A number of mobilization specialists from ODM and the Mobilization Program Advisory Committee participated in the NPA study, which found that an effective program of administrative readiness would make it possible to avoid a future declaration of martial law.[91] Such a program would require steps to ensure coordination among civilian authorities, such as "working out agreements with State and local authorities"; arranging for "support in the way of medical personnel and medical care and equipment"; providing for "regional stockpiles of food and medical supplies"; and planning for the mobilization of trained personnel to areas that had been attacked. Only through "practice and testing in advance," it concluded, "can these varied groups learn to work in combination sufficiently well to forestall the collapse of the effective civil administration of what will be a chaotically disturbed economy."[92]

Operation Alert 1955

Soon after the conclusion of the Kefauver hearings, government preparations to establish authority relations in a future war were put to the test in the 1955 Operation Alert exercise, which ODM conducted in collaboration with the Federal Civil Defense Administration. As FCDA director Val Peterson explained, the exercise was designed to test the ability of "civil defense organizations at the city and State level to meet the problem that was posed for each one of them and to test the availability of resources and the capability to move those resources to meet the problems that were occasioned by the attack."[93] The key "event" in the exercise was a declaration of martial law by President Eisenhower to use the military to aid stricken localities. Eisenhower's declaration set into motion a controversy over the limits of executive authority in a future emergency: Was the centralized authority required for postattack response compatible with core features of the American constitutional system, such as the sovereignty of state governments and the supremacy of civilian authority?

An appraisal of postwar civil defense later noted that the 1955 exercise "prompted broad reexamination of the emergency requirements and functions of government."[94] Some officials defended Eisenhower's declaration, arguing that in a future nuclear emergency, the president would be compelled to take any measures necessary to ensure national survival, whether or not such measures violated key tenets of the Constitution. Mobilization experts in ODM ultimately worked toward a different solution. Building on their program of administrative readiness, they sought to equip the government to manage a postattack emergency within the framework of American constitutional government.

"LIMITED MARTIAL LAW"

Held over three days in June, Operation Alert 1955 (OPAL 55) was structured around the scenario of a devastating nuclear attack: sixty American cities were struck by sixty-one bombs ranging in size from twenty kilotons to five megatons. Nuclear fallout covered sixty-three thousand square miles. The bomb damage pattern envisioned in the exercise generated what Innis Harris called a "shocking number of casualties by World War II standards": more than eight million deaths on the first day, and eight million more within six weeks of the attack. The president, cabinet members, and leaders of executive branch agencies played prominent roles in the exercise; fifteen thousand federal workers were moved to relocation sites.[95]

The scenario depicted a desperate situation for survivors: federal stockpiles could provide basic necessities for only about two and a half million people for three weeks after the attack, while twenty-five million people were in dire need of food and water.[96] Local governments were overwhelmed by the task of providing food, housing, and essential services, and FCDA was ill equipped to help them. Although these elements of the scenario had been planned months in advance, President Eisenhower insisted that aides not brief him. In his view, the point of a war game was to force participants to wrestle with challenges that could not be foreseen ahead of time. "I refused to let them tell me the conditions under which this problem was to be operated," he explained, because "played decisions should be made in the proper atmosphere of emergency."[97] As the exercise began, ODM officials presented Eisenhower with a description of the massive destruction and vast numbers of casualties caused by the attack. "I was suddenly told," he later recounted, "that fifty-three of the major cities of the United States had either been destroyed or so badly damaged that the populations were fleeing; there were uncounted dead; there was great fall-out over the country."[98] The president was also confronted by nearly total destruction of the government apparatus in attacked areas. According to one account of the exercise, "Local government had totally collapsed under the weight of the panic of its citizenry, and political boundaries had been erased by the ensuing chaos."[99]

In response to the first reports of mass devastation, Eisenhower declared a state of "limited martial law" to take control of local emergency response.[100] "As I saw it," he later explained, there was "no recourse except to take charge instantly; because even Congress, dispersed from Washington because of a bomb, would take some hours to meet, to get together, to organize themselves."[101] The text of the martial law declaration laid out the central problem of how to manage resource allocation in a situation in which civilian government has collapsed. Since "civilian authority and control of essential functions has broken down," it began, "immediate restoration of authority in such areas" was needed. Moreover, martial law was required "because of the necessity to mobilize all of the nation's resources at once for war and to transport men and materiel across state lines in order to provide relief for the stricken areas."[102] The declaration of martial law proved highly controversial and brought attention to the question of the authority of the executive in a postattack emergency. In this sense, then, the exercise achieved its aim: it exposed a gap in mobilization planning that required further attention. As Eisenhower noted, the incident "did at least have this benefit: to cause us to study more deeply and in a more analytical fashion our whole history to see what would be the best thing to do under such circumstances."[103]

Eisenhower's declaration apparently came as a surprise to the civil defense and mobilization officials who had designed the exercise. "No one actually had any way of knowing that the decision would be made to declare martial law," FCDA administrator Val Peterson later testified. "This injected a somewhat unexpected note."[104] Similarly, ODM's Innis Harris recounted, "No one was prepared for . . . the martial law action." Such a declaration "wasn't in the plan."[105] As Clinton Rossiter later observed, "One of the remarkable events of that three-day test of our readiness for atomic war was the startled discovery by Mr. Eisenhower and his staff that 'the inherent powers of the Presidency' . . . would be the nation's chief crutch in the aftermath of the ultimate disaster."[106]

Although the text of Eisenhower's martial law declaration laid out the rationale for the decision, it offered little explanation of what exactly "martial law" meant. As a later congressional report put it, careful examination of the proclamation revealed "a confusing and contradictory mixture of decrees."[107] Did it mean military administration of the country, in which civilian officials would take orders from military commanders? Or was the military merely to provide assistance to civilian administrators, who retained their authority? Did it mean a "dictatorship," in which decree rule by the executive replaced legislative and juridical processes? Why was martial law declared in the entire country rather than only in those areas that had been attacked? "Everyone was speculating as to what it meant," Harris later recalled. "The lawyers were running for the books to draft an appropriate proclamation, looking back to see what Abraham Lincoln did."[108]

As we have seen, Collett and Paul's proposal for the postattack chain of command had not anticipated a declaration of martial law. Indeed, they sought to avoid recourse to martial law by planning for civilian control of the allocation of resources. For this reason, ODM director Flemming saw Eisenhower's decision as a "key development" of the OPAL 55 exercise: it "left no doubt that in case of an attack of the magnitude envisioned in our test, the President would call for martial law throughout the country."[109] The implication was that ODM should draft plans for martial law rather than working to develop complex standby powers for coordinating government response, thus cutting "at least one and possibly two years off the work of preparation that ODM must do in planning to counter an attack."[110] "We now know what to expect," Flemming concluded, "and can make definite plans to handle the situation." To begin work on such plans, ODM turned to the interagency Emergency Action Task Force, changing its objective to "developing necessary implementation and procedure" for a future proclamation of martial law.[111]

Members of the Emergency Action Task Force were sharply divided about their new charge. For some, OPAL 55 had demonstrated the inadequacy of the organization of civilian preparedness to manage a nuclear attack on the United States. Admiral Arthur Radford argued that under the scenario envisioned in the exercise, "the requirement would be for action on the spot," something that FCDA could not provide since it "has no national organization and has only an advisory function." ODM representatives maintained that it would be necessary to establish a centralized agency with command authority. The crucial question, General Paul argued, was "how best to win [the] war within [the] structure of orderly government." Similarly, Huglin insisted that the objective was to "keep government strong enough to win the war." "Whether called Martial Law or reinforced civilian control," he argued, "the control agency must have authority to enforce draft of materials, facilities, people and dollars." Huglin proposed that the group draft a standby "emergency document on martial law" that would be ready for immediate issue in a future wartime emergency.[112]

Representatives of the military and FCDA pushed back against Huglin's suggestion to draft a declaration of martial law. Beers argued that "such a document is hotter than an H-bomb"; it would be "terrifically detrimental" to even consider it. Gallagher of FCDA agreed that martial law had "become a political issue," arguing that "martial law is a palliative, a way out." "For 12 years, Beers and I have worked to build up civil defense so [the] Military [will] not need to get into the act." If the military were to take over, he argued, "states and communities are released of responsibility"; it would be an "abandonment of everything we've built up." We should "do our original job," added Beers, "and develop plans for civilian controlled organization."[113]

"A TERRIBLE MISTAKE"

As ODM officials deliberated about the implications of the exercise for their program of readiness planning, a broader controversy arose over Eisenhower's declaration of "limited martial law." Among military and civilian officials participating in the exercise, observed Harris, Eisenhower's decision "caused no end of consternation."[114] Governors rebelled at the idea that they would be under the command of the armed forces in a future emergency and argued that the martial law declaration effectively rendered their preparedness efforts irrelevant. "If the military is taking over the responsibility," reported FCDA administrator Val Peterson, "the Governors see no reason

to continue in the civil defense area."[115] State and local civil defense officials were furious that they had been sidelined. As the Massachusetts Civil
Defense Agency stated, "From a practical point of view it seems most unfor-
tunate that the concept of martial law was introduced. . . . There is it seems,
a dangerous inclination to evade or at least deprecate civil defense preparations on the ground that 'the Army will take over if anything happens.'"[116]

Commentators on the exercise were bewildered and alarmed by the
president's declaration. "Among judges and lawyers, and in the law schools,"
noted Arthur Krock in the *New York Times*, "there was a stir over the President's immediate reaction in favor of martial law in the imagined circumstances."[117] Journalist Douglas B. Cornell concluded that the decision "to
proclaim an imagined state of martial law for the Nation" added up to "some
kind of military dictatorship."[118] What was needed instead, an editorial in the
Washington Post and Times Herald declared, was "the formulation of definite
plans for mobilization of our full national strength, in case of atomic attack,
without resort to martial law."[119]

The controversy over the martial law declaration continued in the
months following the exercise: the Kefauver Subcommittee convened a
special session to discuss the declaration; President Eisenhower ordered
a review of historical precedents; and a range of government officials and
civilian experts weighed in. The controversy came to a head in high-profile
hearings on civil defense conducted in the first half of 1956 by the Military
Operations Subcommittee of the House Committee on Government Operations, chaired by Rep. Chester Holifield. In testimony to the Holifield Committee, administration officials—including ODM director Flemming and
FCDA administrator Peterson—defended the decision. While they allowed
that the declaration was unclear on certain key points, and had led to considerable confusion, they claimed it was never intended to displace civilian
authority. Rather, the goal of the proclamation was simply to provide military assistance to civilian leadership.

FCDA general counsel Raoul Archambault Jr. also defended Eisenhower's declaration. The "existing means, agencies, and processes which Congress had provided and the administrative preparations by some states and
cities," Archambault contended, "proved wholly inadequate to cope with
the situation as simulated during the exercise." Moreover, a legislative solution was impossible. Legislation was necessarily based on past experience,
Archambault argued, and therefore could not anticipate the requirements
of a future crisis. Statutory provisions "cannot envision every difficult situation that might arise nor provide the solution" to unprecedented challenges.

"There is no legalistic formula," claimed Archambault, "nor any fixed set of rules." Echoing the arguments that German jurist Carl Schmitt had advanced thirty years earlier (see introduction), he argued that the determination of what should be done in a national crisis was "the proper exercise of executive authority based on a factual situation that demands action to protect the national interest."[120] The declaration of martial law in OPAL 55 was "an extension of the President's duty to preserve the safety of the Nation in periods of extreme danger."[121]

By contrast, Charles Fairman, a legal scholar from Harvard University who was a leading expert on martial law in the United States, was withering in his criticism of Eisenhower's decision. The proclamation of martial law, he argued, was a disastrous misstep that could have been avoided through careful planning.[122] Operation Alert had "bungled into crude compulsion where insight, administrative skill, and inspiring leadership were needed."[123] Echoing Collett's point that a new organization could not be "superimposed" on the existing structure of government in a crisis, he argued that the decision to declare martial law had ignored the knowledge and capacities of civilian government. "The existing system of administration, run by a host of public servants who know their respective tasks and the characteristics and vagaries of their fellow citizens," Fairman insisted, "has a tremendous going-concern value."[124] It would be a grave mistake to dispense with such practical knowledge in the context of an emergency: "This administrative apparatus would be indispensable in carrying on after an attack. It would be fantastic to think of supplanting it. Rather, we must build it up and prepare it to absorb the weight of the blow."[125]

Fairman emphasized the disjuncture between the detailed preparations that had been made for the exercise, on the one hand, and Eisenhower's apparently ad hoc reaction on the other. The premise of the exercise had been "settled weeks in advance": the destroyed cities, the uncounted dead, the fleeing survivors, the widespread fallout. These were, indeed, the "very assumptions on which the exercise had been set up." In Fairman's view, the president's declaration of martial law was sudden and unreflective. "It was an on-the-spot decision—unstudied, uncoordinated, an improvisation of the moment. I think it was a terrible mistake." For Fairman, Eisenhower's conviction that a test exercise must be experienced as a surprise misconstrued the purpose of preparedness exercises: to enable officials to anticipate and prepare for the completely novel situations they would face in a future war. The point of conducting such exercises, Fairman argued, citing the congressional testimony of army civil defense expert Barnett Beers, was "to obviate an occasion for resort to 'martial law' by 'well prepared plans and implementation of those plans for civil defense.'" When the president spoke of the massive damage due to the attack, he was not describing a circumstance that had not been anticipated. Rather, he "was describing what all must recognize to be the foreseeable consequences of a nuclear attack. Great evils such as these are what is constantly to be planned for, and to be assumed to have occurred in any realistic exercise."[126]

The crucial point for Fairman was that the president's declaration of martial law had ignored an alternative means for a liberal constitutional government to deal with emergency situations. A breakdown of civilian government might well occur "if what was assumed had actually happened," given "our present state of inadequate preparedness." To avoid such a circumstance, "We should work hard to brace the civil administration so that it would not break down." Indeed, the goal of an exercise like OPAL 55—and of governmental preparedness more generally—was to create new experiences and expectations that would guide planners and policymakers in developing the statutes and administrative capacities that would be necessary to address a future emergency within the framework of law, civilian government, and the Constitution. As Fairman put it, "Operation Alert was a training exercise to advance that preparation." Rather than abandon civil authority for military rule, officials should improve the administrative capacities of the civilian government to manage emergency situations.[127]

The report of the Holifield Committee sided with Fairman and against administration officials who had defended Eisenhower's decision to declare martial law. The proclamation had, the Committee argued, "created an entirely new role when it theoretically placed the whole Nation under some form of military control." For this reason, the declaration had generated "a great deal of confusion and uncertainty."[128] Committee members did not question Archambault's claim that declaring martial law was within the president's constitutional power to do whatever was necessary to ensure national survival. As Holifield put it, "what has to be done will be done." But the Committee sided with Fairman in arguing that martial law should be a last resort. Vigorous preparedness measures should be taken to avoid it. "I believe I voice the feeling of the committee," Holifield concluded, that "a civilian organization should be built up which would perform under the functions of an emergency declaration, an emergency power as given by the Congress to the President rather than setting up any other system." Emergency powers should be "provided in advance by the Congress with a method for declaration and a method for termination," rather than relying on "the more rigid and extreme powers of the declaration of martial law."[129]

A Techno-Administrative Solution

The critical reaction to Eisenhower's declaration of martial law led ODM to shift its approach to maintaining civilian authority in a future emergency. As we have seen, ODM director Arthur Flemming's initial response to the results of Operation Alert 1955 was to charge the Emergency Action Task Force to plan for a declaration of martial law. His assumption, based on the outcome of the exercise, was that following a future attack, the president would declare martial law to establish an orderly "chain of command." But given the broad resistance provoked by Eisenhower's declaration, by late 1955 ODM went in a different direction. As an agency memorandum explained, ODM was no longer concerned with "theoretical legal questions but with the development of operational aspects of the problem of maintaining civilian command."[130] Instead of planning for martial law, it sought to enable the complex coordination of actions by civilian agencies in a future emergency. This approach to governmental preparedness—following a familiar pattern of Progressive reform (see chapters 2 and 4)—aimed to equip the executive with the technical and administrative means to manage an emergency without recourse to "exceptional" measures.

EMERGENCY ACTION DOCUMENTS

Work on the "operational aspects" of maintaining civilian government in a future emergency was initially centered in the Emergency Action Task Force. In October 1955, the task force met to discuss how to "maintain civilian authority, preferably without the use of martial law."[131] Members of the task force started from the assumption that the Federal Civil Defense Administration would be the key point of contact with state and local government "in getting the operations job done." Meanwhile, the Office of War Resources—the wartime incarnation of the ODM—would be "responsible for coordinating Federal resource support to meet survival requirements." But several questions remained. On what legal authority and through what organizational means would the federal government act to provide relief in states that had been devastated by an attack? What operating responsibility would be borne by states, given "limitations on pre-planning and variations in strength of organizations"? How would critical resources such as manpower, medical supplies, food, and trained personnel be distributed across state lines? And how, finally, might the armed forces enter this picture? By early 1956, the task force had worked out a basic approach to addressing these questions—through a structure of distributed authority. Governors would "have the responsibility" to utilize all resources within their state "toward effective recovery of survivors within their State from the initial shock of enemy attack." Although states would be responsible to "support the national government and its Chief Executive in all efforts to defend and preserve the nation," they would retain the sovereign power granted to them in the Constitution.[132] The military would enter the picture only in support of civilian authority, if local government was overwhelmed.

To plan for this complex coordination of authority relations, the task force formulated what were called "emergency action documents." These documents were "standby" presidential proclamations or executive orders that would be issued in the event of an enemy attack to set up the preplanned elements of emergency government organization. Such orders—which would be "self-triggering" at the outset of a wartime emergency—functioned to establish legal authority and coordinate actions across the federal government and between state and federal levels of government. Beginning in 1956, mobilization planners produced dozens of these documents as part of their work on Mobilization Plan D-Minus, which laid out the government organization, action steps, and essential functions required to manage resources following an attack on the United States.

An initial set of emergency action documents, drafted in May 1956, established the "basic responsibilities" that would be assumed by key emergency agencies, most centrally the Department of Defense, FCDA, and the Office of War Resources (OWR). One, a draft executive order on "Assigning Major Areas of Responsibilities for Defense," focused on the respective responsibilities of the Department of Defense, FCDA, and OWR in the postattack allocation of resources. FCDA would work with federal agencies and with local and state governments to provide "emergency food, clothing, [and] medical care"; establish "sanitary safeguards"; restore "public facilities and utilities basic to the resumption of commerce and industry"; and run programs to meet "essential consumer requirements." The Department of Defense, after establishing its own requirements for military operations, would make remaining materiel and personnel available to civilian authorities in support of their "emergency relief and rehabilitation functions"—including the "provision of emergency medical, health, and sanitation services" and the "emergency restoration of damaged vital facilities and provision of emergency welfare services." Meanwhile, OWR would mobilize resources and direct the production required to meet both military requirements and "essential civilian requirements." As part of this resource mobilization function, it would adjudicate "conflicting claims for manpower, production, energy, fuel, transportation, communications, housing, food, and health services" made by military and civilian agencies.[133] In short, OWR would serve as the War Production Board of a future nuclear conflict—coordinating the allocation of resources essential to the survival of the population.

Other emergency action documents described the steps that would be taken if civilian government at the local level proved unable, "after devastating attack, to discharge its responsibilities and exercise its authority." For example, a draft presidential proclamation addressed the provision of "assistance to civil authority by the armed forces." The draft proclamation envisioned the "loss of effective means for performance of the lawful duties and functions of local governments" in many areas of the country. Under these circumstances, the armed forces, upon request of civilian authorities, would "give assistance to a state in exercising civil authority in an affected area." Such requests for assistance in maintaining law and order would be made by state governors through regional FCDA directors. Requests concerning the mobilization of vital resources—communications, food, medical personnel and supplies, emergency housing, and "production vital to the defense of the United States"—would be made through regional OWR coordinators.[134]

For members of the Emergency Action Task Force, this proposed structure of assistance to civilian authority was a means of maintaining governmental control in an emergency without resorting to martial law. As General Paul explained, the condition that would trigger military assistance was "the helplessness of civil authority, after devastating attack, to discharge its responsibilities and exercise its authority." The proposed structure maintained in all cases the "supremacy" of civilian government by providing for the "military to respond to requests from the civil authorities for aid, rather than for the military to take the initiative and to supersede local government," and, in the most extreme circumstances, for a civilian agency in the federal government to assume control. The aim was "to keep civil authority in control of government at state and national levels, under D-Minus conditions." "At no time," General Paul emphasized, "is the disputed term martial law used."[135]

In addition to these documents concerning the basic structure of emergency government, the Task Force also prepared emergency action documents for dozens of areas of resource management. It coordinated this work with agencies across the federal government, initially by circulating guidelines on the preparation of emergency action documents. These guidelines instructed agencies that their emergency action documents should address postattack actions that "involve matters of supreme national importance requiring immediate action or execution by the President resulting from attack or threat of attack." As Flemming explained, agencies were to integrate the emergency action documents with other elements of "administrative readiness," such as the lists of emergency actions that would be required to "carry on the essential functions of Government in the event of an attack upon the United States."[136] Once the agencies' emergency action documents were approved, copies would be placed on file in the White House and at agencies with wartime responsibilities. As General Huglin stated in an October 1956 memo, the documents were to be "ready for instant release in the event of attack" so that the initial steps taken by the government would be taken automatically. Agencies would "act under the assumption that the order has been signed and published," whether or not channels of communication were open. By preparing such documents in advance, Huglin argued, "the authority to go at once into a wartime organization will be automatic and development of the organization will proceed without any delay resulting from difficulties in issuance of basic authority."[137]

For example, one such document—an executive order on "Providing for the Mobilization of the Nation's Resources"—would, upon issuance, activate the emergency powers of the federal government and establish emergency agencies. The executive order first established the condition of emergency that justified its issuance, positing that the "survival of the United States as a free nation has been immediately and gravely imperiled by armed attack," and large-scale destruction of life and property had "made impossible the conduct of the national affairs through normal channels."[138] The tasks of "defense and survival" would therefore take precedence over all other demands on the nation's resources, both human and material. To accomplish these ends, the executive order activated emergency powers that had been established in statutes such as the 1950 Defense Production Act: control of materials and establishment of priorities for their use; allocation of production and service facilities to essential uses; requisition of private property and materials; a range of financial and price controls; and controls over manpower. These powers would be exercised by the Office of War Resources, established on the basis of the existing Office of Defense Mobilization, and led by the ODM director. Within OWR, a set of emergency agencies would then be established, drawing on "nuclei" of staff from peacetime agencies. Each emergency agency would be responsible for one of ODM's resource categories: production, communications, energy and natural resources, food, health, housing, and transport.

MOBILIZATION PLAN D-MINUS

ODM's project on emergency action documents was among the final steps in its preparation of Mobilization Plan D-Minus—the blueprint for governing a future war emergency involving an attack on the United States. D-Minus was circulated to executive branch departments in 1957. Three years after the National Security Council had officially shifted its war assumptions, ODM had a mobilization plan in place that took into account the scenario of a nuclear attack. The plan's stated purpose was to "prepare governmental actions which will be required for survival of the nation and its people, the prosecution of a war, the maintenance of free world unity, and the recovery from atomic attacks on the United States." The language of Plan D-Minus echoed that of Progressive reformers who had sought a techno-administrative solution to the challenges that emergencies posed to constitutional democracy. It sought "to provide governmental machinery best suited to meet the unusual demands of such [a] situation." Referring to debates about the constitutionality of emergency measures, Plan D-Minus emphasized that in setting up this governmental machinery, it was essential to "insure segregation of those functions peculiar to the emergency from the normal functions of government, and their rapid and complete liquidation when the emergency no longer exists." Indeed, according to the plan, this was a central purpose of planning for a future emergency. By preparing to rapidly set up an administrative machinery of emergency management within the framework of civilian government, it would be possible to "prevent functions, which are essential in an emergency but unnecessary or repugnant in peacetime, from becoming consolidated with the normal functions of government."[139]

The initial pages of Plan D-Minus laid out the scenario of a catastrophic attack, which resulted in twenty-five million killed and another twenty-five million injured from blast damage and radiation, as well as "almost complete paralysis in the functioning of the economic system in all of its aspects." The scenario described the attack's "immediate severe impact on organized governmental activities," indicating that government control had been "seriously jeopardized" and that federal direction was "virtually non-existent." But the plan suggested that, if adequate preparations were in place, "[r]estoration of the economy and our society will be possible" despite the "magnitude of the catastrophe that has struck the nation."[140] The possibility of such restoration would depend on the existence of a well thought-out technical and administrative plan for relief and recovery. The remainder of D-Minus ran through the elements that had been formulated in ODM's planning process over the previous three years: the emergency organization of government, the essential functions required of emergency agencies, and the specific actions these agencies would take to perform these functions.

The plan first detailed the process for setting up the "governmental machinery" of emergency management.[141] In the immediate aftermath of the attack, a series of presidential emergency action documents would be issued to activate emergency powers and establish the basic structure of emergency government. Among these would be an executive order transforming the Office of Defense Mobilization into the Office of War Resources and charging the new office with management of the nation's resources. Other executive orders would set up emergency agencies located within this new office. Along with existing agencies that bore significant wartime functions, such as the Department of Defense and the Federal Civil Defense Administration, these new emergency agencies would take on the "essential functions" required to organize relief and recovery.

The bulk of D-Minus was composed of long checklists of action steps that ODM had developed in collaboration with executive branch agencies. These actions addressed a dizzying array of problems related to the mobilization and distribution of "all our significant resources" in a future war emergency. The lists were divided into resource categories—manpower, food, housing, communications, transport, power and fuel, and productive facilities—and assigned to specific federal agencies. The Department of the Interior would, among scores of other actions, assess damage to electric power facilities; arrange dispatch of electricity over alternative lines; authorize emergency construction; order curtailments of power usage "with due regard to essential survival items and services"; requisition essential equipment for repairs; and develop a program for the recovery and reconstruction of the power supply, taking into account all requirements for materials, equipment, manpower, and other resources. The Department of Agriculture would, again among scores of other actions, estimate food supplies and requirements based on preattack estimates and "known results of actual attack"; request claims for food and other agricultural products for civilian, industrial, military, and foreign uses; issue freeze orders on stocks, including import items such as sugar, coffee, tea, and spice; issue priorities and antihoarding orders; set up rationing programs; protect livestock and crops from diseases and pests; and institute "plant, pest, and plant disease detection and warning services." The Housing and Home Finance Agency would assess damage to the housing stock; direct aid "in support of immediate survival operations"; procure tents, temporary barracks, trailers, and "improvised community facilities" for emergency housing; evaluate the criticality of housing needs in different parts of the country; and provide for the "planning and construction of needed community facilities in new or expanded communities."[142] Other agencies and commissions across the federal government were responsible for planning or carrying out emergency actions spread across a number of resource categories. Among these were the Departments of State, Defense, Labor, Commerce; the Interstate Commerce Commission; the Atomic Energy Commission; the Federal Power Commission; the Selective Service; and, of course, the Federal Civil Defense Administration and OWR, whose emergency actions spanned virtually every resource area.

In one sense, ODM planners' techno-administrative "solution" to the dilemmas of emergency government in a constitutional democracy followed the template laid out by government reformers of the late New Deal. They sought to equip the executive with administrative tools to deal with a future emergency without recourse to steps that would undermine constitutional order. In another sense, however, the problems of emergency government to which this governmental pattern was applied were entirely different. In Plan D-Minus, "mobilization" was no longer about maximizing production in support of a "total war" effort. Instead, it concerned the survival of the population and the recovery of the economy following a devastating attack. The form of mobilization planning had also changed. Previous mobilization plans were detailed catalogs of the specific military equipment and materiel that would be required to prosecute a war (see chapter 2), serving to guide action in the present to achieve a definite end. By contrast, Plan D-Minus addressed the governmental capacities and actions that would be required to respond to an unprecedented and catastrophic future event. Its purpose was not to fix a set of definite actions, but to guide a program of exercises, through which preparedness measures could be regularly tested and modified. In sum, the mode of governing the future at play in D-Minus was not instrumental action in the present to achieve a definite end, but rather an iterative process of planning for an unprecedented future event whose precise contours could not be known in advance. Although the prospect of nuclear war would gradually recede as the central concern of emergency planning, we can identify in D-Minus the crystallization of an apparatus of emergency government. The diagram initially formulated in the National Security Resources Board's project on postattack rehabilitation (described in chapter 4) was now linked to organizational forms, administrative routines, legal authorities, and techniques of intervention.

6

Survival Resources

In 1958, the US Army Medical Service School published a compendium of lectures that had been delivered as part of the school's ongoing Symposium on the Management of Mass Casualties. The central problem the lectures addressed was how to provide for the medical needs of the population in the aftermath of a catastrophic nuclear attack that would cause millions of casualties and destroy much of the nation's health infrastructure. "After the onset of the megaton bomb era," wrote General Elbert DeCoursey in the foreword to the volume, it was urgent to grapple with the "tremendous potential in numbers of patients, the relation of time to lethality, and the potential shortage of physicians." These factors, he argued, "require plans for organization before the catastrophe."[1] Most of the speakers at the symposium were military doctors, especially surgeons, and their presentations focused on medical topics such as triage, radiation sickness, and the treatment of burns.

One of the presentations, however, had a different emphasis. The statistician H. Burke Horton, director of the National Damage Assessment Center in the Office of Defense Mobilization (ODM), focused not on the use of medical techniques to treat patients but on a problem of resource management: how to calculate the total amount of health resources that would be needed to manage the vast number of casualties to be expected in a thermonuclear war.[2] In the early 1950s, as director of the Interindustry Research Program in the Air Force, Horton had led the development of a novel technique of bomb damage assessment that used electronic computers to model the effects of an atomic attack on the United States. After the program was shut down in 1953, Horton was recruited to ODM, where he established the National Damage Assessment Center. ODM officials initially planned to use the damage assessment technique to identify the vulnerabilities of key industrial sectors to atomic attack. But by the time of Horton's presentation to the Army Medical Service, five years later, the central problem for defense mobilization had shifted. National security strategists no longer saw the ramp-up of the industrial production system in a future war as the major task for mobilization planning. Instead, the advent of thermonuclear weapons and the threat of widespread radioactive fallout meant that planners would have to focus on ensuring the survival of the population and the continued operation of essential services.

Horton began his lecture by pointing to the daunting problem of managing mass casualties in the event of a thermonuclear war. It was "a task so vast that the job appears at first to be impossible." Yet the problem could not be dismissed: "it refuses to go away." In his lecture, Horton suggested a method for addressing it based on his work at the National Damage Assessment Center (NDAC). Using novel computer-modeling techniques, he explained, the center had developed an "experimental approach to the problem" of planning for nuclear war. This approach began with the compilation of an inventory of the nation's resources, with location information for each entry mapped onto a "standard grid." The resource inventory included key industrial facilities, dense population clusters, port facilities, major power plants, rail yards, communication centers, bridges, tunnels, and hospitals. Once these resources were located and recorded on magnetic tape, the center's technicians ran "experimental bombings" to "simulate the effect of attack upon the United States," using a "special electronic computer procedure."[3] Given a set of ground zeros, the computer could summarize the effect of blast, fire, and fallout damage on each resource.

Using these simulations, NDAC analysts had determined that the management of mass casualties was critical to the broader task of maintaining the nation's vital systems in the wake of an attack, since these systems relied on human labor. "The recovery of our transportation, communications, and electric power industries," Horton explained, would rely on the "surviving supply of technically trained manpower." Thus, human survival would be the "limiting factor" to the nation's recovery. The simulations indicated that the loss of "human resources" in a future attack would be "exceedingly high—higher than the losses of most of the other resources." Casualties could be expected in the range of thirty to fifty million, half of whom would die within the first twenty-four hours of an attack. Moreover, unless "great improvements" were made in the availability of shelters and medical care, a "very large proportion" of the population that survived the initial attack would "also be dead after sixty days."[4] Prospects for postattack relief were bleak: assistance was unlikely to come from outside a devastated area after an attack, and initial stocks of supplies would be rapidly exhausted. Existing civil defense capacities were far from adequate to the scale of projected need.

"It seems clear," Horton summarized, that "long-range survival and recovery will be the most massive problems ever encountered by this nation, even assuming that we win a complete military victory." In this situation, "the critical importance of passive defense measures" taken in advance of an attack "is now apparent to all." In developing such measures, Horton argued, the technique of anticipatory damage analysis could play a crucial role. It could identify the highest priority uses of limited resources given vast potential need and reveal "our greatest weaknesses and deficiencies" in advance of an attack. For instance, based on knowledge of probable attack sites and likely fallout patterns, the technique could address questions such as the following: What are the shelter requirements of each part of the United States? What essential stocks are required and how should they be positioned? In using the technique of damage simulation to address these questions, the center sought to transform what Horton called a "problem that is almost beyond comprehension" into an object of rational calculation and management.[5]

This chapter describes how, in the mid- to late 1950s, mobilization planners defined a new objective for emergency government: ensuring the survival of the population in the aftermath of a nuclear attack. Our analysis focuses on a specific area of planning—the stockpiling of essential supplies such as food and medical equipment. As we will see, attempts to build up a sufficient stockpile of survival items were continually frustrated. Here, however, our concern is not with the success or failure of the stockpile policy. Rather, we investigate this seemingly mundane preparedness measure in order to trace a broader transformation in emergency government: from economic management in an ongoing crisis to securing the operation of vital systems in a future catastrophe.

To plan for a stockpile program, federal officials used the tools of mobilization planning to investigate the systems involved in producing, transporting, and storing what they called "essential survival items." Like mobilization planners of World War II and the Korean War, they sought to test the feasibility of their plans by calculating whether supplies of such items would meet requirements in a future emergency. But faced with the prospect of an unprecedented and radically uncertain future event—what Herman Kahn referred to as the "unthinkable" catastrophe of thermonuclear war—officials drew on novel techniques for such feasibility testing. Employing digital computers to simulate the devastation of an enemy attack and to estimate the gap between supplies and requirements of survival resources in its aftermath, they constructed a "balance sheet for survival and recovery."

Planning for a stockpile of survival resources also raised fundamental issues concerning the organization of emergency government. The Federal Civil Defense Administration (FCDA) was in charge of a stockpiling program for medical supplies and equipment. But with its emphasis on local planning, FCDA had neither the capacity nor the authority to anticipate and plan for resource needs at the national scale, and its efforts were increasingly criticized as out of step with evolving assumptions about a future war. Meanwhile, in parallel to FCDA's stockpiling program, mobilization planners took up the problem of planning for the provision of essential survival items from a different perspective. As ODM focused on the widespread damage that would result from a nuclear strike, it began to consider population survival rather than postattack industrial production as the most important question for mobilization. In this context, the distinction between emergency resource management and civil defense—the respective responsibilities of ODM and FCDA—broke down. In 1958 these offices were merged in a single federal emergency planning office that was responsible for preparing to manage survival and recovery in a future war. Emergency government had been redefined as emergency preparedness.

Health Resources Planning

Our account of this shift begins in the late 1940s, within the National Security Resources Board (NSRB), which, as we have seen, was the first federal office responsible for nonmilitary defense planning. As part of its work on civilian mobilization (see chapter 4), NSRB investigated the type and quantity of medical supplies and personnel required to care for the population in the aftermath of a nuclear attack. The Board established a Health Resources Office under the direction of Dr. Norvin Kiefer and charged it with planning for the "allocation, mobilization and utilization of health resources during a national emergency."[6] As Kiefer explained, health resources planning considered "such functions as the maintenance of medical care, public health services, and sanitation measures for civilian populations during wartime."

Kiefer noted that localities would likely be overwhelmed by the population's medical needs in a future war. Following an atomic attack on a city,
local officials would be faced with the "treatment of many thousands of
casualties," thus presenting "enormous and unprecedented administrative
and organizational difficulties."[7]

The Health Resources Office sought to address these "administrative and
organizational difficulties" using the tools of mobilization planning: surveying existing inventories, studying manufacturing potential, and analyzing
the vulnerability of the production system to disruption. According to one
official, such information would enable resource planners to "ascertain any
weaknesses that might hinder prompt distribution of these supplies and to
plan a system which would overcome existing deficiencies."[8] This research
revealed several critical issues. For instance, as Kiefer reported, because of
a lack of reserves in warehouses, on dealers' shelves, and in hospitals, "the
amounts of surgical supplies and equipment would be grossly inadequate
for major wartime civilian disasters."[9] Moreover, since the health supply
industry was already working at near capacity, "any significant increase in
production would require considerable time." As a result, noted Kiefer,
NSRB was "giving serious consideration to methods of building necessary
security reserves of health supplies."[10]

The Health Resources Office provided guidance for the provision of such
"security reserves" in a manual it prepared for the Federal Civil Defense
Administration in late 1950, *Health Services and Special Weapons Defense*. In
most respects, the manual followed the schema for civil defense laid out in
the NSRB Blue Book (see chapter 4): state and local agencies would have
operational responsibility for civil defense, while the federal government
would coordinate local efforts and provide technical guidance. But in one
area—ensuring the supply of health resources after an attack—the manual
recommended that the federal government play a more directive role. The
manual pointed out that a central challenge that would be faced by civil
defense health services in a future atomic attack was "one of sheer volume—
of providing personnel, supplies, and facilities to treat simultaneously the
great numbers of living casualties resulting from each attack." To address
the "difficult and urgent" problem of "the immediate provision of adequate
health supplies following a civilian war disaster," essential medical supplies
would have to be accumulated "through some system which does not now
exist." To put in place a "complete civil-defense program in this country,"
the manual argued, would require "building large stores of consumable

health, particularly surgical, supplies."[11] As in military-industrial mobilization, a federal program of stockpiling essential supplies was a mechanism for increasing the "cushion" of the system of health provision.

This proposal to build a federal stockpile of health supplies raised a number of issues, such as what items should be in the stockpile, where these items would be warehoused, and what agency would be responsible for procuring and storing them. It was "impossible to know on what area or areas an enemy might concentrate attack," the manual explained, and it would be too expensive for each local government "to purchase and store health supplies sufficient to cope with every possible wartime disaster." Therefore, the federal government should accumulate "a common pool of such supplies, available to any attacked city." The manual proposed that the newly established FCDA set up federal stores of "certain health supplies." Several issues had to be considered in locating these stockpiles. It was necessary to place them "far enough from any potential target area to make their loss by enemy attack unlikely," but close enough to urban populations "to ensure reasonable availability to any of the various critical target areas" of the United States. And these reserves should "be located at points where excellent road, rail, and air transportation is available."[12] The manual addressed the type and quantity of items that would have to be stockpiled in a lengthy table, developed by the Office of Medical Services of the Office of the Secretary of Defense, listing the minimum supplies that would be needed to treat one thousand casualties during the first week following an atomic bomb disaster. The table listed roughly eighty separate items—including surgical textiles, essential drugs and antibiotics, anesthetics, and medical instruments—that made up a standard package that could be used for planning purposes. The standard package weighed an estimated 19,218 pounds, could be stored in "537 individual packages, none weighing more than 70 pounds," and "would require about 65 percent of the cubic capacity of one freight car for transportation."[13]

Medical Unpreparedness

Soon after FCDA was established in early 1951, the new agency initiated the medical stockpiling program that NSRB had proposed. John Whitney, director of the FCDA Health Office, explained that the program involved both advising state and local governments on the "establishment of stockpiles of medical and other supplies" and, at the federal level, creating reserves of "equipment and supplies to be furnished to the states and cities when needed."[14] Following the approach laid out by NSRB, the FCDA program was organized around what were called "critical target areas"—the likely sites of a future enemy attack. FCDA also established a reserve of "packaged disaster hospitals," modeled on World War II–era Mobile Army Surgical Hospital (MASH) units. In a wartime emergency, injured civilians would be evacuated to the perimeter of the city. There, they would be treated using stockpiled medical supplies, in preselected buildings—such as schools and churches—where the packaged disaster hospitals were to be set up.

To justify its budget requests to Congress for support of the medical stockpiling program, FCDA had to define its overall stockpiling objective: What would the stockpiled supplies be used for, and how many were necessary? Civil defense planners, however, had neither the mandate nor the tools to calculate the amount of resources that would be required to treat the casualties of a future attack. Instead, in formulating budget proposals, FCDA officials estimated how many casualties could be treated by existing medical personnel—essentially deriving resource "requirements" from the number of doctors and nurses available to deliver medical care.[15] On this basis, FCDA requested funds to stockpile supplies for a "casualty load of about 5 million over a three-week period." The agency estimated that approximately $400 million in medical supplies and equipment would be needed.

Over the next three years, FCDA assembled a significant cache of medical supplies in its stockpile: 9 million doses of penicillin, 33 million capsules of broad-spectrum antibiotics, 2 million whole blood collecting and transfusion sets, 132,000 radiological monitoring instruments, and more than 25 million doses of vaccines and antitoxins for biological war defense and communicable disease control. The supplies were stored in thirty-two sites in small urban communities located within four hours of critical target areas.[16] Yet these supplies amounted to only about one-third of the total that FCDA had requested from Congress, as the agency repeatedly failed to convince legislators to appropriate the funds necessary to meet its stockpile objective. FCDA also fell far short of its goals for the packaged disaster hospital program. By the end of 1955, only a fraction of the six thousand hospital units the agency had requested to manage "the great casualty load anticipated in an all-out attack" had been procured.[17]

In early 1955, FCDA's inability to meet its stated objectives for medical preparedness came to the attention of policymakers and officials. In an investigation of "Health Planning for Total War," a congressional task force found that the nation was "medically unprepared for atomic attack."[18] The task force was "particularly disturbed by the absence of a complete medical plan for survival in the event of atomic war." Although "long-range plans have been discussed," its report noted, "responsibility for integrated direction and control of all available medical resources has not been determined."[19] In conclusion, the task force report called for greater resources and authority for FCDA, an "increased emphasis on Federal stockpiling of medical supplies and equipment," and the assumption of postattack medical responsibilities by cities and organizations outside critical target zones that could come to the aid of attacked cities.

FCDA's planning assumptions came under further scrutiny as information about the effects of thermonuclear weapons became available to national security officials. A major catalyst was the 1954 Castle Bravo H-bomb test in the Marshall Islands and subsequent public outcry over the hazard posed by exposure to radioactive fallout.[20] In February 1955, the chairman of the Atomic Energy Commission, Lewis L. Strauss, released a public statement on "the effects of high-yield nuclear explosions." Strauss warned that in a future atomic war, radioactive contamination would spread well beyond the urban areas that were the focus of civil defense preparations. "The fallout from large nuclear bombs exploded on or near the surface of the earth," he explained, "would create serious hazard to civilian populations outside the target zones."[21] The implication for civil defense strategy was that the geographically circumscribed "critical target area" was no longer an adequate unit for planning.

In light of these concerns about medical preparedness for a future war, civil defense and mobilization planners built the problem of how to manage mass casualties into the scenario of the 1955 Operation Alert exercise. As we saw in chapter 5, one goal of the exercise was to test the structure of governmental authority in a future wartime emergency. But the exercise also tested whether sufficient essential health supplies would be available in the event of a thermonuclear attack accompanied by widespread fallout. In the scenario for the exercise—which depicted an atomic attack in which fourteen American cities "were struck by megaton bombs bursting at ground level"—dangerous fallout spread over "an area of about 63,000 miles."[22] The total number of casualties requiring medical treatment "on a continuing basis" was 8 million, many of them due to exposure to radiation. According to a post-hoc report, the exercise demonstrated that available health supplies would fall far short of anticipated need. Federal stockpiles would be "sufficient to care for only 2.5 million of the injured for 3 weeks." If available medical supplies were "distributed uniformly throughout the entire 8 million requiring medical care," they would last only three days. The exercise also showed that medical requirements would far outstrip the available supply of packaged disaster hospitals. While roughly 9,000 of these units would be needed in such an attack, FCDA had been appropriated funds for only "a total of 532 improvised hospitals." Moreover, federal officials were not prepared to administer the health resources that were on hand. In the event of an attack, stricken localities would require outside help, but the provision of such support was hindered by "the lack of data on total national resources which would be available in an emergency, and of complete plans for making use of these resources."[23]

In the Holifield Committee hearings on civil defense, held the following year, members of Congress pressed civil defense officials to explain why they had not made adequate provision to deal with the massive casualties that would result from a thermonuclear attack. FCDA director Val Peterson testified that his agency had neither the statutory authority nor sufficient resources to address the distinctive problems that radioactive fallout posed for medical preparedness. Peterson distinguished between the responsibility of the Office of Defense Mobilization for ensuring the "broader economic functioning of our society" during a future war, on the one hand, and FCDA's specific task of "getting the things that are necessary to take care of the people" on the other. According to Peterson, the specter of radioactive fallout radically challenged FCDA's capacity to meet this latter responsibility: "This is a totally new problem that we didn't know anything about until a year or so ago." "Nobody in the world—nobody—military or otherwise," he testified, "has any capability to do this job."[24]

One line of questioning in the hearings concerned the agency's method of planning for the postattack treatment of casualties. As we have seen, FCDA had initially limited its total stockpiling objective to the amount of supplies that would be needed to treat five million people over three weeks, given "the lack of medical personnel" required to care for higher numbers of casualties. Herbert Roback, an aid to Congressman Holifield, asked FCDA officials whether they were "doing anything in the way of trying to maximize or increase the resources of the medical profession so you can lift your stockpile-target assumptions." John Whitney, the director of FCDA's Health Office, responded that the question was moot, given that the stockpiling program was still not close to achieving even its initial—now hopelessly outdated—objective. Peterson agreed that there was little point in discussing higher stockpile objectives given the state of the program. "We now have

$160 million" in stockpiled medical supplies, he reported. "Until we get closer to our goal of $410 million, the thing is a little bit academic." Meanwhile, even as congressional support for medical preparedness stalled, the destructive power of atomic weapons continued to increase, and casualty estimates skyrocketed. "This year we ask the congress for $58,400,000 for medical supplies and equipment," Peterson testified, while "current planning assumptions indicate casualties requiring treatment may reach as high as 27½ million"—more than five times the agency's initial estimate.[25]

FCDA officials were also pressed in the hearings on how stockpiled medical supplies would be delivered to those who needed them after an attack. Roback queried Peterson regarding the "warehousing and distribution" of the agency's stockpiled medical supplies. "A spotting of those warehouses on the map indicates that a large number of those are located close to the metropolitan centers." "What," Roback asked, "is your distribution theory?" "There again, we have had to change," Peterson replied. "When you have an improvement in military weapons, increased devastation and better ways of delivering them, we have to shift." Civil defense officials had planned "to locate these stockpiles within 4 hours riding time of these big cities." But this plan had been part of a civil defense strategy that, Peterson acknowledged, was now obsolete. "The theory of Civil Defense was, you kept the people in the city, kept them at the lathe, kept them producing." This approach was now "out the window," since cities would be uninhabitable after a thermonuclear attack. "What knocked it into a cocked hat," Peterson testified, "was the development of this confounded radioactive fallout problem. Some of these warehouses we have located 4 hours from a city appear now to be in what would probably be the fallout pattern."[26]

The problem that fallout posed for FCDA's stockpiling strategy led the Holifield Committee to question the entire federal approach to planning for survival resources, which was based on local responsibility for the provision of relief and assumed a limited federal role. "As the weapons grow larger," the Committee argued in its final report, "so do the requirements of planning and with nationwide fallout as a hazard, the geographic limits on the planning problem are the Nation itself." In other words, ensuring the postattack survival of the population should be treated as a problem of national resource management—the traditional purview of mobilization planning. "It may be," the report concluded, "that the major effort in what is now called mobilization planning should be directed to planning not for the long buildup during war but for coping with the sudden and smashing letdown after the attack."[27]

A "Balance Sheet for Survival"

In fact, by the mid-1950s the office that was responsible for resource planning on the scale of "the Nation itself"—the Office of Defense Mobilization—was coming to a similar conclusion. In a future war, the central problem of emergency resource management would be the survival of the population. Thus, independently of FCDA's medical stockpiling program, ODM officials began to incorporate "essential survival items" into mobilization plans. To address survival as a resource problem, mobilization planners investigated what items—including but not limited to medical supplies—would be essential after an attack and compared supplies and requirements of these items, formulating what ODM Director Arthur Flemming called a "balance sheet for survival and rehabilitation."

FROM INDUSTRIAL RESOURCES TO HUMAN RESOURCES

Soon after the release of the Atomic Energy Commission's report on the effects of high-yield nuclear weapons in early 1955, an ODM subcommittee chaired by damage assessment specialist Burke Horton investigated the implications of radioactive fallout for its most important nonmilitary defense policy: dispersing key industrial facilities outside "critical target areas." The concept of the critical target area had guided ODM planners in identifying both the locations from which key industrial facilities should be dispersed and safe locations for industrial facilities and stockpiles of critical materials (see chapter 4). But the subcommittee argued that with the advent of thermonuclear weapons, this concept could no longer be used to guide nonmilitary defense planning. "The area of fallout from each bomb is so large, and wind variations are sufficiently great," it concluded, "that dispersion offers little promise for dealing with this threat."[28] In search of guidance on the strategic implications of fallout for nonmilitary defense, ODM director Arthur Flemming turned to the Project East River Committee, which a few years earlier had issued an influential report that urged dispersal outside of urban centers to reduce industrial vulnerability. Now, Flemming asked the Committee to revisit these findings in light of advances in weapons technology. In his charge to the Committee, Flemming explained that "our existing dispersion policy of ten miles from the perimeter of the target area is obsolete under the new weapons conditions." There are "serious hazards" other than blast and fire to be addressed, he added, particularly those "caused by fallout."[29]

On the specific question of industrial dispersal, the Project East River Review Committee confirmed ODM's internal conclusions. It noted that advances in weapons and delivery systems had "greatly exceeded the rate of progress in our nonmilitary defenses" and warned that the time was "close at hand when the United States and its potential enemy" would increase the destructive power of weapons "so as to pose the threat of annihilation." In particular, the report argued that with the advent of thermonuclear weapons, "the fall-out of radioactive material has now become a much more serious problem." The spread of "potentially lethal" radiation over "thousands of square miles" undermined the premise of industrial dispersion strategy, which defined a safe distance from likely targets. "The most important consequence" of new weapons capability, the report concluded, was that "the potential disaster area will be larger than any city boundary and will frequently overlap several state boundaries."[30]

Beyond the discussion of dispersal, the Project East River Review Committee also pointed to a broader implication of thermonuclear weapons for mobilization planning. "The principal production problem in the United States in the event of war," it argued, "may well be not the manufacture of military supplies and weapons but the production of those basic items needed for the rehabilitation of the civilian economy and for the reestablishment of our entire economic structure."[31] Such a shift would hold significant implications for the overall strategy of nonmilitary defense. ODM would now have to take into consideration a problem that had previously been understood as a matter of local civil defense planning—the provision of relief to the civilian populace. As the Committee's report put it, the tasks of mobilization and civil defense were now "most intimately related, and to attempt to separate them by subject or by time-phase is artificial and leads to confusion and duplication."[32]

The Mobilization Program Advisory Committee (MPAC) advised ODM on what these findings might mean for resource planning. As we have seen (chapter 5), the Committee argued that ODM should orient its planning to a "'broken back' type of war where survival and restoration assume roles of major importance." Beyond the immediate challenge of survival—providing relief to a stricken population—postattack industrial rehabilitation would depend on an adequate labor supply, without which it would be impossible to restore the industrial economy.[33] Thus, MPAC recommended that ODM consider the protection of "human resources" as a key area of responsibility.[34] To ensure "the manning of industrial plants," argued MPAC member Shaw Livermore, it was also necessary to address "the whole problem of safety and survival of the population," including the provision of "assistance and succor to victims of attack." ODM would thus have to calculate requirements not only for industrial production but also for "what we would have to have in a severely crippled nation to survive and carry on to victory," including "basic things like food, clothing, house-heating, and whatnot."[35]

Livermore pointed to stockpiling as a device for ensuring the availability of these supplies. Items that were essential to postattack survival and recovery, he argued, "ought to be in the stockpile just as much as magnesium or manganese, or any of the things that are actually now in the stockpile." Livermore noted that although "a few of these things FCDA already have in their own little baby stockpile," a national program to store essential survival items would have to go beyond FCDA's work on critical medical supplies. It would entail maintaining reserves of "what you would need in the first six months or a year . . . to carry you over a period after a severe attack" that destroyed the nation's production capacity. In comparison to the stockpiling of industrial materials, Livermore argued, the cost of such a program would be "very slight." But "the thinking involved in what ought to be added to the stockpile is very great."[36]

ODM officials agreed with the direction in which MPAC was pushing mobilization planning. Director Arthur Flemming noted that MPAC's emphasis on a "war of annihilation" was precisely the direction "called for in Plan D-Minus," the mobilization plan on which ODM was "now concentrating." And yet, it was "difficult to orient our thinking in such a way as to give due emphasis to situations where survival becomes a major objective."[37] In response to MPAC's proposals on how to ensure the survival of the population after an attack, staff from ODM's Manpower Division, which was responsible for human resources planning, commented that while this suggestion "represents a broader view of ODM responsibility than has been taken in the past," it was a proposal that "in our view, makes a great deal of sense." They were concerned, however, that the protection of the civilian population was outside ODM's purview. This area had "thus far been considered as clearly an FCDA responsibility," and despite some familiarity with FCDA's work on the topic, ODM manpower specialists did not have "detailed knowledge of where they stand at the present time," nor "any charter for evaluation of FCDA activities."[38]

For Flemming, the problem was not that his office lacked the authority to undertake planning for survival resources. "ODM statutory authority and its delegations from the President in the form of Executive orders are adequate," he maintained, "for coping with any emergency that might arise." Rather, the problem was one of resources and attention. It was necessary to generate "government-wide acceptance" of the need for "regular budgeting" as well as "staffing in the departments" to work on preparedness in this new area.[39]

ESSENTIAL SURVIVAL ITEMS

The task of garnering "government-wide acceptance" for establishing a stockpile of survival items raised a number of practical problems. How would responsibility for this program—which touched on the functions of ODM, FCDA, and a range of other agencies—be distributed? What agency would take charge of overall planning and budgetary requests? As we will see, these questions would soon lead to a significant reorganization of emergency government, in which ODM and FCDA were merged in a single office that oversaw both national resource planning and civilian protection and relief. Beyond such organizational issues, however, planning for a stockpile of survival items raised a basic question about how ODM could, as Flemming put it, "orient our thinking to a war of annihilation": How was it possible to know—in the present—what resources would be needed, and what resources would be available, in a future thermonuclear war?

Flemming outlined this challenge in testimony to the Holifield Committee, which, as described earlier, had been sharply critical of FCDA's stockpiling program. "We must be prepared during the period immediately following the attack," he urged, "to provide the resources which would be essential for survival and rehabilitation." The "items necessary to survival," Flemming explained, "must be planned for in advance, requirements determined, stockpiles built up if necessary, and vulnerability to attack lessened to the extent possible." What was needed was a "supply-requirements balance sheet for survival and rehabilitation." But ODM did not have the basic information that would be needed to formulate such a balance sheet. "We do not have adequate requirements information for a situation involving an attack on this country," Flemming told the Committee, "nor do we have adequate information relative to the resources that would still be available to us following such an attack."[40]

The lack of information about postattack survival requirements that would be required to undertake this new kind of mobilization planning was underscored by the results of the 1956 Operation Alert exercise. According to Innis Harris of the Plans and Readiness Office, the exercise demonstrated the imperative to "take account of [fallout] in all of our emergency planning," but also pointed to the difficulties that mobilization planners faced in doing so. "More basic data on resources were needed," Harris explained, "as well as faster means for estimating damage, subtracting losses from preattack resource estimates, and then estimating the surviving resource capabilities."[41] In a debriefing with representatives from other federal agencies, Flemming similarly pointed to the limited information that planners could draw on in preparing to manage the survival of the population in a future war. The exercise confirmed the "inadequate development of the character of and the order of magnitude of essential survival requirements." It was necessary to determine "what constitutes survival requirements," he argued, and to develop "methods to calculate the probable size of these."[42] The availability of such items had to be ensured well in advance of a future war, since in an emergency "there would not be time to tool up to produce vital medical supplies and other essential items to meet the survival needs of the population."[43]

A first step in developing such methods, according to Flemming, was to identify "those items, the lack of which immediately after attack would be intolerable." To investigate this problem, Flemming constituted an Interagency Committee on Essential Survival Items and charged it with compiling a list of "essential items that would be needed in an immediate postattack period to sustain life at a productive level and to ensure national survival." Once formulated, the list of survival items would serve "as a basis for the development of programs to insure the availability of minimum stocks during the immediate postattack period."[44] The Committee's final list, completed in early 1957, included more than two hundred items and was organized according to six resource categories—health, food, household supplies, energy, water, and shelter. The first resource category, "health supplies and equipment," was the largest and most detailed, containing nearly one hundred items: twenty different types of pharmaceuticals (such as antibiotics and morphine), six kinds of blood-collecting and dispensing supplies, fifteen biologicals (such as antirabies serum and plague vaccine), sixteen kinds of surgical textile, and thirty-seven types of surgical instrument—catheters, forceps, and bone saws, among other items.[45] The category of "sanitation and water supplies" was perhaps most suggestive of how the Committee imagined postattack life: bulldozers, radiation dose-rate calculators, warning signs, and protective masks; water purifiers such as hydrated lime, iodine tablets, and chlorine; general supplies such as chlorinators, pumps with appurtenances, and welding equipment; storage and transport equipment (storage tanks, tank railroad cars); laboratory equipment and supplies; and finally, insect and rodent control items, such as DDT, spraying equipment, and delousing outfits.

According to the influential 1957 report of the Gaither Committee (discussed below), the Interagency Committee's work was "the first methodical, government-wide approach to studying the needs of the surviving civilian population" in a future war.[46] The Interagency Committee's list of essential survival items offered a starting point in formulating a "balance sheet" for the postattack survival of the population. But planners still needed tools for addressing basic questions about postattack supplies and requirements of survival items. How many people would need medicine or other essential items in the wake of an attack? What inventories or stocks of these supplies would remain? And, given anticipated damage to industrial facilities, what items could be rapidly produced to supplement materials already on hand?

Simulating a Future Catastrophe

To address these questions, mobilization planners drew on a technique of computerized damage assessment that ODM had been developing for testing the feasibility of industrial mobilization plans. In congressional hearings, Flemming explained how ODM anticipated using the damage assessment procedure in its essential survival items program. "If the bomb-damage assessment studies indicate that we would be confronted with serious gaps on the supply side," he testified, ODM would "take steps to close those gaps."[47] Flemming's matter-of-fact description belies what was in fact a daunting technical (and, indeed, epistemological) problem: how to generate systematic knowledge about an unprecedented future event so that adequate preparations could be made in the present. Burke Horton, who led ODM's work on damage assessment, compared this problem of anticipating the effects of a future thermonuclear catastrophe to weather forecasting. "What was really required," he argued, "was a mapping of the new hazards in much the same way that temperatures and rainfall maps are prepared for agricultural purposes." The principal difference, he explained, was that whereas such weather maps were based on observed past experience, in the case of hazards such as fallout or blast from a thermonuclear attack, "these maps needed to be prepared before it 'rained' the first time."[48]

THE NATIONAL DAMAGE ASSESSMENT CENTER

The initial impetus for ODM's work on techniques to anticipate future bomb damage was the shift in strategic assumptions in 1953–1954 to a war that involved large-scale attack on the United States. An attack on key industrial facilities, Flemming observed in a 1954 lecture, "would obviously affect our ability to meet requirements" for military production. Therefore, ODM should "develop a program that will anticipate" losses "and try to do everything we possibly can to offset the damage that might be caused by an attack."[49] As production specialist William Lawrence explained, "The existing pattern of requirements" that ODM used as a basis for mobilization would have to be "changed in the light of assumed casualties and loss of productive capacity." An anticipatory bomb damage assessment capability would enable mobilization planners to formulate "new balance sheets between resources and requirements."[50]

At this time, the Air Force's Interindustry Research Program was developing a computerized procedure for simulating attacks on US industrial facilities (see chapter 3). When the Eisenhower administration shut this program down, the Mobilization Program Advisory Committee (MPAC) recommended that ODM adopt the Air Force simulation procedure for use in civilian mobilization planning.[51] The goal, argued MPAC's Edwin George, would be to "make ODM the focal point" in the federal government for "devising statistical methods" to anticipate future bomb damage. Flemming acted on the MPAC recommendation and recruited Burke Horton, who had led the Air Force program, to lead ODM's work on damage assessment. Horton was soon joined by a number of his former Air Force colleagues who had backgrounds in mobilization planning and air intelligence. Among these were Joseph Coker, an economist who had worked in wartime resource evaluation and air targeting; James Pettee, an air intelligence specialist in the Pacific during World War II and, after the war, a target analyst with the Air Force; and General Harold Huglin, a graduate of the Air Corps Tactical School and a bombing commander during World War II, who led the Directorate of Management Analysis in the Air Force, where the Interindustry Research Office had been housed.

Within ODM, Horton served as chief of a new Damage Assessment Division, charged with planning "hypothetical attacks against many targets in the United States" and calculating "the overall physical damage, radiological contamination and casualties that can be expected." ODM anticipated that these studies would be used in its nonmilitary defense programs, directing "attention to sectors requiring corrective action such as changed procurement policies, stockpiling, relocation, and protective construction."[52] The simulation procedure would be especially valuable in guiding preparedness planning across the federal government. Computer simulation could be "so realistic and so convincing," Flemming suggested, "that the results can be taken and utilized by the civilian agencies in dealing with industry and in dealing with our civilian economy."[53]

Horton also chaired a steering group that coordinated ODM's work with damage assessment programs run by the Federal Civil Defense Administration and the Department of Defense. The ODM and FCDA programs had distinct but complementary functions, corresponding to the (increasingly contested) assumption that national resource management and local civilian relief were separate problem areas. A 1957 report on federal damage assessment programs by F. A. Ramsey explained the division of labor between these programs. Given ODM's responsibility for managing "the remaining resources of the Nation" following an attack, its damage assessment program sought to create "an overall picture of the scope of the damage" to vital resources and mobilization capacity. FCDA, meanwhile, was "primarily interested in determining the results of the attack on the population and upon the physical structures within the target areas" with a view toward the provision of local relief.[54]

Initially, damage assessment specialists borrowed time on UNIVAC computers in military research and planning offices to run tests of the simulation procedure.[55] By 1957, ODM's damage assessment unit had acquired its own UNIVAC, which Horton described as "one of the most powerful computers in existence." The computer was installed in a new National Damage Assessment Center located in the government's secret emergency relocation site, buried deep within Mount Weather, Virginia, and staffed by personnel from fourteen federal agencies.[56] According to Joseph Coker, the unit's acquisition of the UNIVAC made possible "the solution of problems so massive and so complex" that before the development of electronic computers, they "would have been beyond the practical limits of human organization and integration."[57] With the aid of the digital computer, NDAC technicians were able to produce a novel form of knowledge about the United States—a detailed picture of the effects of a future nuclear attack on the nation's vital systems.

"A MATHEMATICAL STATISTICAL MODEL OF THE UNITED STATES"

As we saw in chapter 3, the damage simulation procedure that Horton's Interindustry Research Program had developed was made up of three elements. The first was a resource inventory that contained information about the spatial location and significance of critical facilities and essential services that might be damaged in an attack. The second element was an event model that included an assumed pattern of enemy attack, as well as methods for analyzing the spatial distribution and intensity of weapons effects—whether blast, fire, or radiation. The final element of the procedure was the calculation of damage, which was computed based on factors such as the distance of a given resource from a bomb's detonation point, the size of the bomb, and the physical vulnerability of a given resource.

Soon after it was set up, ODM's damage assessment unit conducted a series of test runs of this procedure using available data from civilian and military agencies. Although these tests validated the unit's approach, according to a 1954 review of the program, they also indicated that the procedure was "far from being fully developed" and that a "tremendous job" remained before "really effective use could be made" of the damage assessment technique in preparing for a future attack.[58] To make the procedure operational, over the period from 1954 to 1957 the unit added data on hundreds of thousands of resources to its inventory and developed increasingly sophisticated techniques to simulate attacks. Through this work, NDAC technicians built what Horton described as a "mathematical statistical model of the United States."[59] By running "experimental bombings" using this model, according to Ramsey, mobilization planners could generate a "vast, comprehensive, and composite story of the overall damage to the Nation" that would be caused by a future attack and an assessment of "what remains alive of the population, the economy, and the military."[60]

The resource inventory. Horton described NDAC's resource inventory as a "library of all major resources of the United States." This "vast storehouse of information" was needed, he explained, so that mobilization planners would "know where things are located in order to determine what is left if there ever comes a time when some 'holes' appear on the landscape."[61] Stored on magnetic tapes, the inventory contained information on the "type, exact location, amount, physical vulnerability, and detailed classification of resources," including manufacturing facilities, essential services, and population centers.[62]

The first records included in the resource inventory incorporated data that had been compiled in the government's mobilization planning and air-targeting offices. For example, the inventory contained records on roughly 9,500 industrial plants for which both geographic coordinate data (supplied by the Air Force) and industrial classification and output data (from the Commerce Department's Survey of Manufacturers) were available. Among these were 1,500 key military end-item producers and other plants on the list of "critical industrial facilities" developed by the Industry Evaluation Board (see chapter 4).[63] In expanding this inventory of national resources so that it would be useful for preparedness planning, ODM was particularly concerned with the vulnerability of industrial systems. As Flemming noted, in order to gauge an attack's "effect on actual production," it was necessary to consider damage to "the complex chain of suppliers and subcontractors" that led up to the production of a given end item.[64]

To investigate such vulnerabilities, NDAC drew on a method of "vertical analysis" that employed techniques developed in wartime mobilization planning and air intelligence. Edwin George, who had led the development of "vertical controls" to rationalize the use of resources in the War Production Board during World War II, described the impetus for conducting such vertical analyses. "Our system," he explained, "is actually a fabric of finely interlaced parts and any great holes that would be torn in it would paralyze production far beyond their own diameter." An assessment of the vulnerability of the military-industrial production system would therefore have to "account for vertical end product and weapons systems, embracing flows of materials, components, parts, subassemblies, and essential processing equipment."[65]

Vertical analysis was carried out by the Business and Defense Services Administration (BDSA), a descendent of wartime and postwar offices in the Commerce Department that had worked to generate systematic knowledge about chains of industrial production. As William Truppner, the director of BDSA's Office of Industrial Mobilization, put it, the unit was responsible for assessing the feasibility of mobilization plans by drawing on "industrial knowledge regarding the fabric of our economic structure."[66] In the vertical analysis of a given end product, the agency "made a list of the principal components and materials necessary to manufacture that product."[67] As in World War II–era resource evaluation and target selection, vertical analysis revealed that the most acute vulnerabilities were often hidden deep within production systems. For instance, BDSA found that it was possible to lose an entire truck assembly plant without any loss of output "so long as the flow of materials and components can be diverted to other assembly lines." But "if the only supplier of a specific component" were destroyed, "production could be lost completely" even if assembly plants for end items remained functional.[68]

Vertical analysis dramatically changed the picture of damage that was generated in simulated attacks, pointing to a range of different resources, such as supplies, essential services, and human resources, that would have to be included in NDAC's damage simulations. As Horton reported, in their analyses of "the relationship of the final assemblers to the suppliers of components, subassemblies and materials and the availability of essential services such as electric power," damage assessment specialists discovered that "superficially light to moderate damage estimates become virtual paralysis in many important industrial categories." Vertical analysis also pointed to the potential scarcity of "human resources" as a critical vulnerability. "The recovery of our transportation, communication, and electric power industries," Horton explained, "is dependent upon our ability to use scarce manpower for work out in the open." The ability to conduct such work would be contingent on "residual radiation from fallout and upon the surviving supply of technically trained manpower." Given the vast spread of fallout from a thermonuclear attack, the loss of manpower would be a limiting factor for production. "If we lose ten per cent of our hard resources—power plants, houses, manufacturing plants and similar resources," he reported, "we find that there is likely to be a twenty per cent loss or more of the human resources."[69]

NDAC's damage assessment specialists drew on such results in expanding their resource inventory. As Flemming explained, the unit recorded "information regarding [the] chain of production for selected critical weapons systems and survival requirements" on magnetic tapes, so that ODM would "have a much clearer picture of our actual postattack production capability."[70] By 1957, the center's inventory contained detailed information on roughly 150,000 resources—"virtually all of the important resources of the United States," according to Horton.[71] By 1960, the inventory had grown to 300,000 records.[72] In addition to information on manufacturing facilities, it included details on key nodes in essential services and other critical facilities: "military bases, port facilities, major electric power generating stations, rail classification yards, communication centers, bridges, tunnels, locks, dams, hospitals, medical personnel, livestock, and arable land." The resource inventory also incorporated detailed information on "the people of the United States," as Horton put it, drawing on data assembled by the Federal Civil Defense Administration, organized according to "twenty-five thousand 'standard locations,'" including census tracts, urban places, and in some cases county seats for sparsely settled areas.[73]

The event model. Once information about resources had been "systematically recorded on magnetic tape," the next step in the damage assessment procedure was to "simulate the effect of an attack against the United States" based on a set of "ground zeros" and "weapon yields."[74] Initially, NDAC conducted what Horton described as "one-shot studies."[75] These studies began with a "subjective placement of the weapons" by a "red team" that would choose a set of aiming points and "specify the weapon size and time for each designated point of attack."[76] NDAC would take this set of presumed enemy objectives and "'game' it through the defenses," yielding a "hypothetical set of ground zeros."[77] These ground zeros were then fed into computer models—with names such as BLAST, FLAME (NDAC's early fire model), and DUSTY (for fallout)—that estimated the distribution of thermonuclear weapons effects.

Modeling fallout was a particularly difficult challenge for NDAC technicians as they sought to grasp the scope of the problem of postattack survival of the population.[78] The damage assessment group had at first used a "rough method for determining the effect of fallout on the Nation's resources" that was "based upon 'stylized' patterns," a 1955 report on the program observed. But this approach did not "make sufficient allowance for meteorological variables" that could dramatically affect the intensity of radiation in a particular location.[79] NDAC's computer routine for estimating fallout patterns— DUSTY—addressed this limitation by incorporating "mathematical models of the fallout of radioactive debris from the large mushroom clouds resulting from nuclear detonations."[80] Inputs to DUSTY included a hypothetical attack pattern (detonation points and weapons sizes) and meteorological parameters (such as wind speed and pressure). The latter were derived from frequencies for conditions in a given geographic area, based on five-year weather histories provided by the US Weather Bureau.[81] The outputs of DUSTY were the distribution of radiation from each detonation and the total exposure to fallout at any given geographic location from all the detonations that might affect it.[82] Through such simulations, Horton explained, NDAC could generate "fall-out maps of the entire United States" for any given attack pattern. As we will see, the unit prepared such "one-shot" studies for the 1957 Operation Alert exercise and for the Gaither Committee report (issued the same year), which influenced thinking about nonmilitary defense across the federal government.

But according to Coker, such one-shot attack simulations did not provide an "adequate basis for emergency planning." As damage analyst James Pettee observed, these simulations were "subject to an unlimited amount of variables in design and execution"; the likelihood of any given event could "only be approximated."[83] To make such an approximation, NDAC developed a second type of study that did not simulate a single attack but rather, as Horton put it, "calculated the odds or chances of damage over a wide spectrum of likely attacks."[84] These studies employed an "elaborate gaming procedure"—a "hazard analysis model" called RISK—to account for the large uncertainties about the pattern of attack and the spread of weapons effects.[85]

The central feature of the RISK model was the use of random numbers to generate multiple attack simulations.[86] RISK incorporated a range of factors that would affect the distribution and intensity of weapons effects, such as enemy targeting strategy; weapon selection; abort, attrition, and dud rates; aiming point error; and meteorological variables. Rather than assign set values for these inputs, the RISK model employed the technique of Monte Carlo simulation—developed in the early 1940s for modeling subatomic processes as part of the Manhattan project.[87] Through this procedure, an attack simulation would be run "not once, but many times," so that the "individual peculiarities" of any given run would be "appropriately weighted in the final representation."[88] NDAC analysts estimated that one hundred trials were sufficient to get "statistically satisfactory results." "It is in this way," concluded Pettee, "that reality is simulated in the model."[89]

In sum, the aim of NDAC's simulation procedure was to generate a range of possible futures under varying, randomly generated conditions. Using Monte Carlo techniques, planners created a statistically significant number of such futures, on the basis of which they derived probabilities for particular outcomes. The result of such simulation was a map that represented, as Coker put it, "the *probability* that particular places of interest to us will be exposed to various ranges of blast and radioactive fallout intensities, and also the probabilities of various ranges of resource losses and availabilities."[90] Such maps of weapon effects represented what Horton called a "common denominator of probable hazards"—an expression of vulnerability in terms of the risk of a certain level of exposure.

Damage calculation. The final step in the procedure was to calculate the total damage caused by a simulated attack. As Horton described this step, each resource location was "compared with all nearby ground zeros" to estimate weapons effects at that location. On the basis of this estimation of effects and assumptions about physical vulnerability, each resource would be "assigned to a damage category."[91] In parallel, the computing routine calculated the impact of "blast and other direct effects" on human beings by reference to "curves relating percentage of mortalities and other casualties to distance from ground zero," with automatic adjustments made for weapons yield. Calculations of casualties from fallout were based on what NDAC defined as the "effective biological dose" of radiation for people located at a particular geographic location, with adjustments made for "shielding factors" that would reduce exposure.[92] In a final step, the effects of the simulated attack on each category of resource—whether human or physical—were "summed up and the results printed out on a high-speed printer," indicating the total number of casualties, losses of other resources, and summaries "of what is left for recovery."[93]

By the late 1950s, NDAC had integrated the calculation of damage with other aspects of the simulation procedure in a complex model called JUMBO. The JUMBO model incorporated "data on resources of all important types throughout the country" (from the inventory), information about "weapons effects factors and various attack assumptions" (from RISK), and "fallout and firespread estimates" (from submodels like DUSTY and FLAME).[94] Drawing on these various sources of data, JUMBO could generate various kinds of outputs: facility listings showing blast and fire damage as well as radioactive fallout at each facility; facility damage summaries by sector; resource availability summaries; human casualty summaries, by cause and severity of injury; manpower casualty summaries, broken down by occupation and industry for each metropolitan area; and livestock loss summaries.[95] These damage summaries could then be compared with estimated postattack requirements, thus laying the groundwork, as Horton put it, for "rational steps to reduce or minimize risks in an economic manner."[96]

"The Survival and Resource Problems Become One"

By early 1957, ODM had put in place two key elements that would enable it to approach postattack survival as a problem of resource management. Using the interagency committee's list of two hundred "essential survival items" and the computer simulation technique, they could investigate whether supplies would be sufficient to meet the population's needs in a postattack situation. By this time, national security strategists had come to understand population survival as a central task for nuclear preparedness planning. But this task seemed overwhelming—"almost beyond comprehension," as Burke Horton put it. In this context, ODM's tools for analyzing survival as a resource management problem offered national security experts and government officials a way to grasp the scope and contour of postattack survival—and to imagine the measures that might be taken to ensure it. The previously distinct functions of managing military-industrial production and organizing relief to the civilian population—the respective tasks of mobilization and civil defense organizations—were thus merged into a single project of national resource management.

PLAN D-MINUS AND OPERATION ALERT 1957

In May 1957, ODM released Plan D-Minus, the first mobilization plan based on the assumption of a large-scale nuclear attack on the United States. As we have seen (chapter 5), D-Minus was the culmination of ODM's effort to plan for a crippling attack that would cause millions of casualties, paralyze government, and devastate the national economy. According to the Plan, the key challenges to be addressed in this scenario would involve the resources and infrastructural systems essential to human survival. In the immediate aftermath of a future attack, governments at all levels would have to provide medicine, food, shelter, and water for the remaining population. After this initial "survival" period, the central task of government would be to rebuild the infrastructure of economic and social life. Even as radiation contamination diminished and reconstruction could begin, the recovery process would be "limited primarily by manpower shortages": the capacity for economic recovery would depend both on "the extent to which necessary services— power, transportation, communications, etc.—can be provided" and on "the number and types of workers available for production purposes."[97] The fact that manpower was identified as a "limiting factor" for recovery made it all the more critical—from the perspective of nonmilitary defense—to address the postattack survival of the population.[98]

In the 1957 Operation Alert exercise, which was based on Plan D-Minus assumptions, mobilization planners incorporated the damage simulation technique into preparedness planning. In the exercise, NDAC technicians used the damage assessment procedure to generate information about survival requirements (based on the size of the surviving population, the number of casualties, etc.) and available supplies (based on an inventory of existing resources and an assessment of damage). Specifically, NDAC sought to calculate "the supply-requirements situation which would probably exist in a postattack period with respect to approximately 50 survival items."[99] This was a "feasibility test" of a certain kind. Now, instead of testing whether existing resources were sufficient for a given industrial mobilization plan, it was a test of whether selected survival items would be available in sufficient quantities to meet the population's requirements.

The damage simulation conducted as part of Operation Alert 1957 revealed not only the direct effects of blast, firestorms, and radiation, but also the vulnerability of the resources that were essential for population survival. As Innis Harris of ODM later reported, "approximately 50 percent of the hypothetical casualties" from the simulated attacks "occurred through lack of shelter and lack of critical survival items."[100] For example, ODM planners estimated that, prior to the simulated attack, stocks of penicillin held by manufacturers and wholesalers amounted to roughly 150 trillion units. Although the planners considered this amount sufficient to match overall requirements given casualty projections, the NDAC analysis showed that much of the inventory was located in target zones, so that "more than half the supply would have been destroyed in the attack." Thus, the exercise demonstrated that "hypothetically many people would have died solely because existing stocks of penicillin were not maintained in dispersed locations."[101] In this way, the exercise confirmed ODM's arguments for instituting preparedness measures focused on essential survival items, such as planning stockpiles based on likely patterns of attack.

THE GAITHER COMMITTEE

The release of Plan D-Minus established population survival as a primary concern for mobilization planning. In the late 1950s, ODM's approach began to shape broader discussions in the national security establishment about planning for a future war. Specifically, NDAC's damage simulation technique provided a way to conceptualize an apparently unmanageable problem—to turn a problem that was "almost beyond comprehension," as Horton put it, into a manageable one—and to formulate concrete measures to address it. One venue in which ODM's techniques were taken up by leading national security experts and advisors was the Gaither Committee, established in April 1957. Named after its first chairman, Rowan Gaither, the Committee has served as a landmark in Cold War historiography, seen as a key site of debate around Cold War nuclear strategy. While historians have mainly focused on the Committee's warnings concerning US military vulnerability, its initial charge was to investigate proposed passive defense measures linked to population survival, such as a nationwide shelter system and an expansive program to stockpile essential survival items.[102]

The Committee was established in the context of a high-profile debate over proposals for a massive shelter program to protect the civilian population from radioactive fallout. With the advent of thermonuclear weapons, conceptions of civilian protection based on the assumption of evacuation from urban target areas had become increasingly untenable. By 1957, a growing number of strategists saw mass shelter as the "ultimate solution" to the problem of population survival in a thermonuclear war.[103] Early that year, FCDA submitted a report to the National Security Council recommending the construction of a nationwide fallout shelter system, at an estimated cost of $32 billion.[104] Meanwhile, the House Committee on Governmental Operations submitted legislation proposing $20 billion in spending on fallout shelters.[105] The Eisenhower administration opposed such a major government investment in shelters, raising questions about both cost and efficacy. In this context, the ODM Science Advisory Committee established the Gaither Committee, charging it to "form a broad-brush opinion of the relative value of various active and passive measures to protect the civil population in case of nuclear attack and its aftermath."[106] The Committee was thus asked to conduct a cost-benefit analysis of measures to address a future event that had no precedent in human experience.

The Gaither Committee's work was divided into study groups, one of which focused on passive defense measures.[107] This group included a number of specialists in industrial mobilization and resource management, including Edwin George and Burke Horton, who were both closely involved with ODM's damage assessment program. Herman Kahn served as an informal advisor to the group.[108] The group used a "series of hypothetical attack situations" developed by ODM's damage assessment unit to analyze the post-attack balance of critical survival items. Referring to the use of the simulation technique to estimate casualty levels, one member of the group, radiation specialist Walmer Strope, later reflected that the "real innovation" of the Gaither study was "the consequence of the developing computer technology" and its application to the problem of survival planning. He recalled that the results of the damage simulation procedure, "the first of this kind," were "rather convincing."[109]

The computer simulations indicated that survival in the immediate post-attack period would depend on having sufficient stocks of survival items on hand or near their points of use. With respect to food stocks, the passive defense study found that, given the country's vast agricultural areas, "the provision of adequate nutritional and caloric diet" would still be possible "even after a massive nuclear attack."[110] The provision of housing was also not "expected to be a critical problem," although life would be "uncomfortable due to displacement and overcrowding." The availability of sufficient water supplies, however, was identified as a critical vulnerability: urban areas would likely experience severe shortages due to equipment damage and the disruption of distribution systems. As for the postattack medical situation, the study found that the prospect of providing emergency care for massive numbers of casualties at first "staggers the imagination." But "if proper sorting of the injured can be done," there was "a reasonable chance of phasing the limited medical care so that it will do the most good for the greatest number of people." However, storage and distribution of the necessary medical supplies remained a challenge, as "current bulk storage in some 42 warehouses does not appear practical," especially "when one considers the urgency of time in medical treatment."[111]

The Gaither passive defense group also drew on NDAC's damage simulation technique to analyze the nation's capacity for "producing the goods and services its citizens need" once stocks of survival items had been depleted. The simulation indicated that three months after an attack, in a scenario that assumed "strategic surprise," 49 percent of facilities for the production of transportation equipment, 40 percent of facilities for petroleum and coal products, and 58 percent of capacity for chemical production would survive. But a different picture appeared when specific industrial subsectors were examined in detail. "These indications which rely on general statistical measures can be misleading," the study noted, "for within a given industry certain segments may be highly vulnerable." Here, the study applied the technique of "vertical analysis" to systems for producing the basic items required to sustain human life. In the case of the chemical industry, it found that a substantial percentage of facilities might be expected to survive an attack, "yet, within that industry, production of certain vital medicine and drug items" was "highly concentrated and vulnerable." For example, 75 percent of sulfa drug production was located in the metropolitan New York area, and 100 percent of terramycin production was located in the "highly-vulnerable" Groton–New London area. When industries were examined "by segment or individual product," the study concluded, "survival probabilities look much dimmer than they do when over-all industry statistics are examined."[112]

In its study of the vulnerability of essential services, the passive defense group again used the technique of vertical analysis. The study focused on the transportation system, since the distribution of supplies such as medicines and foods would be essential to postattack survival. The damage simulations suggested various factors that could severely limit postattack transportation. One was the availability of personnel: "unless a shelter program is adopted," the study found, "manpower" would be the "limiting factor" to the recovery of the transportation system. Transportation would also be curtailed by the interruption of fuel supplies resulting from the destruction of major petroleum refineries and from "severe damage" to pipelines. The recovery of the transportation system would also be limited by damage to vital nodes. For instance, the group found that "under one of the attack plans studied, all but one of the bridges across the Mississippi River were destroyed."[113]

The passive defense group presented its findings to the NSC in the second volume of the Gaither Committee report. The introduction to the volume explained that the postattack environment would "not resemble life as we know it today," and that the "survival problems" would be "ugly and formidable." In each of the attack patterns the Committee examined, the survival problems were "of such magnitude that they are not easy to comprehend." But these problems could be overcome by identifying the most acute scarcities in postattack survival resources. "In all of the Attack Plans studied," the group explained, "manpower is the limiting factor in every case, unless shelter protection is provided for the labor force." The group therefore called for large-scale investment in a mass shelter program to protect the population. This shelter program, it argued, should be "designed to interact successfully with pre-attack civilian warning and communication, with storage and stockpiling plans, with post-attack industrial recuperation and with measures to insure government continuity and recovery." The report emphasized the need to specifically plan for survival items and essential services. Securing the postattack survival of the population and the rehabilitation of the economy would be "extremely difficult, if not highly improbable" unless preparations were made in advance to ensure that "adequate stocks of survival items are available in survivor's hands, in protected distribution channels or in government stockpiles to meet survival needs after attack."[114] The computerized damage assessment technique enabled the group to quantify this problem and to thereby address the Committee's charge to estimate the costs of an adequate stockpiling program.[115]

A NEW ORGANIZATION FOR EMERGENCY PREPAREDNESS

The classified report from the Gaither Committee was circulated to government officials in late 1957. Along with the launch of Sputnik, which indicated that the Soviets possessed advanced ballistic missile capability, historians regard the report as a watershed of sorts in the history of the Cold War. According to the historian Philip Funigiello, although the threat of massive atomic attack had been discussed for several years in high-level planning bodies, the Gaither report forced "official Washington" to take it seriously for the first time.[116] The report's recommendations were not universally supported. For example, the Eisenhower administration rejected its proposal for massive expenditures on shelters and a stockpile of survival items. But the Gaither report's framing of the centrality of resource management to postattack survival—which the Committee's passive defense group had inherited from ODM mobilization planners—was widely accepted in the late 1950s.

For many officials and experts, this orientation to postattack survival implied that the entire approach to emergency preparedness for a future attack had to be rethought, and the federal approach to nonmilitary defense had to be reorganized. Thus, the Holifield Committee argued that "mobilization in the sense of preparing production, manpower, and other resources for great wars with long periods of buildup, after which the productive and military might of America is brought to bear for final victory is outmoded." Therefore, it recommended that civil defense and defense mobilization be integrated into a "unified and continuous concept" of nonmilitary defense, encompassing "all those tasks necessary to prepare the Nation to withstand and overcome the ravages of enemy attack."[117] In this "unified" conception, nonmilitary defense would aim to "strengthen our capacity to substantially withstand attack, our national resiliency, by insuring the continuity of civil government and the protection of civilian life." If an aggressor were to strike, the Committee concluded, such a program would be "the indispensable means to national survival."[118]

The Eisenhower administration acted on these recommendations in 1958, presenting Congress with a plan to merge the Office of Defense Mobilization with the Federal Civil Defense Administration in a new Office of Civil and Defense Mobilization. In a message that accompanied the administration's plan for the merger, Eisenhower argued that the existing organization of nonmilitary defense was "out of date." Specifically, technical advance in weapons and delivery systems had "led to a serious overlap among agencies carrying on these leadership and planning functions."[119] William Finan of the Bureau of the Budget elaborated this point in his testimony on the merger to the House Subcommittee on Government Operations. When laws such as the National Security Act (1947), the Federal Civil Defense Act (1950), and the Defense Production Act (1950) were enacted, Finan told the Subcommittee, "hydrogen bombs, radiological fallout, and missile delivery were not immediate factors in defense planning." The possibility of an attack "which might affect vast areas of the country and require the use of all available resources, human and material, in the survival effort" made it "vital" that the "inseparability of defense mobilization and civil defense be recognized."[120]

Gordon Gray, who had taken over the directorship of ODM from Arthur Flemming, similarly explained that the "early delineations of functions and assignments to delegate agencies"—with FCDA in charge of "local survival operations" and ODM responsible for "overall Government management of resources"—were based on "a concept of limited and localized attack." But with the development of thermonuclear weapons and the resulting problem of fallout, it had "become apparent," Gray testified, "that the survival and resource problems become one."[121]

A report produced by McKinsey and Company for the Bureau of the Budget described the rationale for a merged office and the functions it would perform. "The tin hat and sand bucket defense measures of former years are outdated," the report argued, as was "heavy reliance on defense mobilization concepts in which it was assumed that the Nation would have time after hostilities to assemble, control, and allocate its resources." The devastation that would result from a nuclear attack "makes manifest that the Nation's total energies and resources would be dedicated to human survival." The central function of a new office of "civilian mobilization" would be to plan for the aftermath of an attack. This office would establish "realistic assumptions" as to the types of situations that would be faced by civilian authorities in a future war and would conduct "supply-requirements studies to determine the availability of critical resources." It would develop emergency operating plans and organize exercises so that federal agencies and local and state governments could test their own plans. The office would also direct readiness programs to ensure adequate supplies of "health and food items required in the survival period following an attack."[122] These programs would include planning for the strategic stockpile and the use of Defense Production Act powers to ensure the production and supply of survival items. In short, the office's task would be to maintain a national "balance sheet" of survival resources.

From Industrial Mobilization to Emergency Preparedness

In late 1941, amid the nation's preparation for entry into World War II, mobilization specialists working in federal emergency agencies had labored to prepare the statistical tools required to run a war economy. They conducted massive surveys of American industry; constructed a "balance sheet" of supplies and requirements; and ran feasibility tests to gauge whether a program of military production could be carried out, given available resources and the overall productive capacity of the American economy (see chapter 2).

In the late 1950s, many of the same specialists, now working in the newly constituted Office of Civil and Defense Mobilization (OCDM) and other federal mobilization planning offices, faced a different problem of emergency government. Anticipating the unprecedented event of thermonuclear attack, they employed the tools of mobilization planning—national resource surveys, balance sheets, and feasibility analyses—to plan for national survival in a future war.

SUPPLY-REQUIREMENTS ANALYSIS

By 1957 mobilization planners had identified the resources that would be essential for survival of the population following an attack on the United States. And they had begun to use NDAC's simulation technique—in Operation Alert 1957 and in the Gaither passive defense study, for example—to examine the availability of selected survival resources in a future war. These simulations yielded rough national estimates for fifty out of the two hundred items that had been identified by the Committee on Essential Survival Items. But planners did not have the comprehensive data required to envision and plan for the full range of postattack survival needs or to anticipate whether sufficient supplies would be available to meet requirements in specific geographic locations. Thus, the Department of Commerce, which was responsible for compiling information on survival resources, reported that the OPAL 57 exercise had demonstrated "the lack of statistical data on production, capacity, and inventories of survival items." Without such data, there could "be no survival items program which will be meaningful and of real value in the event of an attack upon the Nation."[123] While planners had "knowledge and tools to handle the traditional job of mobilizing military production," Innis Harris concluded in 1958, they were still "not equipped with the statistical tools to mobilize critical items for human survival, under attack and postattack conditions."[124]

To remedy this lack of information about postattack survival resources, mobilization planners investigated what Gordon Gray described as "the Nation's capacity to produce" survival items, "the likely effects of enemy attack on that capacity, and the estimated requirements for those items."[125] This work was overseen by the Business and Defense Services Administration (BDSA), which was charged to produce a "supply-requirements balance sheet" that could be used to plan programs to "insure the availability of these essential survival items" in a future war.[126] In this obscure agency within the Department of Commerce, government planners concretely addressed the problem of "thinking about the unthinkable" effects of a future catastrophe. As BDSA administrator H. B. McCoy framed it, the challenge of estimating available supplies of survival resources was "of enormous proportions." Indeed, there was a "serious question as to whether human beings" had the "capacity to think through the awesome consequences of nuclear attack and to develop the most effective means of dealing with them." Yet such information was "absolutely necessary for the total consideration of the survival items program."[127] Until the postattack supply of these items could be estimated and "matched against estimated requirements for these same items under assumed attack conditions," planners would not know "the extent to which our postattack needs can be met nor will we be able to decide upon remedies to meet potential short supply situations."[128]

BDSA's specific charge was to investigate the postattack supply of survival items in different parts of the country: What quantity of each survival item would be available in stocks or inventories, and how much could be produced? To address these questions, BDSA staff applied the vertical analysis technique they had used to study industrial chains of production to investigate the postattack supply of survival items. They oversaw "extensive statistical studies on the production, production capacity, and national inventories" of specific survival items that had been "identified and agreed upon for planning purposes."[129] As in the agency's work on industrial production, a key question was what resources and facilities should be included in studies of the production and distribution of survival items. As William Truppner, director of BDSA's Office of Industrial Mobilization, described this issue, "We have created the most complex interdependent industrial complex in the history of the world and we have achieved great efficiency in production substantially by a concentration of our efforts and dependence on other people for their concentration on other efforts." This efficiency and interdependence, however, was also a source of vulnerability. BDSA's analyses suggested that "the biggest single impact" on the nation's ability to produce all items, including survival items, would come "not from the impact of the damage but rather from the breaking of the production mesh." When the "mesh is broken at any place," Truppner warned, "the thing spreads out."[130] Drawing on data that had been gathered through industrial surveys, BDSA conducted vertical analyses to identify "links in the chain of production which, if broken by attack, would create critical bottlenecks" in the production of survival items.[131]

Once the facilities involved in the production, storage, and distribution of a given survival item had been identified, BDSA worked with various federal departments to survey "production and manufacturers' inventories of those survival items" under their respective jurisdictions.[132] The Department of Commerce, for example, circulated questionnaires to between "10,000 and 15,000 manufacturers to obtain complete production capacity and inventory data" that covered one hundred survival items for which it bore planning responsibility.[133] The Department of Health, Education, and Welfare surveyed available stocks of medical supplies in major US hospitals to identify wholesale and retail inventories. The Department of the Interior conducted surveys of "electrical power, petroleum, and coal," while the Department of Agriculture surveyed production and storage facilities for grain and "other food groups on the survival items list" as well as for "cold storage and meat."[134]

In a final step, data collected in the surveys were transferred to electronic tapes that could be fed into the National Damage Assessment Center's simulation model to calculate the postattack supply of essential survival items.[135] Available postattack supplies were then compared with requirements, which were also estimated using NDAC's simulation technique. Based on such analysis, according to Truppner, the director of OCDM would know "how much reliance he can place on our productive facilities in the event of an attack, so that he has an entry in his statistical balance sheet in calculating the size and nature of a stockpile program for survival items." The director could also use these results to justify expenditure requests to Congress. "The Government does have a few hundred million dollars tied up in this," Truppner explained, "and, when one goes to the Congress and asks for money, one must have some basis for the size of the program for which the funds are being requested."[136]

BDSA work on the supply-requirements balance of survival items began with an analysis of essential medical supplies, which had been identified as a critical gap in national preparedness. Production specialist William Lawrence argued that such analysis was "a prerequisite for a proposed stockpile program." In principle, such a calculation was "a relatively simple task." But it was "tedious and very detailed," and a study of this kind "had never been done before in the history of the Government."[137] First, BDSA estimated postattack requirements—what the remaining population would need to assure its survival.[138] Here, the agency drew on data generated by NDAC using the RISK model, which estimated that an attack would result in 60 million casualties—both dead and wounded—out of a total population of 180 million. BDSA then applied dosage factors computed by the Public

Health Service against this casualty pattern. The result was an estimate of the "total gross requirement for medical survival items" to care for the remaining population.[139] The next step was the supply analysis, based on BDSA's studies. Once BDSA and the Public Health Service had "surveyed stocks at producer, wholesale, hospital, and retail drugstore levels," analysts "applied the RISK I damage pattern to them on a geographical basis" to estimate what volumes of survival resources would be available in different locations throughout the country.

The findings of this assessment were alarming. Postattack medical supplies, Lawrence reported, would cover "our 180-day needs for only 21 of the 119 items on the list." Lawrence emphasized the enormous size of these requirements, which were estimated to be "more than three times the normal capacity output of United States companies without any damage." A medical supply stockpile program appeared, therefore, to be "the only means of survival." Gaps in the supply of essential survival items suggested a "total requirement for stockpiling of $574 million" for the six-month "survival period," and an additional $150 million for "a number of 200-bed" packaged disaster hospitals, bringing the total cost for medical stockpiling to $724 million.[140]

SURVIVAL

By 1960, NDAC had been renamed the National Resource Evaluation Center (NREC). This new name reflected the origins of many of the center's practices in wartime mobilization planning and air intelligence. The career trajectory of economist Joseph Coker, who succeeded Burke Horton as the Center's director, exemplifies these connections to the midcentury scenes of resource evaluation and air target selection. Coker first encountered economic vulnerability analysis in the early 1940s as a resource evaluation specialist in the War Production Board. He then moved into the world of air intelligence, serving in the Economic Vulnerability Section of the Joint Target Group and, after the war, in the Strategic Vulnerability Branch of the Air Intelligence Directorate. During this period, he completed a doctoral dissertation on "the economics of strategic target selection," which laid out the basic principles of industrial vulnerability analysis and introduced the concept of "resilience" to indicate a system's capacity for adaptive transformation in response to an external shock.[141] In the mid-1950s, Coker was among the air intelligence specialists whom Horton recruited to the Office of Defense Mobilization as it turned from economic management of a military-industrial economy to planning for survival following a thermonuclear attack.

Under Coker's direction, one of the Center's projects was to build a computational model, called SURVIVAL, which combined the different elements of a balance sheet analysis in a single automated procedure.[142] According to NREC statistician Wallace B. Oliver, this model was "designed to compare estimated post-attack survival item stocks with the requirements of the surviving population" so that mobilization planners "could determine, by geographic area and time period, the surplus or deficit of supply for individual survival items."[143] One module of the program—described in the lefthand side of figure 6.1—analyzed survival requirements. This analysis incorporated estimates of the likely number of casualties from JUMBO—the damage model. Casualty estimates were phased over five periods: the first, second, and third months after the attack; the next quarter; and then the remainder of the year. The program then incorporated estimates of "stated daily requirement per thousand population" for eight different casualty classes, each of which would have different requirements for food, shelter, medicine, and other items. The supply analysis—described on the right side of figure 6.1—assessed both available stocks and production capacity. SURVIVAL calculated postattack inventories of essential items by feeding data about all facilities that held stocks, including "production, wholesale, distribution point, and public utilities," into the JUMBO damage assessment model.[144] For each facility, the simulation procedure estimated damage and radiation levels to determine the availability of stocks in different postattack periods. Postattack production capacity was assessed by calculating damage to industrial facilities, the availability of inputs (including raw materials and essential services), and the condition of the local labor force, which would be essential to run production after an attack.

As Coker explained, the final outputs of SURVIVAL were "estimates of surpluses or deficits, item by item, period by period, and area by area." These results could guide mobilization preparedness, providing "a rational basis for decisions on stockpiling survival items." Specifically, they could guide "preattack preparations" for "controlling, conserving, producing, allocating, transporting, and distributing critical survival items under various emergency contingencies."[145] Significantly, the model allowed for "parametric control," which enabled planners to adjust assumptions about damage and other factors, thus making it possible to test the efficacy of measures such

THE SURVIVAL MODEL

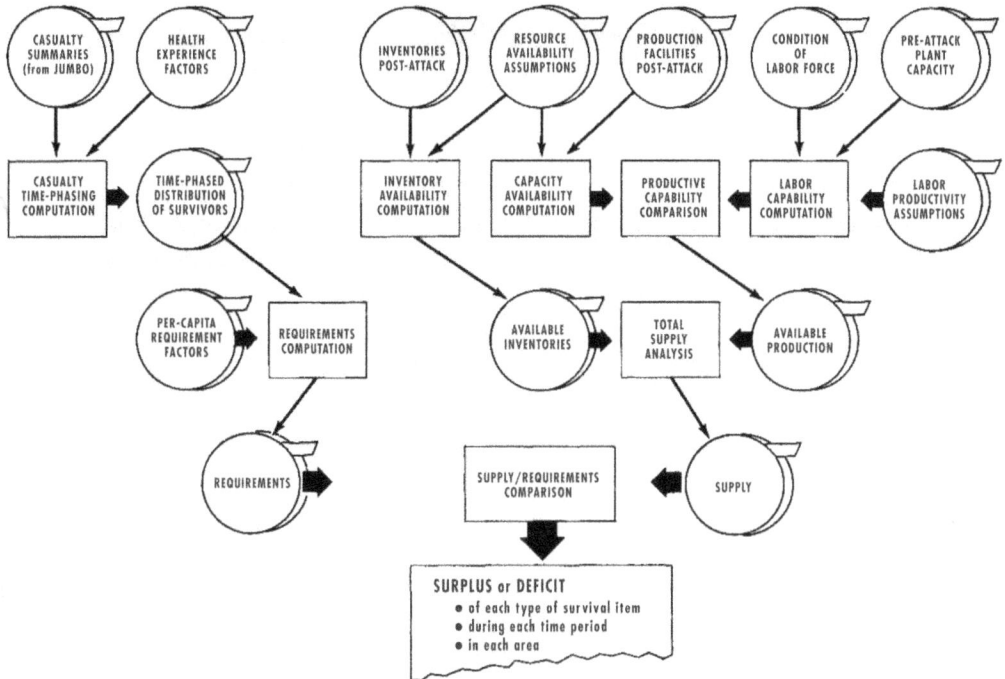

CASUALTY SUMMARIES (from JUMBO)

HEALTH EXPERIENCE FACTORS

INVENTORIES POST-ATTACK

RESOURCE AVAILABILITY ASSUMPTIONS

PRODUCTION FACILITIES POST-ATTACK

CONDITION OF LABOR FORCE

PRE-ATTACK PLANT CAPACITY

CASUALTY TIME-PHASING COMPUTATION

TIME-PHASED DISTRIBUTION OF SURVIVORS

INVENTORY AVAILABILITY COMPUTATION

CAPACITY AVAILABILITY COMPUTATION

PRODUCTIVE CAPABILITY COMPARISON

LABOR CAPABILITY COMPUTATION

LABOR PRODUCTIVITY ASSUMPTIONS

PER-CAPITA REQUIREMENT FACTORS

REQUIREMENTS COMPUTATION

AVAILABLE INVENTORIES

TOTAL SUPPLY ANALYSIS

AVAILABLE PRODUCTION

REQUIREMENTS

SUPPLY/REQUIREMENTS COMPARISON

SUPPLY

SURPLUS or DEFICIT
• of each type of survival item
• during each time period
• in each area

FIGURE 6.1. The SURVIVAL model. The National Damage Assessment Center developed an automated procedure, SURVIVAL, for calculating the relation between requirements for essential items and available supplies in a future postattack situation. The type of balance sheet analysis that had required thousands of human calculators and battalions of tabulating machines during World War II could now be carried out on a digital computer. *Source*: Joseph D. Coker, "Application of Statistics to the Resource Management Program," Proceedings of the Social Statistics Section, American Statistical Association Annual Meeting, December 27–30, 1964, Chicago, IL, http://www.asasrms.org/Proceedings/y1964/y1964.html.

as "facility hardening," decontamination, and the rapid repair of damaged facilities.[146]

Mobilization planners used the SURVIVAL model to perform the same kind of analysis that had been conducted as part of World War II mobilization: surveying resources and production capacity, balancing supplies and requirements, and testing whether critical resources would be available in a future emergency. But in adapting this technique to address the survival items "balance sheet" following a nuclear catastrophe, they also incorporated a range of practices drawn from other domains we have examined in this book. From the field of resource evaluation, they adopted tools for analyzing specific points of vulnerability in vital production systems. From the world of air targeting, planners took up ways of modeling the "event" of a future attack, and for anticipating its effects on facilities and populations. And from planning for postattack relief operations, they incorporated information about the resources needed to sustain the population; about existing inventories of these resources, as they were dispersed in tens of thousands of stores, hospitals, and warehouses; about the health infrastructure of hospitals, beds, doctors, and nurses; and about the production and transportation systems required to produce and deliver them. In drawing together these disparate elements into a simulation procedure, mobilization planners stabilized a new object of knowledge and target of government intervention: the resource systems on which the survival of the population depends.

From Nuclear War to Climate Change

In one sense, the late 1950s marked a kind of apotheosis for the mobilizers. Their framing of nuclear preparedness as a matter of resource planning for national survival was accepted as a basis for organizing nonmilitary defense programs; they had access to the most advanced computers in existence; and they employed sophisticated simulation techniques to prepare for a future war. The type of balance sheet analysis that had taken months of labor for mobilization planners like Stacy May, Robert Nathan, and Simon Kuznets at the outset of World War II, and that had enlisted thousands of workers, battalions of tabulating machines, and endless flows of punch cards in the War Production Board, could now be carried out by an automated computing routine. The mobilizers were confident that their calculative techniques made it possible to rationally plan for the otherwise "unthinkable" problem of national survival in a future war. They believed that planning for the post-attack management of survival resources would be, as H. B. McCoy testified in 1958 to the Joint Committee on Defense Production, "of inestimable benefit to the people of this nation."[1]

But from a broader perspective, we can see that the period from the late 1950s to the early 1960s was the twilight of the mobilizers' influence in the world of national security strategy. Amid increasing skepticism about the very premise of preparing for nuclear war and a proliferation of competing demands for resources and attention, their models did not exert the hold on decisions about resource allocation that feasibility testing did during World War II and the Korean War. "In OCDM we seem to suffer from some kind of phobia," observed industrial production specialist William Lawrence, who had worked in mobilization offices during both wars and was involved in postattack survival planning in the late 1950s. "I feel," he lamented, "that the past three Operations Alert have demonstrated, and conclusively so, that, if we were to suffer a nuclear attack on this country in the near future, we would have the utmost chaos." Yet "no lesson learned in one Operation is ever considered again, and no decision is ever made on a vital question."[2]

Indeed, mobilization and civil defense planners struggled to convince legislators to support stockpiling and shelter plans, as is evidenced by their faltering efforts to acquire the packaged disaster hospitals that were the nucleus of the medical stockpile. By 1957, federal emergency planning agencies had procured only around 1,800 of the hospitals, far short of the 6,000 units that they estimated would be required to treat "the great casualty load anticipated in an all-out attack."[3] Meanwhile, casualty estimates continued to skyrocket, by 1958 reaching 60 million in the event of an all-out nuclear exchange. In response, officials increased their estimates of the mobile hospitals needed to 9,500 and—given that "fallout and disruption of transportation and communication" would limit access to essential supplies—sought to add to the stocks of survival items.[4] In 1962, during a brief period of government support for civil defense, 750 packaged disaster hospitals were added to the stockpile, bringing the total to roughly 2,500. But this support was not sustained. Out-of-date drugs were not replaced, and the safety and sterility of equipment was not ensured. The stockpile became increasingly obsolete and too expensive to upgrade. By 1969, hundreds of millions of dollars in unused medical supplies were gradually deteriorating in thirty-two storage facilities around the country. In 1973, as a cost savings measure, the federal government decided to close down its medical stockpiling program and dispose of the stored supplies. Government auditors had concluded that much of it was "unfit for use," and the rest could be "given to state agencies or sold to private bidders."[5]

For our purposes, the significance of the mobilizers' work on planning for national survival does not lie in its practical influence on government policy or in the success (or failure) of the programs they proposed. Instead, this work exemplifies a novel diagram of emergency government—a distinctive way of understanding and managing threats to the health and well-being of the population, and, thus, a new chapter of biopolitical modernity. Prior to the late eighteenth century, as Michel Foucault argued in his lectures on bio-politics, calamities such as famine, epidemics, and drought were attributed either to fate or to divine punishment. These events were thus understood (and governed) as "inevitable misfortune," external visitations that broke the flow and regularity of normal time. Political authorities sought to man-age such events through a "system of legality and a system of regulations," such as disciplinary controls over the grain trade or the confinements of the plague town.[6] By contrast, Foucault argued, the techniques of population security introduced in the late eighteenth and early nineteenth centuries made it possible to approach these events in an entirely different way—not as external visitations but as the effects of a new register of reality, the social milieu. Thus, early political economists linked food scarcity to the calculated choices of growers, distributors, retailers, and consumers, and a nascent field of public health analyzed disease incidence in terms of risks associated with the living conditions of the population—unclean water supplies, cramped living quarters, and the lack of centralized sanitation. This new way of under-standing the problems of collective existence and of constituting targets of intervention was made possible by the collection of massive amounts of information on living conditions and on the incidence of afflictions such as disease and poverty. By drawing on this vast archive of experience, govern-ment agencies brought the previously external and unaccountable "event" into a framework of statistical calculation. Adapting a phrase from Ian Hack-ing, the chance event was thus "tamed," as techniques of population security were used either to distribute risks over a population (through social insur-ance, for example) or to regularize vital flows by constructing systems of sanitation, water supply, electrical power, and so forth.[7]

From this perspective, the mid-twentieth-century consolidation of vital systems security can be understood as a return of the "chance" event at the limits of population security. In anticipating the postattack world for which they had to prepare, nonmilitary defense planners could not draw on prior historical experience. Recall Burke Horton's comment that the development of computerized techniques of attack simulation made it pos-sible to map new hazards, such as radioactive fallout, "much the same way that temperatures and rainfall maps are prepared for agricultural purposes," with the key difference that the new maps "needed to be prepared *before it 'rained' the first time.*" There was no archive of historical incidence that would enable planners to calculate either the probability or the likely effects of a nuclear attack. Moreover, the vast consequences of such an attack could not be distributed—since they would affect such a large percentage of the population and the economy—or regularized, since the very systems built to manage the vicissitudes of food supply, disease, weather, or other contingencies would be disrupted. Indeed, planners identified this complex of vital systems as precisely the site where a future event would make its effects most painfully felt, as the immediate effects of bombing would be multiplied by the disruption of systems for the delivery of medicine, food, or drinkable water supplies. In its work on postattack survival, ODM sought to tame this kind of unprecedented event. Its procedure for simulating future attacks created a different kind of "archive"—an archive of *possible future events*—that would guide preparedness planning by transforming the radical uncertainty of a nuclear attack into a foreseeable contingency for which preparations could be made.

This new concern with the vulnerability of vital systems was first reflected in law when the 1947 National Security Act charged the National Security Resources Board with ensuring the "continuous operation" of "industries, services, Government and economic activities" that were essential to "the Nation's security" (see chapter 4). When the legislation was enacted, the vital systems in question were still limited to systems of industrial production. Gordon Gray's comment, a decade later, that the survival problem and the resource problem had "become one" marked a shift in the object of mobilization planning: from military-industrial production to the essential services and critical systems that underpinned biopolitical modernity. Thus, the 1958 merger of civil defense and mobilization planning functions in a new Office of Civil and Defense Mobilization (OCDM)—an apparently minor organizational reshuffling—can be understood as a point of inflection in the genealogy of American emergency government. The apparatus of emergency government that had initially been assembled to manage the economy for total war was redirected to a new problem: ensuring the continuous functioning of vital systems in the face of potentially catastrophic disruption.

Vital Systems Security as Political Rationality

The establishment of the Office of Civil and Defense Mobilization did not resolve the problem of how to organize emergency preparedness within the federal government.[8] In 1961, President John F. Kennedy used reorganization authority to move civil defense functions to an Office of Civil Defense, located in the Army. Meanwhile, he placed responsibility for emergency resource management in a new Office of Emergency Planning. This latter office was renamed the Office of Emergency Preparedness in 1968, only to be abolished in 1973, when President Richard Nixon signed an executive order that dispersed its functions to various federal agencies, including the Treasury, the General Services Administration, and the Department of Housing and Urban Development.[9] This diffusion of responsibility led to an increasingly complex federal organization of emergency preparedness and response in which, as historian Scott Knowles writes, a "staggering amount of agencies" were performing "different, but sometimes overlapping functions."[10] In 1979, President Jimmy Carter acted to reassemble these functions in a new Federal Emergency Management Agency (FEMA), once again joining emergency resource management at the national scale with preparedness planning for local emergency relief. In turn, FEMA was incorporated into a new Department of Homeland Security after the attacks of September 11, 2001, as part of (yet another) broad reshuffling of domestic security organizations.[11]

Over the course of these successive reorganizations, the purview of emergency government gradually expanded to address a wide range of emergencies, from natural disasters to environmental catastrophes and energy crises.[12] This shift was already reflected in Kennedy's charge to OCDM's successor, the Office of Emergency Planning, to "advise and assist the President" in the "determination of policy for the emergency plans and preparedness assignments of the Federal departments and agencies," and to "make possible at Federal, State, and local levels the mobilization of the human, natural, and industrial resources of the nation to meet all conditions of national emergency." Nonmilitary defense planning to manage the consequences of a nuclear attack has persisted as one part of federal emergency management. But in practical terms, the organizations that inherited the functions of the Office of Emergency Planning have been more centrally concerned with events that have actually occurred: a series of natural disasters in the early 1960s, culminating with Hurricane Betsy in 1965; terrorist attacks and oil shocks in the 1970s; a renewed wave of natural disasters in the early 1990s; the attacks of September 11, 2001; and a succession of more recent disasters, from Hurricanes Katrina and Sandy to the massive wildfires of 2018 and 2020 in California and the coronavirus pandemic.

Congress and the executive branch have responded to such events by charging federal emergency management organizations with new responsibilities for emergency preparedness, post-disaster relief, and hazard mitigation. In addressing these disparate responsibilities, the emergency management organizations that descended from the mobilization and civil defense planning agencies of the 1950s have been beset by many of the tensions that initially arose in the early Cold War. Among these are the balance between the functions of emergency relief and emergency resource management; the ambiguous lines of federal versus local authority in preparing for and responding to emergencies; the complex coordination of emergency preparedness across multiple agencies in the federal government; and the relation between executive authority and congressional prerogative. These persistent problems can be traced to fundamental features of the US government, such as its federal structure and its diffused pattern of sovereignty. But they are also the product of historical contingencies. There is no particular reason, for example, why powers and functions of emergency resource management should primarily be federal, given the devolution of so many other aspects of emergency government to states and localities. In this sense, the genealogy we have presented offers insight into the specific circumstances in which characteristic problems and tensions of American emergency management arose, and why, perhaps, they have proven to be so persistent.

But the significance of the genealogy presented in this book does not only concern the specific domain of federal emergency management. The knowledge practices, administrative and legal forms, and technical instruments initially devised to prepare for thermonuclear war have been taken up in other federal government agencies with responsibility for health, finance, transportation, and energy, as well as by state and local governments; by private sector organizations, such as reinsurance and catastrophe modeling companies; and by international organizations working in areas such as humanitarian relief, disaster response, and global health. The techniques invented for anticipating and managing a nuclear attack have often been generalized as they have been adopted in new settings. Thus, bomb damage assessment evolved into catastrophe modeling, a tool that is now applied to a range of potential events including natural disasters and disease outbreaks in fields from urban planning to insurance. The scenario-based exercise, which migrated from the military to civilian government in the early 1950s, has become a standard planning technique in various domains of civilian government and in strategic planning in corporations. Planning for the continuity of "essential functions" now underpins policies for critical infrastructure protection and governmental continuity at the federal, state, and local levels.[13]

What links these diverse domains, despite their heterogeneity, is a common concern with the breakdown of vital systems that can turn a manageable disaster into a devastating catastrophe: the collapse of levees, the failure of emergency communications systems, the blockage of road networks needed for evacuation in a wildfire, or the collapse of a health infrastructure, short of resources and overwhelmed by patients in need of care. We also find across these various domains a strategic orientation that was initially formulated by mobilization planners working on postattack survival in the late 1950s: ensuring the provision of resources that are essential to saving lives and sustaining the operation of vital systems in a future emergency. In sum, vital systems security is among the dominant political rationalities of the contemporary world. Its emergence—alongside and in relation to apparatuses of population security and sovereign state security—marks a "reflexive" moment in biopolitical modernity. The very systems that were invented to ensure health and well-being, and to manage the vicissitudes of life, have themselves become sources of vulnerability to catastrophic disruption.

COVID-19: SECURING ESSENTIAL FUNCTIONS

In early spring 2020, as Covid-19 began to spread across the United States, public health officials proposed a strategy for managing the pandemic. They argued that nonpharmaceutical interventions, such as banning public gatherings, closing schools, and ordering people to stay in their homes for all but essential activities, should be used to "flatten the curve" of disease incidence. This approach drew on an old technique for managing infectious disease—physical separation—but directed it toward a novel objective: preventing the health infrastructure from being overwhelmed.[14] Pointing to epidemiological models, health officials argued that unless stringent measures were taken, there would not be enough hospital beds, ventilators, or medical personnel to treat the expected surge of cases. In overcrowded hospitals, the disease would likely spread to health care workers and uninfected patients, exacerbating the crisis. Care would have to be rationed, death rates would rise, and other medical emergencies might go unaddressed. From this perspective, the key question for public officials was the balance between the availability of health resources and the requirements for care.[15]

As health experts and officials focused on ensuring the continued operation of health care infrastructure, obscure parts of the system for producing, allocating, and delivering health resources came to the center of national attention. Diagnostic testing was limited by bottlenecks in the supply chain and by poor coordination between federal and local agencies. The resulting lack of basic epidemiological data made it nearly impossible to know how widespread the disease was and where to target interventions.

It was discovered that the Strategic National Stockpile held only 12,700 ventilators—many not in working condition.[16] At the same time, the overseas sourcing of critical supplies such as ventilators and N95 masks made it difficult for domestic manufacturers to ramp up production. Debate arose over the use of a little-known piece of legislation, the 1950 Defense Production Act, to increase the supply of essential items. It was not clear which governmental organization—the Department of Health and Human Services, the Centers for Disease Control, the Department of Homeland Security, or the Executive Office of the President—had authority over the federal response, or how to adjudicate among competing claims from states for scarce medical resources. There was no formal system for establishing priorities in the allocation of essential resources, and governors competed to acquire critical supplies.

Meanwhile, officials scrambled to organize shelter-in-place policies that would reduce the rate of disease transmission while allowing "essential functions" to continue operating during the crisis. In March, the Cybersecurity and Infrastructure Security Agency within the Department of Homeland Security released a guidance document for state and local officials on "the identification of essential critical infrastructure workers during the Covid-19 response." During the pandemic emergency, the guidance stated, certain industries had "a special responsibility to continue operations" even in the face of stay-at-home orders from public health authorities. "In the modern economy," it explained, "reliance on technology and on just-in-time supply chains means that certain workers must be able to access certain sites, facilities, and assets in order to ensure continuity of functions."[17] The document identified dozens of specific occupations in which workers should remain physically on site, classified according to sixteen infrastructure sectors, including food and agriculture, healthcare and public health, transportation, and commercial facilities.

In sum, while the US pandemic response in spring 2020 was confused and contested in many ways, it is striking that public health officials nonetheless generally agreed on the basic aims of government response: preventing the health care infrastructure from being overwhelmed, providing an adequate supply of critical medical resources required to treat patients, and ensuring that, amid shutdowns designed to reduce disease transmission, the essential functions of social and economic life would continue. As we have shown in this book, this governmental aim—securing the operation of vital systems—was first identified in the middle of the twentieth century, in efforts to prepare for nuclear war. The elements marshaled to meet this objective (however successfully or unsuccessfully) in the case of Covid-19 had also initially been assembled in the context of war and Cold War: stockpiles of vital resources; balance sheets of supplies and requirements; measures to allocate resources among competing priorities; lists of essential functions that would have to continue under emergency conditions; and arrangements for coordinating emergency response, both among different parts of the federal government and among national, state, and municipal governments.

The case of the US government's response to Covid-19 helps to clarify our argument that vital systems security—along with sovereign state security and population security—is among the dominant rationalities of contemporary government. The elements of response that are assembled amid rapidly unfolding emergency situations draw on techniques of intervention, knowledge practices, and forms of governmental organization that make up a coherent diagram of government—a way of thinking about and governing collective existence as a complex of vital and vulnerable systems. In the early twenty-first century, this way of governing collective existence is not necessarily articulated in an explicit policy framework, but rather is embedded in practices, institutions, laws, and forms of expertise in diverse domains, from public health to local emergency preparedness planning and practices of critical infrastructure protection. These various elements make it possible to constitute a range of novel "events" in a particular way—as threats to the functioning of vital systems—that experts and officials can act on through concrete interventions.

CLIMATE EMERGENCIES

As the coronavirus pandemic intensified across the United States, a number of commentators drew an initially surprising connection to another, seemingly very different threat. Bill Gates wrote that Covid-19—which caused a "tragic number of deaths" and "economic hardship not seen in many generations"—offered a preview of "another global crisis: climate change."[18] An article by three McKinsey consultants also pointed to "fundamental similarities" between the widespread effects of Covid-19 and those of climate change. "The current pandemic," it argued, "provides us perhaps with a foretaste of what a full-fledged climate crisis could entail in terms of simultaneous exogenous shocks to supply and demand, disruption of supply chains, and global transmission and amplification mechanisms."[19] Physicist Adam Frank, meanwhile, suggested that in "fostering a recognition" that the health infrastructure and other vital systems are "a whole lot more fragile than we thought," the outbreak could be seen as a "fire drill" for climate change. "When most of us think about climate change," he explained, "we visualize changes to the planet: soaring temperatures, rising oceans, melting ice caps." But to understand the "powerful connection" between the experience of the pandemic and the future of climate change, one had to take a different perspective, approaching "modern civilization" as a "series of networks laid on top of each other." We will encounter climate change, Frank concluded, as "one emergency after another, year after year, as heat waves, floods, fire and storms blow cascades of failures through our systems."[20]

Such analyses suggest that climate change is today being framed in the now-familiar terms of system vulnerability thinking: as a series of punctuated events that threaten to disrupt the vital systems on which modern society depends. They also indicate that, as climate change unfolds, vital systems security is becoming more central to contemporary politics and to the practical operation of government. The knowledge practices and tools of intervention developed in the Cold War United States as part of nuclear preparedness—catastrophe modeling, vulnerability assessment, stockpiling of essential supplies, and emergency readiness—are now brought to bear to address present and future disasters related to climate change. And the form of distributed governmental responsibility that arose in the 1950s for responding to a nuclear attack is the organizational form through which climate-related events such as floods, hurricanes, and wildfires are now being managed. In turn, tensions that were identified in Cold War nuclear preparedness confront today's urban planners, disaster managers, and economic advisors as they deal with the anticipated effects of climate change. Like preparedness planners of the early Cold War, they face the challenge of understanding and governing the elements of collective existence—whether a city, an economic activity, an infrastructure, or a community—from the perspective of a catastrophic future. How should we evaluate the risk of such disasters? How should we calculate the present value of vulnerability reduction and preparedness measures to manage uncertain future events? How should we distribute the burden of paying for such measures? How should we weigh the need to reduce vulnerability against the imperatives of economic growth or welfare? Are these aims in tension or complementary?

In the 1950s, discussion of such questions was mostly confined to a small circle of mobilization and civil defense planners in the US federal government. Today, they are confronted in cities and localities around the country—and around the world—as governments, NGOs, and multilateral organizations address future climate change and manage its already-unfolding effects. For example, the question that was posed in Cold War discussions of industrial dispersal—regarding the tensions between vulnerability reduction and existing economic and metropolitan organization—today recurs in discussions of relocating populations in the face of climate risk. How should the freedom to choose where to live—or to continue living in a particular vulnerable location—be weighed against the costs that exposure to disasters imposes on broader collectivities? Should people living in areas that are becoming ever-more risk prone due to climate change be made to bear the costs of those risks? If not, who should?

There is another way in which current discussions of how to address climate change connect to the genealogy of emergency government presented in this book. In the last several years, advocates for more robust measures to reduce greenhouse gas emissions have begun to discuss the linked ideas of climate "emergency" and climate "mobilization." Here, the reference is not to what legal scholar Kim Lane Scheppele calls the "small emergencies" of floods, blackouts, heat waves, and other events that will occur with ever greater frequency and intensity as the climate changes.[21] The reference, rather, is to "big" emergencies such as war and economic depression that are understood to pose an existential threat to national life. The Democratic Party's 2016 platform thus proclaimed its commitment to "a national mobilization" to reduce greenhouse gases and to a "global effort to mobilize nations to address this threat on a scale not seen since World War II."[22] Similarly, the advocacy group Climate Mobilization has released a climate "Victory Plan"—modeled on the plan that Stacy May and Robert Nathan produced in late 1941. The group calls for an "emergency restructuring" of the economy that would imply "nothing less than a government-coordinated social and industrial revolution."[23]

The idiom of "climate emergency" has proliferated with remarkable speed. By early 2021, almost fourteen thousand scientists had signed a "Warning of a Climate Emergency" circulated by the Alliance of World Scientists; nearly two thousand government jurisdictions, representing over 800 million people worldwide, had issued climate emergency declarations.[24] These declarations emphasize the severity of the threat presented by climate change, and the massive, urgent response required to address it. They also generally proclaim their democratic character: the International Climate Emergency Forum thus declares its "commitment to democratic principles," and Climate Mobilization emphasizes that, as a "grassroots" democratic movement, it seeks to be "as inclusive as possible while unwaveringly demanding WWII-scale climate mobilization."[25] But a number of commentators have pointed out that it is far from clear how—and whether—liberal democracies can organize "emergency" measures to mitigate and adapt to climate change, or whether "government-coordinated social and industrial revolution" would prove compatible with democratic norms. For example, philosopher Dale Jamieson points out that climate change exposes key vulnerabilities of modern democracies: the short-term time horizons of elected representatives, the large numbers of veto points (particularly in the American political system), and the subservience of experts to "the voice of the people." Meanwhile, historian Nils Gilman argues that the rhetoric of "catastrophism" often found in claims of climate emergency may well open the door to antidemocratic responses, whether in radical economic measures to reduce emissions, in responses to ever more intense and destructive disasters, or in more autarchic forms of politics that arise in an increasingly resource-constrained world.[26] Gilman thus suggests the possibility that efforts to address a "climate emergency" might lead to what Ulrich Beck called a "totalitarianism of hazard prevention" in which an "exceptional condition" threatens to "become the norm."[27]

As we have shown in this book, the ability of constitutional democracy to address crisis situations was a central issue in the succession of mid-twentieth-century crises during which American emergency government first took shape. Progressive reformers thought that crisis situations—which, they argued, required powerful executive authority based on technical knowledge about rapidly unfolding situations—presented a fundamental challenge to American democracy. The prospect that government response to emergencies would undermine American democratic institutions arose repeatedly during the Great Depression, World War II, and the early Cold War, from accusations about Roosevelt's "dictatorial" measures in the New Deal to Eisenhower's declaration of martial law during a nuclear preparedness exercise. Midcentury reformers, officials, and experts sought to address these dilemmas not by amending the Constitution but rather by assembling "devices and techniques" that, as Clinton Rossiter put it, would make American emergency government "strong enough to maintain its own existence without at the same time being so strong as to subvert the liberties of the people it has been instituted to defend."[28] As the effects of climate change unfold, this foundational problem of American emergency government is still with us: how to invent an "administrative machinery" that can address future crises within the frame of democratic government.

NOTES

Preface: A Vulnerable World

1. See, for example, Comfort, Boin, and Demchak, *Designing Resilience*.

2. In critical social science, this argument was classically advanced by Ulrich Beck in *Risk Society*. For a more recent discussion, see S. Graham, *Disrupted Cities*.

3. See, for example, Cooper, *Disaster*.

4. For an account, see Collier and Lakoff, "Problem of Securing Health"; and Lakoff, *Unprepared*.

5. Publications from this early phase of the project include Collier, "Enacting Catastrophe"; Collier and Lakoff, "Distributed Preparedness"; Collier and Lakoff, "Vulnerability of Vital Systems"; Collier and Lakoff, "Problem of Securing Health"; and Lakoff, "Preparing for the Next Emergency."

6. In fact, civil defense and other forms of emergency preparedness and response were connected more closely in the 1950s and 1960s than has been generally recognized (see S. Knowles, *Disaster Experts*). Nonetheless, "preparedness" took shape as a mode of civil defense planning for nuclear attack.

7. For example, S. Knowles, *Disaster Experts*; and Quarantelli and Dynes, "Response to Social Crisis and Disaster."

8. This office was initially named the Office of Emergency Planning when it was created by executive order in 1961. It was subsequently named the Office of Emergency Preparedness, until its dissolution in 1973.

9. Rossiter, "Constitutional Significance," 1213.

10. Katie Rogers, Maggie Haberman, and Ana Swanson, "Trump Resists Pressure to Use Wartime Law to Mobilize Industry in Virus Response," *New York Times*, March 20, 2020.

11. Andrew Jacobs, "Despite Claims, Trump Rarely Uses Wartime Law in Battle against Covid," *New York Times*, September 22, 2020.

12. Isaac Stanley-Becker, "Biden Harnesses Defense Production Act to Speed Vaccinations and Production of Protective Equipment," *Washington Post*, February 5, 2021.

Introduction: The New Normalcy

1. Industrial College of the Armed Forces, *Emergency Management of the National Economy*. The college was created in 1924—in the aftermath of the chaotic mobilization of World War I—to train officers in wartime procurement and industrial production. After a brief hiatus during World War II, when college officers were dispatched to support the industrial mobilization effort, the ICAF resumed its activities in 1943. Much of the material in *Emergency Management of the National Economy* (1954) had been written earlier.

2. Fesler served during World War II on the War Production Board and, prior to that, on the National Resources Planning Board and on the President's Committee on Administrative Management (see chapters 1, 2).

3. The offices include the National Resources Planning Board and its postwar successor, the National Security Resources Board; the President's Committee on Administrative Management, which addressed governmental reorganization in the late 1930s; the civilian War Production Board and an array of military mobilization planning and procurement offices established during WWII; the postwar Office of Defense Mobilization; the US Army's Office of the Provost Marshal General; the US Air Force's Strategic Vulnerability Branch; and the Federal Civil Defense Administration, among many others.

4. Flemming, foreword, iii, iv.

5. McDermott, *Office for Emergency Planning*, 6, 13. The executive order to which McDermott refers must have been in draft form at the time of his speech, as it was only promulgated the following year. McDermott's quotations, however, are identical to the final text.

6. Federal Emergency Management Administration, *National Preparedness Goal.*

7. Cybersecurity and Infrastructure Security Agency, "Sector Risk Management Agencies," accessed April 3, 2021, https://www.cisa.gov/sector-risk-management-agencies.

8. Federal Emergency Management Administration, *National Preparedness Goal*, 11, 13.

9. See Foucault, "Nietzsche, Genealogy, History"; and Foucault, "Polemics, Politics, and Problematizations."

10. Steinberg, *Acts of God.*

11. S. Knowles, *Disaster Experts*, 7.

12. Rose, "Governing 'Advanced' Liberal Democracies," 42.

13. Ibid.

14. Hacking, "Historical Ontology," 94. Hacking adds, drawing a contrast with Kuhn's *Structure of Scientific Revolutions*, that "crisis is not offered as the explanation of change. . . . Foucault's explanations of change are complex and programmatic."

15. On this approach to genealogy that emphasizes tracking "recombinatorial" processes, through which contingent elements are linked in historically specific situations, see Koopman, *Genealogy as Critique*; Walters, *Governmentality*; Collier, "Topologies of Power"; and Foucault's remarks in *Birth of Biopolitics.*

16. A reference to the term "vulnerability specialist" appears in 1954 in *The Bomb Damage Problem*, a special report prepared for the Directorate of Management Analysis, DCR/Comptroller Headquarters, US Air Force (see chapter 3).

17. Kupperman, *Technological Advances*, 1.

18. Ibid., 2–3.

19. See Beck, *Risk Society.*

20. Lovins and Lovins, *Brittle Power*, is another prominent example from this period of an understanding of modern society as dependent on the continuous functioning of vulnerable systems.

21. "Description of position: Chief, Systems Evaluation Division, NRAC, OEP (1968)" and "Recommendation for Robert H. Kupperman as Chief, Systems Evaluation Division," Collection Number 95022, box 3, Robert H. Kupperman Papers, Hoover Institution Library and Archives, Stanford, CA. The Systems Evaluation Division, whose activities fall outside the scope of this book, was located within the National Resource Analysis Center, which was the successor to the National Damage Assessment Center and the National Resource Evaluation Center, discussed in chapter 6. The Systems Evaluation Division is one unit that links mobilization preparedness of the 1950s to the more familiar concerns of emergency management as they were articulated in the 1970s.

22. Other classic accounts have described a similar shift in the objects, aims, and logics of government, ranging from Karl Polanyi's history of the birth of market society in the nineteenth

century (*Great Transformation*) to T. H. Marshall's theses on social citizenship in the twentieth century ("Citizenship and Social Class").

23. Foucault, *Security, Territory, Population*, 64, 70, 339.

24. See, for example, Ewald, "Return of Descartes's Malicious Demon"; Foucault, *Security, Territory, Population* and *Birth of Biopolitics*; Rabinow, *Anthropos Today*; Donzelot, *Policing of Families*; Hacking, *Taming of Chance*; Poovey, *History of the Modern Fact*; and Rose, *Powers of Freedom*.

25. Hacking, *Taming of Chance*, 5.

26. Ewald, "Insurance and Risk"; Rose, *Powers of Freedom*.

27. Foucault, *Security, Territory, Population*, 79.

28. On the genealogy of resilience, see Walker and Cooper, "Genealogies of Resilience." Our account differs from Walker and Cooper in a significant respect. Whereas they trace the genealogy of resilience to laissez-faire economics, we show that resilience was conceptualized and operationalized in domains such as wartime mobilization planning and nuclear preparedness planning—the areas in which we find some of the most comprehensive and "totalitarian" forms of economic intervention in American history. For a more open-ended approach to the genealogy and contemporary forms of resilience, see B. Anderson, "What Kind of Thing Is Resilience?"

29. The body of work that has investigated vital systems security in various domains includes Langely, *Liquidity Lost*; Silvast, *Making Electricity Resilient*; Özgöde, "Governing the Economy"; Whitington, "Modernist Infrastructure"; Dunn Cavelty and Giroux, "Good, the Bad, and the Sometimes Ugly"; Collier and Lakoff, "Vital Systems Security"; Lakoff, *Unprepared*; Folkers, "Existential Provisions"; Fearnley, "Epidemic Intelligence"; and Boyle and Speed, "From Protection to Coordinated Preparedness."

30. Similarly, contemporary vital systems security remains closely tied to sovereign state security, as in critical infrastructure protection policies, for example.

31. Vital systems security is thus connected to techniques of what has been called "possibilistic" knowledge. See, for example, Clarke, *Worst Cases*; De Goede, "Beyond Risk"; and Amoore, *Politics of Possibility*. In some cases, however, mechanisms such as catastrophe modeling may be incorporated into "probabilistic" forms of knowledge (see Collier, "Enacting Catastrophe"; and the discussion of the Monte Carlo simulation in chapter 6).

32. On infrastructure and social complexity, see, most classically, Wittfogel, *Oriental Despotism*. On infrastructure in modernity, see Edwards, "Infrastructure and Modernity." As Ashley Carse ("Keyword: Infrastructure") has recently documented, the use of the term "infrastructure" dates to discussions of military logistics systems in the middle of the twentieth century.

33. On the role of communications in classical monarchy, see P. Anderson, *Lineages of the Absolutist State*; and Mann, "Autonomous Power of the State."

34. In this sense, our account complements genealogical work in recent decades on other dimensions of modern political rationality. See, for example: Donzelot, *Policing of Families*; Escobar, *Encountering Development*; Ewald, "Return of Descartes's Malicious Demon"; Foucault, *Security, Territory, Population*; Hacking, *Taming of Chance*; T. Mitchell, *Carbon Democracy*; Rabinow, *French Modern*; and Rose and Miller, *Governing the Present*.

35. At this time, governments began to develop and manage infrastructures (whether power systems, railroads, or urban utilities) to foster economic growth or social welfare in projects of urban reform and national development. See Graham and Marvin, *Splintering Urbanism*. On nineteenth-century urban reform specifically, see Rabinow, *French Modern*. On national developmentalism and infrastructure (particularly in the Russian case) see Collier, *Post-Soviet Social*.

36. The mirrored problems of labor mobilization and government response to strikes were also important early sites for reflection on vital systems and their vulnerability. As Timothy Mitchell (*Carbon Democracy*, 22) has shown, for example, labor organizations recognized that "a relatively minor malfunction, mistiming, or interruption" could have "widespread effects" on

interdependent systems of production and targeted strike activity at vulnerable nodes of circulatory systems. Correspondingly, government measures were formulated to block strike activity on
the grounds that it threatened the supply and distribution of goods that were vital to urban life
and industrial activity. For a similar emphasis, see Neocleous, *Critique of Security*.

37. Quoted in Hansell, *Strategic Air War*, 14.

38. Quoted in Gentile, *How Effective Is Strategic Bombing?*, 16.

39. Franklin D. Roosevelt, "Message to Congress on the Use of Our National Resources," January 24, 1935, in Gerhard Peters and John T. Woolley, American Presidency Project, University of
California, Santa Barbara, https://www.presidency.ucsb.edu/documents/message-congress-the
-use-our-national-resources.

40. The term is taken from the title of the 1949 dissertation of Joseph Coker ("Economics
of Strategic Target Selection"), one of the economists who moved from wartime mobilization
planning and air targeting into postwar vulnerability analysis. See chapter 3.

41. On the connection between mobilization planning and systems analysis, see Novick, *Origin
and History of Program Budgeting*; and Erickson et al., "Bounded Rationality."

42. Directorate of Management Analysis, DCR/Comptroller Headquarters, US Air Force,
Bomb Damage Problem, 14.

43. These techniques evolved into today's practices of catastrophe modeling and scenario-
based exercises. On catastrophe modeling, see, for example, Collier, "Enacting Catastrophe";
and Cabantous and Dupont-Courtade, "What Is a Catastrophe Model Worth?" On exercises,
see Lakoff, "Generic Biothreat"; Samimian-Darash, "Practicing Uncertainty"; and Anderson and
Adey, "Affect and Security."

44. Landis Dauber, *Sympathetic State*.

45. On the treatment of natural disasters as "acts of god" and the significance of this understanding for government policy, see Steinberg, *Acts of God*; and Landis Dauber, *Sympathetic State*.
For a striking account of the ad hoc character of American emergency response prior to the
middle of the twentieth century, see the description of the great Mississippi flood of 1927 in
J. M. Barry, *Rising Tide*.

46. National Security Act of 1947, Pub. L. No. 253, 61 Stat. 496 (July 26, 1947).

47. See, for example, Sebald, *Natural History of Destruction*.

48. Rossiter, *Constitutional Dictatorship*, 1, 3.

49. Gulick, "Conclusion," 139.

50. Merriam, "Government and Society," 1501.

51. Merriam, *New Aspects of Politics*, 17. On Merriam's idea of the "lag" between the development of political institutions and the changing situations these institutions had to address, see
Reagan, *Designing a New America*.

52. This rapid expansion in federal functions was due "in part to the accelerated emergence
of the positive state, in part to the crisis that provided most of this acceleration." Rossiter, "War,
Depression, and the Presidency," 417.

53. Waldo (*Administrative State*) referred to the federal government that emerged from World
War II as the "administrative state." On the expansion of the Executive Office of the President
(EOP) in particular, see Relyea, *Executive Office*. As one indicator of the growth of this apparatus
of executive rule, President Hoover's staff consisted of thirty-three people; today, more than two
thousand people are on the EOP's staff.

54. Rossiter, "War, Depression, and the Presidency," 417.

55. Fesler served on the President's Committee on Administrative Management in the late
1930s, which formulated proposals for the Reorganization Act of 1939, discussed later in this
introduction and in chapter 2. See Fesler, "Brownlow Committee."

56. Fesler, *Principles of Administration*, 82.

57. Tilly, *Coercion, Capital, and European States*. Katznelson, *Fear Itself*, is a prominent recent example of this literature. Among others, we have drawn on Yergin, *Shattered Peace*; Hogan, *Cross of Iron*; Leuchtenberg, "New Deal and the Analogue of War"; and Brinkley, *End of Reform*. For an account that touches on many specific themes we address here, see Curley, "Emergency-War Machine."

58. This point about the shift in the circumstances in which emergency powers were used was made by contemporary observers such as Rossiter and Schmitt, and has been made by today's commentators as well, though interpretations of this shift vary. Scheuerman, *Liberal Democracy*, follows Schmitt and American Progressive reformers in emphasizing the increasing prevalence of social and economic crises. By contrast, Mark Neocleous, in his *Critique of Security*, argues that this shift was driven by an attempt to suppress domestic unrest, and particularly to discipline labor organizations.

59. Scheuerman, *Liberal Democracy*, 113.

60. H. L. Platt, "World War I," 128.

61. Our analysis builds on the work of two political theorists who have provided incisive analyses of Schmitt's work. John McCormick (*Carl Schmitt's Critique*) has emphasized that Schmitt's early work, particularly as presented in *Dictatorship*, published in 1921, presents arguments that are dramatically different from those found in Schmitt's later and better-known work. William Scheuerman (*Liberal Democracy*, 256n3), meanwhile, has highlighted the centrality of economic and financial states of emergency to Schmitt's work, an emphasis that, Scheuerman argues, is neglected in most recent scholarship on Schmitt and on emergency government.

62. On temporality, law, and emergency, see Scheuerman, *Liberal Democracy*; and Opitz and Tellman, "Future Emergencies."

63. Schmitt, *Dictatorship*, 8. For Schmitt, this rational-technical characteristic of dictatorship specifically distinguishes it from despotism: "Any dictatorship that does not make itself dependent on pursuing a concrete result, even if one that corresponds to a normative ideal . . . is an arbitrary despotism. In order to achieve a concrete result, one has to interfere in the causal order of things using means whose justification is given by their degree of appropriateness and depends exclusively on the actual contexts of this causal pattern. . . . There is an unfettering of means from law itself" (xlii). In cases of emergency, Schmitt explained, a ruler "is entitled to do everything" that is appropriate in the actual circumstances"; "only the goal governs" (8).

64. Schmitt, *Dictatorship*, 5, 7, 9. This emphasis on the rational-technical quality of emergency dictatorship contrasts with the positions articulated in Schmitt's later, better-known work. McCormick, *Carl Schmitt's Critique*, 152, places Schmitt's argument on this point in the broader context of the post-Weberian concern in German social thought with the relationship between scientific truth and constitutional order.

65. Merriam, *New Democracy*, 231, 236–37.

66. Rossiter, *Constitutional Dictatorship*, 14.

67. This debate focused on the famous article 48 of the Weimar constitution, which provided for the declaration of states of emergency. It read, in part, "If the public safety and order in the German Reich are seriously disturbed or endangered, the President of the Reich may take the measures necessary to the restoration of the public safety and order, and may if necessary intervene with the armed forces," adding that to this end the president could "suspend in whole or in part" fundamental rights provided in other parts of the constitution. For Schmitt, who came to be a critic of liberal constitutionalism, this provision pointed to the fundamental contradiction at the heart of liberal constitutional regimes. For commentators like Rossiter, by contrast, it merely

showed how poorly elaborated these emergency powers were. Rossiter observed that the last sentence of article 48, which indicated that "a national law shall prescribe the details" of the powers it described, "was never acted upon, and throughout the history of the Republic, Article 48 was the only written law on constitutional emergency powers." As a consequence, article 48 became the "foundation for all sorts and degrees of constitutional dictatorship." See Rossiter, *Constitutional Dictatorship*, 31–32.

68. Rossiter, *Constitutional Dictatorship*, 70. A similar question might be raised about critical analyses of emergency powers in the early 21st century. See discussion below.

69. Ibid., 68.

70. Scheppele, "Small Emergencies," 861–62. Scheppele points to how American emergency government has suffused throughout normal politics, a development she finds troubling. Our analysis views these institutions from a somewhat different perspective—the middle of the twentieth century, when many observers doubted that democratic governments could manage recurrent crises of economic depression and war.

71. Delegated authorities that were important in the run-up to World War II included those employed in the lend-lease and stockpiling programs. As Rossiter points out, along with the "immense delegations of discretionary powers" that Congress made to the presidency, Roosevelt made extensive use of the "inherent" constitutional powers of the presidency as he saw them, often in contradiction of existing law or other constitutional provisions. Rossiter offers a catalog of these actions taken on the basis of emergency powers in *Constitutional Dictatorship*, 66–67.

72. Rossiter described the creation of Executive Office of the President as an attempt to adjust the American executive "in all its ramifications to the mounting stresses of a protean, outward-looking, industrial society." In Rossiter, "Constitutional Significance," 1213.

73. McReynolds, "Office for Emergency Management," 138.

74. Scheuerman, *Liberal Democracy*, 122. Every legal rule "codifies a series of expectations drawn from the experiences of legislators, and past history is necessarily used to draw up general norms intended to function as a guide to the future."

75. E. George, "Management of a Wartime Economy," 6.

76. Calhoun, "World of Emergencies."

77. Recent analysis of the ubiquity of "emergency" declarations include Scheppele, "Small Emergencies"; and Scheuerman, "Time to Look Abroad?"

78. A key example of such risk management in Beck's analysis is insurance—whether public or private—which operates by spreading relatively localized risks over a broader population, based on an assessment of the past incidence of loss. On insurance as an exemplary technology of modern government of "the social," see Ewald, "Insurance and Risk."

79. Beck, *Risk Society*, 51.

80. In addition to Beck, see Ewald, "Return of Descartes's Malicious Demon"; Jasanoff, "Technologies of Humility"; Perrow, *Normal Accidents*; and Callon, Lascoumes, and Barthe, *Acting in an Uncertain World*.

81. Beck, *Risk Society*, 5. To return to his key example of insurance, Beck argues that conventional practices of actuarial assessment are incapable of addressing incalculable, unbounded, and potentially catastrophic threats. "In the afflictions they produce [the risks of second modernity] are no longer tied to their places of origin—the industrial plant. The normative bases of their calculation—the concept of accident and insurance, medical precautions, and so on—do not fit the basic dimensions of these modern threats" (22). Beck's specific thesis on insurance—which is a lynchpin of the argument in *Risk Society*—has been disputed by some authors (e.g., Ericson and Doyle, "Catastrophe Risk"; Collier, "Enacting Catastrophe"). Beck responds to these criticisms in *World at Risk*.

82. Beck, *Risk Society*, 29. On the status of experts, see also Eyal, *Crisis of Expertise*.

83. Beck, *Risk Society*, 79.

84. Ewald, "Return of Descartes's Malicious Demon," 274.

85. On the precautionary avoidance of risk, see various contributions to Baker and Simon, *Embracing Risk*. On the prospects for a democratized science and expertise, see Callon, Lascoumes, and Barth, *Acting in an Uncertain World* and Sheila Jasanoff, "Technologies of Humility."

86. Jasanoff, *Science and Public Reason*, 168.

87. In *Risk Society*, Beck develops this argument in reference to a "scientific and bureaucratic authoritarianism" in areas such as environmental regulation and control but later identifies a similar tendency in counterterror policies following the attacks of September 11, 2001, in the United States. See Beck, "Terrorist Threat."

88. For a review of scholars who have advanced this or similar arguments, see Neocleous, *Critique of Security*. See also the exchange in the *Georgia Law Review* 40, no. 3 (2006), including Levinson, "Constitutional Norms"; Scheppele, "Small Emergencies"; and Scheuerman, "Time to Look Abroad?" See also Unger, *Emergency State*; and Alford, *Permanent State of Emergency*.

89. Agamben, *State of Exception*, 18. Agamben draws on Walter Benjamin's claim that "the 'state of exception' in which we live is the rule" (67).

90. Neocleous, *Critique of Security*, 68. Neocleous points out that the history of emergency powers is closely tied to class relations in industrial capitalism: "not only have emergency powers been crucial to the consolidation of capitalist modernity, almost always introduced to deal with a form of resistance by the oppressed, but also that these powers have become exercised as a permanent feature of oppression" (66). Fairman complicates this story, noting that martial law has been deployed on both sides of the struggle between labor and capital. Fairman, "Law of Martial Rule."

91. Neocleous, *Critique of Security*, 58.

92. Taureck, "Securitization Theory," 54–55.

93. Our approach might be referred to as "second-order observation" of crisis, emergency, and catastrophe (see Luhmann, *Observations on Modernity*). Rather than seeking to diagnose the implication of increasingly pervasive crises or emergencies, second-order organization seeks to understand the different ways in which these terms have been taken up in governmental practice, and the kind of political and discursive work that they perform. For example, Calhoun, "World of Emergencies"; Roitman, *Anti-Crisis*; Aradau and van Munster, *Politics of Catastrophe* and "Exceptionalism and the 'War on Terror'"; Anderson and Adey, "Affect and Security"; Adey, Anderson, and Graham, "Governing Emergencies"; and Amoore, *Politics of Possibility*.

94. Beck, *Risk Society*.

95. As one indicator, polling by the Pew Research Center shows that the greatest bipartisan consensus in the United States about the proper role of government relates to addressing terrorism and natural disasters. Meanwhile, strong partisan divides appear in areas of social and economic policy. See Pew Research Center, *Beyond Distrust*.

96. For similar arguments, see Adey, Anderson, and Graham, "Governing Emergencies," 4; and McCormick, *Carl Schmitt's Critique*. Writing explicitly about Carl Schmitt—but implicitly referring to contemporary interpretations of Schmitt, like Agamben—McCormick (*Carl Schmitt's Critique*, 153) writes, "The exception does not reveal anything, except perhaps that eighteenth and nineteenth century liberals were politically naïve about constitutional emergencies; and perhaps that constitutions and their framers are not omniscient. It offers no more existentially profound truth than that."

97. For similar arguments, see Collier, Lakoff, and Rabinow, "Biosecurity"; and Collier and Lakoff, "Distributed Preparedness." For a parallel critique of "securitization," see Barnett, "On the Milieu of Security."

98. Documentary analysis of relatively obscure texts is a hallmark of genealogy. Thus, Jana Sawicki ("Heidegger and Foucault," 156) writes that Foucault's documents tend to be

"administrative treatises, architectural plans, case studies, hospital records," rather than "canonical texts."

99. A classic statement of this position is Weber, *Protestant Ethic*. For a more contemporary account, see Jasanoff, "Technological Risk."

100. Hacking, *Historical Ontology*, 11.

101. Dreyfus and Rabinow, *Michel Foucault*, have referred to a focus on the "serious speech acts" of technical experts and other purveyors of specialized knowledge.

102. "Problematization doesn't mean the representation of a pre-existent object, nor the creation through discourse of an object that doesn't exist. It's the set of discursive or non-discursive practices that makes something enter into the play of the true and false, and constitutes it as an object for thought (whether under the form of moral reflection, scientific knowledge, political analysis, etc.)." Foucault, "Concern for Truth," 257.

103. Rabinow, *Anthropos Today*, 54. Rabinow continues: "The apparatus is a specific response to a historical problem. It is, however, a dominating strategic response. . . . What may have begun, for example, as a pressing problem of urban policing may turn into a set of diverse techniques applicable to other populations, at other times and in other places: the apparatus can be turned into a technology. Given a specific strategic objective and the attempt to develop a successful response, as one might expect, diverse and unplanned effects can and do result. These too can play a role in extending the network of the apparatus."

104. Edwards, *Vast Machine*, 17.

105. Our approach is in conversation with approaches to "technopolitics" in fields such as science and technology studies, anthropology, and sociology. Much of this literature has examined how technical experts gain political power, how technical rationality is invoked to justify political interests, and how technical rationality is shaped by political programs. In many cases, the point of this literature is to show how political interests are cloaked by purportedly neutral technocracy. See, in particular, Hecht, *Radiance of France*; Hecht and Edwards, *Technopolitics of Cold War*; and T. Mitchell, *Rule of Experts*. Although these questions are not unimportant to our analysis, our central focus is on the mutual constitution of the technical and the political. New objects of technical understanding make possible new kinds of politics, and urgent governmental problems shape the questions that technical experts seek to specify and resolve. See, for example, Barry, *Political Machines*.

106. McDermott, *Office for Emergency Planning*, 3.

107. Foucault contrasts a form of critique that seeks to "reject all possible solutions except for the valid one" with a method of "problematization," which he defines as "the development of a domain of acts, practices, and thoughts that seem to me to pose problems for politics." Foucault, "Polemics, Politics, and Problematizations," 114.

108. Garrison, *Bracing for Armageddon*; Krugler, *This Is Only a Test*; Oakes, *Imaginary War*; Davis, *Stages of Emergency*; S. Knowles, *Disaster Experts*.

109. The Office of Defense Mobilization was created in 1950 as an operational agency for Korean War mobilization and was then merged with the National Security Resources Board in 1953, retaining its name.

110. Edwards, *Closed World*; Erickson et al., *How Reason Almost Lost Its Mind*; Hughes and Hughes, *Systems, Experts, and Computers*; Amadae, *Rationalizing Capitalist Democracy*.

111. Kahn, *On Thermonuclear War* and *Thinking about the Unthinkable*; Ghamari-Tabrizi, *Worlds of Herman Kahn*.

Chapter 1. Vital Systems

1. Sherman, *Air Warfare*, 193.
2. Ibid., 19, 195.
3. Ibid., 195, 197, 200.

4. See Rogers, *Atlantic Crossings*.

5. Committee on Regional Plan of New York and Its Environs, *Regional Plan of New York and Its Environs*, 125; Haig and McCrea, *Major Economic Factors*, 20. For a discussion of the role of organic metaphors in the history of urban planning, see Rabinow, *French Modern*.

6. Fishman, "Regional Plan"; Scott, *American City Planning*.

7. H. L. Platt, "World War I," 132.

8. Kantor, "Charles Dyer Norton," 163–82. Delano worked on railroad logistics and served on the Federal Reserve under Woodrow Wilson. Patrick Reagan (*Designing a New America*) refers to Delano as a "father" of New Deal planning.

9. Friedmann and Weaver, *Territory and Function*, 56.

10. Scott, *American City Planning*, 105.

11. H. L. Platt, "World War I," 134.

12. Ibid., 130, 136, 128.

13. Fishman, "Regional Plan," 113. World War I was also crucial for the integration of regional power systems to deal with local shortages caused by war production, both in the United States and in other countries (see Hughes, *Networks of Power*).

14. Scott, *American City Planning*, 193.

15. H. L. Platt, "World War I," 140.

16. Quoted in Scott, *American City Planning*, 173.

17. M. Knowles, "Engineering Problems of Regional Planning," 119–20.

18. H. L. Platt, "World War I," 137. The war provided "a cram course in learning how to reorganize the complex parts of a badly overloaded infrastructure in order to meet industry's skyrocketing need for more energy" (137).

19. Hughes, *Networks of Power*, 296.

20. Scott, *American City Planning*, 109.

21. T. Adams, preface, x.

22. Fishman, "Regional Plan," 111.

23. T. Adams, preface, x.

24. Scott, *American City Planning*, 123.

25. Lewis, *Transit and Transportation Problem*, 5.

26. Ibid., 17–18. The lessons of the war were explicit in the transportation volume of the *Survey*, which noted that the "difficulties encountered in the transportation of both troops and goods during the World War emphasized the need for better belt line communications" (17–18).

27. Haig and McCrea, *Major Economic Factors*, 3.

28. Fishman, "Regional Plan," 112.

29. T. Adams, preface, xvi.

30. Haig and McCrea, *Major Economic Factors*, 18, 43.

31. Warken, *History of the National Resources Planning Board*, 43. It was originally called the Emergency Administration for Public Works, and then renamed the PWA.

32. National Resources Committee, *Structure of the American Economy*, 136. The Capital Parks Commission was a late-1920s example of regional planning; among other prominent figures, Frederic Delano also played a central role.

33. As Brinkley notes, over the 1930s, this planning board "moved through several incarnations," each of which moved closer to a vision originally articulated by Merriam, of an "agency committed to broad concepts of research and planning with influence at the highest levels of government." Brinkley, *End of Reform*, 175. Economists like Wesley Mitchell had also been pushing for some kind of planning board by the late 1920s. See Özgöde, "Governing the Economy."

34. Brinkley, "National Resources Planning Board," 175. Despite constant changes in names and status, the leading figures in these organizations changed little, and the same vision animated all of them. See Warken, *History of the National Resources Planning Board*.

35. New Deal Planners considered their mission to be an "enlarged version of the same kind of resources and development planning that Eliot, Delano, and many others had grown accustomed to considering for cities and regions." Brinkley, *End of Reform*, 174. See also Clark, *Strategic Factors in Business Cycles*, 12.

36. Ickes, *Back to Work*, 60.

37. On the invention of the national economy, see T. Mitchell, *Carbon Democracy*; Özgöde, "Governing the Economy," 68–75; Breslau, "Economics Invents the Economy," 379–411; and Eyal and Levy, "Economic Indicators as Public Interventions," 220–53.

38. Brinkley, "New Deal and the Idea of the State," 106.

39. National Resources Planning Board, *Federal Emergency Administration*, 84, 42.

40. Ickes, *Back to Work*, 84–85, 125–26.

41. Roosevelt, *Year of Crisis*, 122.

42. Ickes, *Back to Work*, 4, 92. Two-thirds of the emergency expenditures for managing the economic crisis during the New Deal were directed to public works programs; Smith, *Building New Deal Liberalism*, 1–2.

43. National Resources Planning Board, *Federal Emergency Administration*, 49.

44. Clark, *Strategic Factors in Business Cycles*, 88. These developments are usually associated with the development of American macroeconomics, and with such aggregate concepts as general demand and fiscal multipliers. Our account emphasizes how these macroeconomic concepts were built on analyses of substantive flows, which were equally central to New Deal economics.

45. Metcalf, "Secretary Hoover," 60.

46. Hoover, *Report of the President's Commission on Unemployment*, 23.

47. Metcalf, "Secretary Hoover," 70.

48. Committee on Elimination of Waste in Industry, *Waste in Industry*.

49. Metcalf, "Secretary Hoover," 66, 69.

50. Notably, Foucault used the same term—"regularization"—to describe the general norm of apparatuses of biopolitical government. Foucault, *Society Must Be Defended*, 247.

51. For this reason, Hoover actively promoted the creation of an American Construction Council, which was formed in May 1922 under the chairmanship of future president Franklin Delano Roosevelt.

52. Metcalf, "Secretary Hoover," 69, 75.

53. The continuities between Hoover-era thinking and that of the early New Deal can be grasped by comparing Metcalf, "Secretary Hoover," to the discussion in the 1934 "A Plan for Planning" in the National Planning Board's *Report on National Planning*. The themes and style of argumentation are nearly identical.

54. National Resources Board, *Report on National Planning*, 48.

55. Ibid., 48, 84. On the arguments of economists in favor of a more interventionist stance, see also Özgöde, "Governing the Economy," 75–87.

56. Ickes, *Back to Work*, 84.

57. National Resources Planning Board, *Federal Emergency Administration*, 49. The question was whether the stimulus created by spending outweighed the counter-stimulus of greater taxation. In this sense, there were always two sides of the equation: the multiplier of government spending and the taxpayer's propensity to save (or consume). Debt-financed expenditure was a different question.

58. Clark's report was one of four studies on the topic. The others were Black, *Criteria and Planning for Public Works*; Gayer, *Public Works in Prosperity and Depression*; and "Government Organization for Public Works Planning," by F. W. Powell (see National Resources Board, *A Report on National Planning and Public Works*).

59. Clark, *Economics of Planning Public Works*, 15, 13, 83.

60. See Özgöde, "Institutionalism in Action."

61. National Resources Planning Board, *Federal Emergency Administration*, 30.

62. National Resources Planning Board, *Report on National Planning*, 76.

63. Roosevelt, *Public Papers and Addresses*, 60.

64. Here, our story is distinct from Timothy Mitchell's account in *Carbon Democracy*, which argues that the national economy, understood as a macroeconomic aggregate, served to obscure what Karl Polanyi referred to as the "substantive economy." In fact, it might be argued that what Polanyi—an admirer of the American New Deal who wrote his seminal book, *The Great Transformation*, in 1944—called the substantive economy had only been constituted by the research and data-gathering efforts of the New Deal (and their equivalents in other countries) during the first decades of the twentieth century. A similar point is made in Özgöde, "Governing the Economy."

65. National Resources Committee, *Structure of the American Economy*, 1. Also see Didier, "Cunning Observation," for his account of agricultural statistics, and Özgöde, "Institutionalism in Action," 319–21.

66. Duncan and Shelton, *Revolution in United States Government Statistics*, 31. "Prior to 1933, statistical research units in the Federal Government were few and small. Research was not thought of as a governmental activity; it was done by universities, foundations, and large corporations. By the mid-1930s, it became standard operating procedure in setting up a new Federal Agency to have as part of the initial organization a division of research and statistics. The few research units which already existed in Federal agencies were greatly expanded" (31).

67. Riefler, "Government and the Statistician," 5.

68. Carson, "History of the United States National Income," 153–81. On the invention of the national income framework, see also Duncan and Shelton, *Revolution in United States Government Statistics*.

69. See Perlman, "Political Purpose"; and Carson, "History of the United States National Income."

70. See Kuznets, *Commodity Flow and Capital Formation*; and Duncan and Shelton, *Revolution in United States Government Statistics*, 89. At the same time, the economist Wassily Leontief, who also worked at NBER before taking a position at Harvard, began his work on input-output modeling, which characterized the total national economy as an integrated system comprising interdependent activities. Leontief's framework was not fully taken up in the government until after World War II, in part due to the volume of data and data-processing capacity that it required. But government economists were aware of and influenced by his work, and Leontief's first input-output table for the US economy (initially published in 1936) was reprinted as an appendix of the National Resource Committee's landmark 1939 study *Structure of the American Economy*.

71. Perlman, "Political Purpose," 141–42. Kuznets' national income framework was both retrospective and relatively static, but economists in the Department of Commerce worked to make it more dynamic and prospective. Under Nathan's direction, the national income statistics were revised to measure "consumer purchasing power" as the "means to economic recovery" and to provide a "standardized report meant to mirror at short intervals . . . the economy as it was actually operating" (141–42). Also within the Commerce Department (run at this point by Harry Hopkins), a new Industry Survey was headed by William Truppner, and a newly formed Industrial Economics Division "had a heavy current policy orientation. The division was in the forefront in applying Keynesian techniques to the analysis of the required magnitudes of public works expenditures, estimates of full employment capacity, and several other areas" (Carson, "History of the United States National Income," 166). Truppner, who later worked in the War Production Board and on postattack survival planning in postwar preparedness agencies, appears again in chapter 6.

72. Sweezy, "Government's Responsibility for Full Employment," 19–26. Currie taught at Harvard in the early 1930s and entered government as part of the "brain trust" in the Department of Treasury assembled by Jacob Viner. He then followed Treasury Secretary Mariner Eccles to the Federal Reserve, which became the center for Keynesian economics in the federal government. Currie came to be regarded as the leading architect of New Deal economic policy. See Sandilands, *Life and Political Economy of Lauchlin Currie.*

73. Sandilands, *Life and Political Economy of Lauchlin Currie*, 69. This group reached the height of its influence amid a renewed downturn in 1937. Currie and his associates had anticipated the downturn, arguing that it would result from fiscal and monetary retrenchment by the federal government. Over the course of 1938, Roosevelt was convinced to pursue the new avenues of response formulated by Currie's group at the Federal Reserve.

74. Early New Deal programs were generally aimed at particularly problem-stricken areas of the country, such as localities with high rates of unemployment. One frequently cited example of the contradictory effects of programs of the early New Deal is that they were funded through taxation rather than debt finance. In the view of the New Deal Keynesians, the government was simultaneously adding to and subtracting from aggregate demand. See Sandilands, *Life and Political Economy of Lauchlin Currie.*

75. Galbraith, *Economic Effects of the Federal Public Works Expenditures*, 23.

76. He qualified this observation somewhat, speculating that the tendency to fill orders out of stocks would be different at different phases of the business cycle.

77. Galbraith, *Economic Effects of the Public Works Expenditures*, 31, 34–35.

78. Bernstein, *Perilous Progress*, 74. Only the mobilization for war provided the level of fiscal stimulus that was required: "In point of fact, America's greatest depression was not brought to an end by inspired policy choices. Far from it. World War II achieved what the New Deal could not" (74).

79. Riefler, "Our Economic Contribution," 95.

80. Jones, "Role of Keynesians," 125. In fact, Currie was talking about a postdefense slump, but the meaning is the same.

81. Carson, "History of the United States National Income," 167. Bassie "traced the effects of defense expenditures on (1) plant and equipment expansion required for defense production, (2) expansion of consumer outlays and residential construction resulting from a higher level of income, and (3) the private plant and equipment expansion required for the increases under items (1) and (2). To do this, defense expenditures, net exports, and change in inventories were considered the possible initiating factors that bring about changes in total gross national product"—the latter calculated by drawing on data collected by Kuznetz and the Department of Commerce.

82. The New Deal economists proposed to go even further in the vision of compensatory spending than Keynes thought possible. Byrd L. Jones reports on a meeting with Keynes in the summer of 1941, in which "New Deal economists defended their continued advocacy of pushing effective demand in excess of existing levels of output," while Keynes argued in more orthodox terms for "government action to limit any potential inflationary gap 'by heavy taxation, a high pressure savings campaign, or rationing on a wide scale.'" The New Dealers countered that "financing defense through increased taxation in 1940" was equivalent to taking "two steps forward and then one step back." Jones, "Role of the Keynesians," 126.

83. Duncan and Shelton, *Revolution in United States Government Statistics*, 84.

84. Civilian Production Administration, *Industrial Mobilization for War*, 38.

85. H. George, "An Inquiry into the Subject 'War,'" 42.

86. Ibid., 42–43.

87. Ibid., 43.

88. Quoted in Rostow, *Concept and Controversy*, 31.

89. Quoted in Biddle, *Rhetoric and Reality in Air Warfare*, 38.

90. Ibid., 139.

91. Douhet, *Command of the Air*. According to Douhet, air war should focus on the civilian population and on industrial centers, using urban terror bombing to crush the enemy's will to resist.

92. W. Mitchell, "Airplanes in National Defense," 38–42.

93. Hansell, *Strategic Air War*, 12.

94. Ibid., 12, 14.

95. For ACTS theorists, this approach appeared feasible thanks to technological innovations of the early 1930s. The long-range bomber and a more accurate bombsight, they argued, made such precision bombing increasingly practical as a strategy. The experience of World War II—where cloudy weather and high rates of attrition made daylight precision bombing nearly impracticable until the end of the war—showed that this optimism was not justified. See Katz, *Foreign Intelligence*.

96. Wilson, "Origin of a Theory of Air Strategy," 19.

97. Air Corps Tactical School, "Air Force," in Air Warfare section, February 1, 1938, 248.101-1, Air Force Historical Research Agency, Air University.

98. Wilson, "Origin of a Theory for Air Strategy," 19.

99. Quoted in Eden, *Whole World on Fire*, 70.

100. Hansell, *Strategic Air War*, 22.

101. Ibid., 12.

102. Fairchild, "New York Industrial Area," Lecture, ACTS, Maxwell AFB, Ala., 6 April 1939, 14, HRA file 248.2019A-12.

103. Such systems would later be called infrastructures. See Carse, "Keyword: Infrastructure."

104. Fairchild, "New York Industrial Area," 20.

105. Hansell, *Strategic Air War*, 12.

106. Fairchild, "National Economic Structure," Lecture, Air Corps Tactical School, Maxwell AFB, Ala., 1939–1940, 22, HRA file 248.2021A-7. Citing a 1936 report from the Federal Power Commission, he noted that power from public utilities ran all aircraft factories, aircraft engine factories, instrument manufacturing plants, bombsight plants, and, indeed, "practically all of the assembling and finishing plants of every kind producing articles of every description" (22).

107. "The situation might be even more acute than that which existed during the World War, when in many districts electric service had to be denied to domestic and commercial consumers and non-essential industries to meet war needs for power." Ibid., 23.

108. Ibid., 27.

109. Ibid., 26, 28, 31.

110. Hansell, *Strategic Air War*, 22.

111. Ibid., 21.

112. Ibid., 22.

113. Ibid., 21.

114. Richard D. Hughes, "Memoirs," unpublished manuscript, n.d., 5, call number 520.056-234, Air Force Historical Research Agency, Air University.

115. Cited in Gentile, *How Effective Is Strategic Bombing?*, 20.

116. Hansell, *Air Plan that Defeated Hitler*, 84.

117. McFarland, *America's Pursuit of Precision Bombing*, 102.

118. R. D. Hughes, "Memoirs," 7–8.

119. Hansell, *Air Plan that Defeated Hitler*, 102. A revised but similar air war plan was developed in August 1942 (AWPD-42), which defined the strategic aim as "undermining and destroying the capability and will of Germany to wage war by destroying war-supporting industries and the industries upon which the war industries and the civilian economy of Germany depended." Culbertson, *Air Intelligence and the Search for the Center of Gravity*, 6.

120. R. D. Hughes, "Memoirs," 9.

121. Katz, *Foreign Intelligence*, 114.

122. Quoted in Gentile, *How Effective Is Strategic Bombing?*, 18.

123. Joseph Coker, a veteran of wartime air-targeting and mobilization planning, used this phrase in his 1949 dissertation, which described the analysis conducted by economists working in air target planning during World War II (Coker, "Economics of Strategic Target Selection"). The term "economics of target selection" was subsequently used by the economist Mancur Olson, in an early essay written when he was serving as a lieutenant in the US Air Force. Olson, "Economics of Target Selection," 308–14.

124. Katz, *Foreign Intelligence*, 17.

125. Bernstein, *Perilous Progress*, 80; Guglielmo, "Contribution of Economists," 113.

126. Guglielmo, "Contribution of Economists," 109.

127. Rostow, *Division of Europe*, 29.

128. "Based on the fragmentary data available, the OSS economists were called upon to derive the capacities of the Russian rail lines to deliver the supply requirements of the invading German armies in order to calculate the earliest date at which the German offensive could resume in 1942." Guglielmo, "Contribution of Economists," 119.

129. Carl Kaysen, interview by Marc Trachtenberg, David Rosenberg, and Stephen Van Evera, 1988, transcript, Security Studies Program, Massachusetts Institute of Technology.

130. Katz, *Foreign Intelligence*, 104. Despres wrote, "The invasion of the Continent in 1943 can succeed only if German strength has in the meantime been considerably reduced." Cited in ibid., 104.

131. See ibid., 104.

132. Guglielmo, "Contribution of Economists."

133. Kindleberger, *Life of an Economist*, 190.

134. Katz, *Foreign Intelligence*, 111.

135. R. D. Hughes, "Memoirs," 21.

136. Guglielmo, "Contribution of Economists," 132.

137. Katz, *Foreign Intelligence*, 115.

138. R. D. Hughes, "Memoirs," 27.

139. R. D. Hughes, "Memoirs," 27.

140. Katz, *Foreign Intelligence*, 115–16; Guglielmo, "Contribution of Economists," 132. To produce these reports, EOU staff looked at interrogation transcripts of German prisoners, aerial photographs, and ground reports. Their first report, on Siemens Cable Works, identified the points that were most vulnerable to bomb damage and most costly to repair, and whose destruction would therefore disrupt the manufacturing process for the longest period (see Katz, *Foreign Intelligence*, 115).

141. Rostow, *Pre-Invasion*, 22.

142. Salant's memo is reproduced in ibid.

143. Katz, *Foreign Intelligence*, 116. "The document, which laid the intellectual foundations not only for the EOU 'Party Line' but for a so-called philosophy of air power that would survive well into the nuclear age, reduced the problematic of strategic bombing to three independent and exhaustive questions: (1) Can you hit it, and at what cost? (2) Can you damage it if you hit it? (3) Will damage to the target hurt the enemy?"

144. Rostow, *Pre-Invasion*, 100. Kaysen commented on how the economic concept of substitution was taken up in thinking about vulnerability: "The most important idea that as an economist I had, all of us economists had, which wasn't in the ordinary military thinking, was the idea of substitution. If you digress and take the whole strategic materials business, we were among the first people to see through the strategic materials business and say this is going to be crap. They'll find ways around it." In other words, noneconomists make the mistake of assuming that a particular material is critical, without asking whether substitutes can be found. Kaysen interview, 2.

145. Rostow, *Pre-Invasion*, 103. Guglielmo summarizes the approach: "The most effective enemy targets would be those industries producing some component vital to armaments production that were geographically concentrated, used specialized equipment, held small inventories, and were working at capacity." Guglielmo, "Contribution of Economists," 134.

146. Katz, *Foreign Intelligence*, 118.

147. Kindleberger, "Some Economic Lessons," 183. In the US Strategic Bombing Survey, this EOU analysis of the power system was cast into doubt. As Joseph Coker later wrote in his dissertation, "The Economics of Target Selection," citing an EOU target potentiality report: "Germany's electric power system would have been an excellent strategic target system if the Allies had possessed the means of destroying a large portion of German electrical capacity. There is reason to believe that the United States Air Forces and the Royal Air Force did possess this capability" (134).

148. Kindleberger, "Some Economic Lessons," 181.

149. Rostow, *Pre-Invasion*, 31. Rostow's discussion of oil illustrates EOU's analytical approach. "A critical part of EOU's analysis related to the estimated level of German oil stocks and rates of oil consumption. These variables would determine, along with damage actually inflicted on German oil production and minimum necessary allocation to the German economy, the timing of the effect on German military operations of the proposed oil offensive" (33).

150. Ibid., 22.

151. R. D. Hughes, "Memoirs," 27.

152. USSBS, *Summary Report*.

153. For a recent appraisal of this shift in the bombing campaign, see Gladwell, *Bomber Mafia*.

154. Gentile, *How Effective Is Strategic Bombing?*, 46.

155. Paul Nitze interview, June 11 and June 17, 1975, Paul H. Nitze Oral History Interviews, Harry S. Truman Library, Independence, MO.

156. USSBS, *Summary Report*, 37.

157. Rostow, *Pre-Invasion*.

158. Nitze interview.

159. Baldwin, "Oil Strategy," 10–11. Oil was a topic of controversy in air war planning, since EOU had recommended it as a target system in advance of the invasion; but the recommendation was ignored in favor of a strategy of attacks on marshaling yards.

160. USSBS, *Summary Report*, September 30, 1945, 2.

161. Gentile, *How Effective Is Strategic Bombing?*, 72, 2, 16.

162. USSBS, *Summary Report*, 30.

163. US Strategic Bombing Survey, *The Effects of Atomic Bombs on Hiroshima and Nagasaki*, 38, 41.

Chapter 2. Emergency Government

1. Franklin D. Roosevelt, "Establishing the Office for Emergency Management in the Executive Office of the President and Prescribing Regulations Governing Its Activities," May 25, 1940. *Code of Federal Regulations: Title 3—The President. 1938–1943 Compilation* (Washington, DC: Government Printing Office, 1968).

2. Emmerich, "Administrative Legacy," 126–27. Clinton Rossiter similarly referred to OEM as the "administrative sky hook" from which Roosevelt hung the "vast hodgepodge of wartime agencies" that he created to manage the domestic war effort; Rossiter, "Constitutional Significance," 1209.

3. Shaw, *Field Organization*.

4. Novick, Anshen, and Truppner, *Wartime Production Controls*, 167.

5. McReynolds, "Office for Emergency Management," 131.

6. Ibid., 138.

7. Ibid., 141.

8. Exec. Order No. 8629, 6 C.F.R. 191 (1941).

9. Rossiter, "Constitutional Dictatorship in the Atomic Age," 410.

10. Tyler Curley makes a similar point: "Mobilization was managed through the creation of an innovative institutional system in the Office for Emergency Management. This structure functioned as an emergency-war machine, designed to coordinate the activities of local, state, and regional actors. It did not require consolidated powers but rather facilitated a collaborative, dynamic federalist approach to statebuilding." Curley, "Models of Emergency Statebuilding," 707.

11. A similar argument has been made by Katznelson: "The New Deal's rearrangement of values and institutions, and its support for the Western liberal political tradition" answered the challenge that emergency situations posed to democratic institutions. Our account is distinct in its emphasis on the centrality of technical and administrative arrangements rather than "values" and "institutions." Katznelson, *Fear Itself*, 5.

12. On the different strands of the Progressive movement and the resulting confusion in discussions of Progressivism among historians, see Kennedy, *Progressivism*.

13. O. L. Graham, *Encore for Reform*. See also Link, "What Happened to the Progressive Movement?"

14. See Karl, *Executive Reorganization*, 182; and Reagan, *Designing a New America*, 66.

15. Huthmacher ("Urban Liberalism," 79) writes that urban reform "established patterns and precedents for the further evolution of American liberalism, an evolution whose later milestones would bear the markers 'New Deal' and 'New Frontier.'" Reagan, citing the historian Stephen Diner, emphasizes that Charles Merriam in particular sought to establish "a new *system* of public administration in the city government." Reagan, *Designing a New America*, 61.

16. The historian Barry Dean Karl has traced the connection between Progressivism and administrative reform in the New Deal through the career trajectories of Merriam, Brownlow, and Gulick. Karl, *Executive Reorganization*.

17. Reagan, *Designing a New America*, 69.

18. Ibid., 53. Brinkley writes that Merriam "spent his life trying to fit his deep faith in planning into the structure of democratic politics." Brinkley, "National Resources Planning Board," 174.

19. Quoted in Reagan, *Designing a New America*, 60.

20. Merriam, *New Aspects of Politics*, 68.

21. Reagan, *Designing a New America*, 59.

22. See, in particular, Karl, *Executive Reorganization*; and Reagan, *Designing a New America* on the shift from local government planning to the federal government.

23. Reagan, *Designing a New America*, 77. Reagan refers to Merriam's contribution as a summary of "the evolution of his own thinking of the last decade."

24. Merriam, "Government and Society," 1488, 1501, 1493, 1489, 1499.

25. Reagan, *Designing a New America*, 77. According to Reagan, the Committee's work "altered the direction and use of social science research in the United States" by proposing a permanent role for technical experts in the ongoing administration of government, particularly national government.

26. Reagan, *Designing a New America*, 186.

27. The quotations in this paragraph are taken from the influential final section, drafted by Merriam, entitled "A Plan for Planning." On the composition of this section of the report and Merriam's central role, see Reagan, *Designing a New America*, 187–88.

28. National Planning Board, *National Planning Board: Final Report*, 13, 22, 28.

29. Emmerich, "Administrative Legacy," 122. See also Polenberg, *Reorganizing Roosevelt's Government*, 191.

30. Sources on the prior pattern of wartime mobilization include Koistinen, *Arsenal of World War II*; Rossiter, *Constitutional Dictatorship*; and Sherry, *Preparing for the Next War*.

31. On the "bold new departure" of WWI mobilization, see Leuchtenberg, "New Deal and the Analogue of War," 84.

32. Ibid., 85. In some cases, Congress created new emergency executive authorities, as in the economic controls of the Lever Act of August 1917 and the reorganization authority in the Overman Act of May 1918; Rossiter, *Constitutional Dictatorship*, 242. Rossiter deemed the "delegatory statute," which transferred congressional powers to the president, to be "the most important single emergency device" of the US federal government during World War I. See also Emmerich, *Federal Organization and Administrative Management*, 42.

33. Polenberg, *Reorganizing Roosevelt's Government*, 194. Polenberg notes that this retrenchment followed a longstanding pattern. "When Roosevelt's critics charged that he sought to establish an executive dictatorship they were following an old tradition. Every strong President had faced a similar accusation." See also Rossiter, *Constitutional Dictatorship*, 254.

34. Leuchtenberg, "New Deal and the Analogue of War," 83, 109. "There was scarcely a New Deal act or agency that did not owe something to the experience of World War I."

35. Rossiter, *Constitutional Dictatorship*, 260. According to Rossiter, legislative statutes enacted during Roosevelt's first one hundred days amounted to the "largest single instance of delegated power in American history." Among these delegations was broad reorganization authority, but Roosevelt did not use this authority, fearing, according to Dickinson, that "politically controversial reorganization plans would jeopardize congressional support for his economic recovery policies." M. J. Dickinson, *Bitter Harvest*, 78.

36. Cited in Leuchtenberg, "New Deal and the Analogue of War," 125, 140. See also Emmerich, "Administrative Legacy," 122. A 1934 review of legislation noted "a veritable deluge of declarations of emergency" in enactments of Congress. See J. P. Clark, "Emergencies and the Law," 270.

37. Rossiter, *Constitutional Dictatorship*, 261. Every statute passed in the early New Deal that delegated congressional power to the president included provisions limiting its duration to the period of the emergency. Roosevelt therefore complicated matters by repeatedly declaring that the economic crisis had been successfully managed. See Karl, *Executive Reorganization*, 26.

38. Karl, *Executive Reorganization*, 193–94. For a discussion of congressional resistance and the court rulings, see Brinkley, *End of Reform*, 18.

39. M. J. Dickinson, *Bitter Harvest*, 82, 46.

40. Franklin Delano Roosevelt, "Message from the President of the United States," in President's Committee on Administrative Management, *Report of the Committee*, iii. On the proposal for such a committee, see Merriam "National Resources Planning Board," 1084.

41. Emmerich, *Federal Organization and Administrative Management*, 50.

42. Karl, *Executive Reorganization*, 209.

43. Emmerich, *Federal Organization and Administrative Management*, 46.

44. Karl, *Executive Reorganization*, 225.

45. President's Committee on Administrative Management, *Report of the Committee*, 1–2.

46. See Polenberg, *Reorganizing Roosevelt's Government*, 191; and Emmerich, *Federal Organization and Administrative Management*, 46.

47. Karl, *Executive Reorganization*, 166. The Committee argued that Congress used its "legislative power to create an executive branch of the government which was only partly under the control of the executive designated by the people."

48. President's Committee on Administrative Management, *Report of the Committee*, 2.

49. Ibid., 29.

50. Ibid., 36.

51. Karl, "Executive Reorganization," 21. According to Karl, reorganization belonged to a "new tradition of executive inventions designed to cope with difficulties in adjusting the Constitution to the needs of modern government without revising the structure."

52. Polenberg, *Reorganizing Roosevelt's Government*, 7–8. The Committee's treatment of reorganization marked, according to Polenberg, a "sharp break" with traditional reorganization. See also Karl, *Executive Reorganization*, 189; and Hobbs, *Behind the President*, 15.

53. Quoted in Polenberg, *Reorganizing Roosevelt's Government*, 8.

54. Franklin D. Roosevelt, "Message to Congress Recommending Reorganization of the Executive Branch," January 12, 1937, American Presidency Project, University of California, Santa Barbara.

55. Emmerich, *Federal Organization and Administrative Management*, 56.

56. Patterson, *Congressional Conservatism and the New Deal*, 218–19. The new legislation "sought to avoid the 'dictator' issue," according to Patterson, by changing "the method of putting plans into effect." Reorganization plans would not go into effect "until sixty days after their submission to Congress and could be vetoed by simple concurrent resolution of both houses." Historians differ on the significance of these revisions. Brinkley refers to the new bill as "so emasculated that it moved quickly through Congress." Brinkley, *End of Reform*, 22–23. By contrast, Dickinson writes that the Reorganization Act, along with subsequent legislation, "strengthened the Executive Office of the President in a way that largely comported with FDR's views as expressed in the Brownlow Report." M. J. Dickinson, *Bitter Harvest*, 112. The reformers who authored the Act saw it as part of an epoch-making change in American government.

57. Franklin D. Roosevelt, "Message to Congress on the Reorganization Act," April 25, 1939, American Presidency Project, University of California, Santa Barbara.

58. Exec. Order No. 8248, 4 C.F.R. 3864, 3 C.F.R. (1938–1943).

59. Koistinen, *Arsenal of World War II*, 15. An initial draft of the executive order included a full description of this "office for emergency management," but it was expunged at the insistence of the Army and Navy, which anticipated (correctly) that such an office would threaten their traditional authority over military procurement.

60. See Scheuerman, "Economic State of Emergency," 1888. In the mid-1930s Carl Schmitt pointed to the repudiation of Roosevelt's attempt to wield emergency authority in the early New Deal as evidence that the concentrated executive power and expert authority required to manage crisis situations was incompatible with the institutions of liberal democracy.

61. Gulick, "Conclusion," 139.

62. Rossiter, *American Presidency*, 134. For a similar contemporary assessment, see Hobbs, *Behind the President*, 11.

63. Emmerich, "Administrative Legacy," 126. Emmerich underscored that the 1939 Reorganization Act and Executive Order 8248 had equipped Roosevelt with "new tools and powers that no previous president had been given."

64. For a parallel account of how expert assessment was used to advance specific interests in the New Deal, see Porter, *Trust in Numbers*, which focuses particularly on the development of benefit-cost analysis for public works projects.

65. Nathan, "GNP and Military Mobilization," 16. In this emphasis on the limits of the economy, and the extent to which fiscal stimulus would affect these limits or lead to destructive economic disruption, New Deal economists and statisticians approached military-industrial mobilization "as an extension of the war against the Great Depression." See also Jones, "Role of Keynesians," 125.

66. Among these were Walter Salant, who later worked in the Enemy Objectives Unit, and several prominent economists who moved into the mobilization program. See Jones, "Role of Keynesians," 126; and Carson, "History of the United States National Income."

67. Warken, *History of the National Resources Planning Board*, 141–42.

68. Following the German invasion of Poland in 1939, Roosevelt had declared a "limited national emergency" along with his issuance of Executive Order 8248. As Rossiter notes, "The idea of a 'limited' national emergency as declared in September 1939 was the President's own; it is completely unrecognized by statute or constitutional practice." Rossiter, *Constitutional Dictatorship*, 267.

69. Koistinen, *Arsenal of World War II*, 15–17. The picture of OEM that emerges in secondary literature is complicated. The office itself—and particularly its first director, William McReynolds—seem to have been relatively inefficacious. But as Emmerich argues, OEM provided a mechanism through which Roosevelt could push economic mobilization, particularly in the period between January 1, 1941—when the Reorganization Act expired—and the passage of a War Powers Act in December 1941. Due to the existence of OEM, he argues, a whole series of defense agencies "could be and were created by executive order as parts of the Office for Emergency Management." Emmerich, "Administrative Legacy," 127.

70. Henderson was also a former official at the Russell Sage Foundation.

71. Nelson, *Arsenal of Democracy*, 129. May moved to NDAC from the Social Science Research Council in June 1940.

72. Civilian Production Administration, *Industrial Mobilization for War*, 38. In NDAC, Mason ran the Economics Research Section.

73. Other prominent recruits to NDAC included Morris Copland from the Bureau of the Budget and V. Lewis Bassie of the Division of Industrial Economics in Commerce.

74. Lacey, *Keep from All Thoughtful Men*, 7n25.

75. Civilian Production Administration, *Industrial Mobilization for War*, 38.

76. Koistinen, *Arsenal of World War II*, 13. Koistinen emphasizes that "New Deal professionals also led in devising the regulatory devices essential for maximizing war output while maintaining economic stability."

77. Nelson, *Arsenal of Democracy*, 125–26.

78. Ibid., 12, 125–26. On the conservative mobilization alliance, see Koistinen, *Arsenal of World War II*, 54; and Civilian Production Administration, *Industrial Mobilization for War*, 152.

79. Nathan, "GNP and Military Mobilization," 7. According to Nathan, New Deal specialists focused on estimating production at full employment because they did not have detailed information about military requirements and could not conduct analyses in terms of specific materials.

80. Robert R. Nathan, interview by Niel M. Johnson, June 22, 1989, Oral History Interviews, Harry S. Truman Library, Independence, MO.

81. Koistinen, *Arsenal of World War II*, 24. Koistinen records that Leon Henderson "led or actively participated in . . . devising policies for antitrust, rapid amortization, priorities, and the initial phases of economic warfare."

82. Ibid., 53, 57. For an overall assessment of the primary role of government programs in wartime expansion, see ibid., 296.

83. Ibid., 116–17. According to Koistinen, maintaining stocks was a long-time priority of the Army-Navy Munitions Board, which was largely frustrated.

84. Civilian Production Administration, *Industrial Mobilization for War*, 75. Only $70 million was appropriated into the stockpile, and acquisitions fell fall short of NDAC recommendations.

85. Koistinen, *Arsenal of World War II*, 64–65.

86. According to Nelson (*Arsenal of Democracy,* 120), British pressure was particularly important in debates in the Roosevelt administration about an expanded production program.

87. Ibid., 61.

88. Brigante, *Feasibility Dispute*, 17.

89. Novick, Anshen, and Truppner, *Wartime Production Controls*, 41. The priorities system described here was in place until mid-1941, when mobilization planners set up the Production Requirements Plan.

90. Koistinen, *Arsenal of World War II*, 65, 66. Koistinen notes that in this negotiation, the civilian officials in NDAC largely ceded control to military procurement officers.

91. Novick, Anshen, and Truppner, *Wartime Production Controls*, 42–43.

92. Nelson, *Arsenal of Democracy*, 127–28.

93. Novick, Anshen, and Truppner, *Wartime Production Controls*, 76.

94. Koistinen, *Arsenal of World War II*, 120.

95. Brigante, *Feasibility Dispute*, 17.

96. See Koistinen, *Arsenal of World War II*, 88.

97. Koistinen, *Arsenal of World War II*, 99. Nonetheless, as Koistinen notes, the civilian experts laid the groundwork for "full-blown mobilization" by collecting data, developing production control systems, and refining regulatory devices.

98. Brigante, *Feasibility Dispute*, 15. Military requirements were not based, as Brigante puts it, "on a strategy of world-wide deployments of American and allied arms and men."

99. Civilian Production Administration, *Industrial Mobilization for War*, 134.

100. Nelson, *Arsenal of Democracy*, 131.

101. George had worked in the Department of Commerce under Hoover and then in the National Recovery Administration. See *Hearings on First Supplemental National Defense Appropriation Bill for 1942, 2nd session, Part 2, Before the Subcomm. On Defense, Comm. On Appropriations*, 77th Cong., 2nd sess. (1941) (testimony of Stacy May), 347.

102. Brigante, *Feasibility Dispute*, 21. By late March 1941, "a usable summary had been made of the defense program in the United States—defined as the total of Army, Navy, Maritime Commission, British, and Lend-Lease requirements for 1942."

103. Nelson, *Arsenal of Democracy*, 133.

104. Ibid., 132–33. See also Civilian Production Administration, *Industrial Mobilization for War*, 139.

105. Nelson, *Arsenal of Democracy*, 133, 137.

106. Brigante, *Feasibility Dispute*, 21.

107. Nathan, "Keynesian Revolution," 139.

108. Civilian Production Administration, *Industrial Mobilization for War*, 140.

109. Brigante, *Feasibility Dispute*, 25.

110. Nathan and May analyzed the Victory Program in terms of GDP but also checked it in relation to critical materials—steel, aluminum, and copper—measured in both tons and dollars.

111. Civilian Production Administration, *Industrial Mobilization for War*, 140.

112. Koistinen, *Arsenal of World War II*, 188.

113. Brigante, *Feasibility Dispute*, 27–28.

114. Ibid., 30.

115. Koistinen, *Arsenal of World War II*, 206.

116. Brigante, *Feasibility Dispute*, 35.

117. Ibid., 28. See also Jones, "Role of Keynesians," 128.

118. Brigante, *Feasibility Dispute*, 28.

119. Cuff, "From the Controlled Materials Plan," 1.

120. Exec. Order No. 9024, Establishing the War Production Board (January 16, 1942).

121. For a list of resource divisions and other operational units, see US War Production Board, *Catalog of War Production*.

122. Quotation from General Administrative Order No. 22, reprinted in US Planning Committee of the War Production Board, *Meeting Minutes*, March 3, 1942.

123. Nitze interview. Nitze also describes the central role that Searls played on the Strategic Bombing Survey.

124. Koistinen, *Arsenal of World War II*, 304.

125. US Planning Committee of the War Production Board, *Meeting Minutes*, February 20, 1942, 2.

126. US Planning Committee of the War Production Board, *Meeting Minutes*, March 10, 1942, 16.

127. US Planning Committee of the War Production Board, *Meeting Minutes*, March 13, 1942, 18, 17.

128. Brigante, *Feasibility Dispute*, 34.

129. Civilian Production Administration, *Industrial Mobilization for War*, 10.

130. Brigante, *Feasibility Dispute*, 40.

131. US Planning Committee of the War Production Board, *Meeting Minutes*, March 16, 1942. The Planning Committee proposed a 35 percent reduction in the total program.

132. Koistinen, *Arsenal of World War II*, 312, 307–8.

133. Ibid., 306–7.

134. Ibid., 303. The New Dealers' concept of feasibility, they feared, would "empower outsiders, impractical theorists, those not sharing the views and values of industry and the military." This dispute can be compared to the postwar struggles between military offices and technical experts as described in Kaplan, *Wizards of Armageddon*.

135. "The official record is barren about how the settlement was devised. No doubt Roosevelt directed what was to be done." Koistinen, *Arsenal of World War II*, 312.

136. Ibid., 208.

137. US Planning Committee of the War Production Board, *Meeting Minutes*, April 13, 1942.

138. Novick, *Origin and History of Program Budgeting*.

139. Novick, Anshen, and Truppner, *Wartime Production Controls*, 94.

140. The Production Requirements Plan was set up by the Supply Priorities and Allocations Board, one of the offices that Roosevelt created in 1941 in his attempts to concentrate planning power in the hands of New Deal officials and specialists.

141. Civilian Production Administration, *Industrial Mobilization for War*, 431.

142. Novick and Steiner, "War Production Board's Statistical Reporting Experience," 217.

143. Novick, "Statistical Materials," 131–34.

144. Novick and Steiner, "War Production Board's Statistical Reporting Experience," 221.

145. Novick, Anshen, and Truppner, *Wartime Production Controls*, 105. The PRP was made mandatory for all uses of controlled metals for the third quarter of 1942 after Priorities Regulation No. 11 of June 1942 brought "all manufacturing users of significant quantities of scarce metals under the Production Requirements Plan."

146. Cuff, "From the Controlled Materials Plan," 1.

147. On the limited use of systematic accounting mechanisms in the military prior to World War II, see Lass, "Into a Wild New Yonder."

148. R. E. Smith, *Army and Economic Mobilization*, 579.

149. Ibid., 579, 580.

150. Koistinen, *Arsenal of World War II*, 317. "Of incomparable value to WPB was the fact that PRP would provide the board with nearly complete information on wartime industrial load essential for maintaining economic stability." See also US Planning Committee of the War Production Board, *Meeting Minutes*, May 5, 1942, which notes that since "most plants have already prepared detailed production schedules," the "assemblage of the data needed for scheduled programs should not be overly burdensome." This was not true at the beginning of the year.

151. Quoted in Cuff, "From the Controlled Materials Plan," 1. See also Brigante, *Feasibility Dispute*, 35.

152. Koistinen, *Arsenal of World War II*, 319.

153. Civilian Production Administration, *Industrial Mobilization for War*, 461.

154. On these alternative systems of control, see US Planning Committee of the War Production Board, *Meeting Minutes*, September 3, 1942, in which the Planning Committee opts for a vertical system of control, and Novick, Anshen, and Truppner, *Wartime Production Controls*, 138.

155. Civilian Production Administration, *Industrial Mobilization for War*, 475.

156. Cuff, "Organizational Capabilities," 104. The Committee recognized that it was impossible to rapidly design and implement this new system and therefore advised Nelson to make the PRP's system of horizontal controls mandatory for all industrial producers—but only for the third quarter of the year, after which vertical controls would be implemented.

157. See Koistinen, *Arsenal of World War II*, 321. Novick later observed that the CMP was the first "program budget"—the approach to budgeting based on systems analysis—which was later adopted by the Defense Department, under Hitch's direction, in the Kennedy administration; Novick, "Origin and History of Program Budgeting."

158. Novick, Anshen, and Truppner, *Wartime Production Controls*, 167, 166, 169.

159. Ibid., 167.

160. The focus on strategic points also related to a more familiar economic argument against total planning, which was that the task of gathering and processing information would be overwhelming, and that a more decentralized system would therefore be preferable. See US Planning Committee of the War Production Board, *Meeting Minutes*, April 13, 1942.

161. Koistinen, *Arsenal of World War II*, 313–14.

162. On the pattern of postwar retrenchment, see Rossiter, *Constitutional Dictatorship*, 254.

163. Quoted in Cuff, "From the Controlled Materials Plan," 3. Clifford, an advisor to President Truman, later served as secretary of defense.

164. On the push for a permanent mobilization after World War II, see Sherry, *Preparing for the Next War*; Hogan, *Cross of Iron*; and Yergin, *Shattered Peace*.

165. Quoted in Sherry, *Preparing for the Next War*, 200, 112.

166. Fesler, *Principles of Administration*, 69.

167. Sherry, *Preparing for the Next War*, 235. Sherry referred to a new "ideology of preparedness": "Believing that American weakness had encouraged Axis ambitions in the 1930s, strategic planners thought that powerful military forces could deter or subdue future troublemakers. Pearl Harbor and the new weapons developed after it demonstrated the nation's nakedness to sudden attack and its need for unprecedented forces-in-being to ward off the coming blitzkrieg."

168. Eberstadt, *Unification*, 18.

169. Cuff, "Ferdinand Eberstadt," 39.

170. Dorwart, *Eberstadt and Forrestal*, 95, 97.

171. Quoted in Katznelson, *Fear Itself*, 8.

172. Dorwart, *Eberstadt and Forrestal*, 103.

173. See Dorwart, *Eberstadt and Forrestal*; and Stuart, "Present at the Legislation."

174. Eberstadt, *Unification*, 20–21.

175. Ibid., 85–86.

176. Ibid., 20.

177. Ibid., 84. Eberstadt was influenced by his experience on the Army-Navy Munitions Board, whose detailed mobilization plans were ignored by civilian leadership. To avoid such waste of planning effort, he proposed to place mobilization planning in a civilian office. See Cuff, "Ferdinand Eberstadt," 40.

178. Eberstadt, *Unification*, 9.

179. Stuart, "Present at the Legislation," 16. The National Security Act "bore a striking resemblance to the recommendations which were put forth by Eberstadt and his team."

180. Renamed the Department of Defense in 1949, the National Military Establishment included the armed forces as well as influential new planning boards, including the Research and Development Board and the Munitions Board.

181. The Act renewed for postwar purposes the Stockpiling Act of 1939.

182. Eberstadt, *Unification*, 21. The unification study noted that "as national military policy keeps step with scientific development, broader questions relating to the location of industrial plants might arise. In the years ahead a substantial degree of industrial relocation may be required if the present concentration of industry proved unduly vulnerable to attack." But it quickly added that the report would not "attempt to forecast such dangers at this time," adding that "the best insurance of national security for the future in these terms . . . resides in a close working relationship between industry and the military services." As we see in the next two chapters, others in the military also saw work on strategic relocation and other vulnerability reduction measures as an urgent priority.

183. Jordan, *U.S. Civil Defense before 1950*, 86. NSRB staff, in their own attempts to glean the intent of Congress in creating the Board, concluded that the aim had been to establish an organization that would carry out the functions of the wartime civilian mobilization agencies—first and foremost, the War Production Board—as well as the "civilian requirements claimant agencies," such as the Petroleum Administration for War, the War Food Administration, the Office of Price Administration, and others.

184. Yoshpe, *National Security Resources Board*, 77.

185. Cuff, "From the Controlled Materials Plan," 3. Mobilization planners who later worked in NSRB included Ralph J. Watkins, Oscar Endler, Ernest Tupper, and William N. Lawrence. Another group of planners, including Shaw Livermore, David Novick, and Edwin George, worked outside of government in advisory positions, where they influenced mobilization planning, as discussed in chapter 5.

186. Yoshpe, *National Security Resources Board*, 102.

187. If feasibility testing had initially been a focal point for fractious debates over mobilization, then in the later years of WWII and in the initial years after the war, it was accepted by both civilian and military leadership as an authoritative technique to assess mobilization plans. See Perlman, "Political Purpose and the National Accounts," 143; and Bernstein, *Perilous Progress*, 79.

188. Yoshpe, *National Security Resources Board*, 103.

189. Ralph J. Watkins, "The National Security Resources Board" (Lecture delivered to the Industrial College of the Armed Forces, Washington, DC, October 15, 1948, National Defense University Archives).

190. Yoshpe, *National Security Resources Board*, 109.

191. Ibid. See also Cuff, "From the Controlled Materials Plan," 3.

192. Yoshpe, *National Security Resources Board*, 99–100. This work was based on a military Industrial Mobilization Plan, which outlined a wartime organization that included an Office of War Mobilization modeled on the War Production Board.

193. For example, in a study titled "Policies and Programs for the Coordination of Military, Industrial, and Civilian Mobilization." See Yoshpe, *National Security Resources Board*.

194. Yoshpe, *National Security Resources Board*, 101.

195. Pierpaoli, "Truman's Other War," 16.

196. Hogan, *Cross of Iron*, 341–44.

197. Cuff, "From the Controlled Materials Plan," 3.

198. Ibid., 4–5.

199. As Hogan put it, Truman sought to "achieve the goals of NSC 68/4 without creating a command economy." Hogan, *Cross of Iron*, 344. See also Yoshpe, *National Security Resources Board*, 145.

200. Yoshpe, *National Security Resources Board*, 163, 147.

201. Pierpaoli, "Truman's Other War," 16–17.

202. Hogan, *Cross of Iron*, 345.

203. Yoshpe, *National Security Resources Board*, 156.

204. Pierpaoli, "Truman's Other War," 145.

205. Edward T. Dickinson, "Analytical Review of NSRB's Planning Experience" (Lecture delivered to the Industrial College of the Armed Forces, Washington, DC, May 23, 1952, National Defense University Archives), 7.

206. US Department of Commerce, National Production Authority, Historical Reports on Defense Production, *Production and Distribution Controls*, Report No. 5, cited in Cuff, "From the Controlled Materials Plan," 4.

207. E. T. Dickinson, "Analytical Review," 7. Dickinson's view is not shared by all historians, some of whom see NSRB's experience with mobilization planning as a failure. For example, Jordan (*U.S. Civil Defense before 1950*, 118) argues that the strain of the Korean War mobilization "broke" NSRB in fall 1951. The Board's struggles relate to a fundamental tension in the conception of NSRB: Was it a strictly advisory body, or did it have operational functions? In some sense, NSRB was asked in 1950 to solve problems it had never been equipped to solve, and Truman, for reasons indicated, was never willing to transform NSRB into a powerful mobilization agency, instead choosing to create a powerful emergency agency—the Office of Defense Mobilization—based on the presumption that it would be eliminated at the end of the war. In fact, ODM persisted as a peacetime planning agency after it was merged with NSRB in 1953 and retained a range of powers that NSRB never possessed, such as the authority to redelegate presidential powers to federal agencies through Defense Mobilization Orders, as discussed in chapter 5. For our purposes, the important point is that NSRB developed a model of mobilization preparedness that planners in both NSRB and ODM later applied to a different problem: reducing the vulnerability of American vital systems to large-scale attack on the United States.

208. Rossiter, "War, Depression, and the Presidency," 219, 237, 219, 235.

209. Brinkley, "National Resources Planning Board," 180.

210. Rossiter, "War, Depression, and the Presidency," 236.

211. Brinkley, *End of Reform*, 186.

212. Hogan, *Cross of Iron*, 363–65.

213. Hobbs, *Behind the President*, 9, 181.

214. Cuff, "From the Controlled Materials Plan," 3.

215. Ibid., 181.

Chapter 3. Vulnerability

1. Kaysen, "Vulnerability of the United States," 190, 192.

2. Ibid., 196.

3. Industrial analysis would require an understanding of the structure of a future war economy that was anticipated in American industrial mobilization plans; thus mobilization planning, once more, went hand in hand with vulnerability analysis at this stage in its development: it was always the vulnerability of the mobilization economy that was in question.

4. Kaysen, "Vulnerability of the United States," 193.

5. Ibid., 195.

6. Ibid.

7. Directorate of Management Analysis, DCR/Comptroller Headquarters, US Air Force, *Bomb Damage Problem*, 6.

8. Following Paul Edwards, this knowledge infrastructure served to "monitor features of interest, model complex systems to find and test causal relationships, and record data in memory systems to track change over time." Edwards, "Knowledge Infrastructures for the Anthropocene," 36.

9. Referring to the period of the Revolutionary War, Douglas Southall Freeman described the Hudson River as "the jugular of America, the severance of which meant death." Freeman, *George Washington*, 500.

10. Cooling, "US Army Support of Civil Defense," 7. For an early reference to the term "passive defense," see also Conn, Engelman, and Fairchild, *Guarding the United States*. On the evolution in thinking about continental defense, see Schaffel, "Genesis of the Air Defense Mission."

11. "Over-all Civilian Defense in Japan," in "Exhibit 'C'" of Office of the Provost Marshal General, *Defense against Enemy Action Directed at Civilians*, 1946, 25, Study 3B-1, Historical Manuscripts Collection, Office of Chief of Military History Manuscripts, US Army Center of Military History. A postwar assessment of wartime civil defense concluded that, while the effort had "accomplished the greatest example of mass mobilization, organization, and training ever voluntarily undertaken by the citizens of the United States," it had failed to adequately prepare the nation for an enemy attack, due to the "absence of unified command and of authority," the amount of energy spent on activities "extraneous to actual defense," and the "total lack of advanced planning which found this nation completely devoid of technically qualified and adequate leadership, as well as personnel, and which necessitated complete improvisation throughout the emergency" (9).

12. Exec. Order No. 8972, 6 Fed. Reg. 6420 (signed December 12, 1941, received December 16, 1941).

13. The Office of the Provost Marshal General was a part of the Army Service Forces that, among other things, trained and fielded military police.

14. Engelhart, "Coordinating Wartime Internal-Security Programs," 54. Notably, internal security did not include defense against direct military attack on the United States.

15. See Morgan, *FDR*, 6. The FBI later investigated Gullion for his role in forming an organization "to save America from FDR, radical labor, the Communists, the Jews, and the colored race."

16. See Collier and Lakoff, "Vulnerability of Vital Systems," and Lakoff and Klinenberg, "Of Risk and Pork," for descriptions of recent uses of facilities ratings.

17. Engelhart, "Coordinating Wartime Internal-Security Programs," 56. Engelhart observed that procurement officers did not include "anything but prime contract plants"—those from which they directly procured military equipment and supplies—in the early inspection responsibility lists; Memorandum for Brigadier General Edward S. Greenbaum, "Critical War Production Capacity Bottlenecks," February 28, 1945, in "Exhibit 'K'" of Office of the Provost Marshal General, *Defense against Enemy Action*, 157. By contrast, resource evaluation specialists argued that not a single prime contractor should be assigned the highest security rating. In 1943 Engelhart was head of the Internal Security Division's Federal Coordination Branch, a consequential role since it meant that he was directly engaged with the civilian mobilization agencies.

18. Ibid., 56.

19. Ibid.

20. Ibid.

21. General Administrative Order, No. 2-38 (May 8, 1942), in "Exhibit 'K'" of Office of the Provost Marshal General, *Defense against Enemy Action*, 144.

22. "Scope and Application," in "Exhibit 'K'" of Office of the Provost Marshal General, *Defense against Enemy Action*, 188, 190. Initially, Stacy May, then WPB's director of statistics, appointed a member of his staff to lead a Resources Analysis Section. In August 1943, these functions were reorganized and expanded when Charles Wilson, the new WPB chair, created the Resources Protection Division, which housed the Resources Analysis Branch.

23. Ibid., 189.

24. See Malvern C. Talbert, memorandum to Secretary of Defense James V. Forrestal, August 25, 1947, Industrial Dispersion Program folder, box 11, ODM Central Files, 1950–1951, Record Group (RG) 304, Records of the Office of Civil and Defense Mobilization, National Archives, College Park, MD.

25. "Scope and Application," 190, 191. In some categories of facilities, such as aircraft, more detailed ratings were employed.

26. These memorandums were part of a correspondence between Colonel Engelhart and Brigadier General Edward S. Greenbaum, a leading Army procurement officer, in the closing months of the war. Early in 1945 Greenbaum had asked Provost Marshal General Archer L. Lerch to provide information on "critical war capacity bottlenecks"—that is, bottlenecks in American military-industrial production—for the ongoing war in Japan. Lerch had succeeded Major General Allen Gullion the prior year. See memorandum for Brigadier General Edward S. Greenbaum, "Critical War Production Capacity Bottlenecks," February 28, 1945, in "Exhibit 'K'" of Office of the Provost Marshal General, *Defense against Enemy Action.*

27. Ibid., 153.

28. "Basis of Resources Protection Board's Ratings," in "Exhibit 'K'" of Office of the Provost Marshal General, *Defense against Enemy Action*, 164.

29. "Summary of Tabs C to I, Inclusive," in "Exhibit 'K'" of Office of the Provost Marshal General, *Defense against Enemy Action*, 157. "The most serious bottlenecks are not, and have not been, in 'table of organization equipment,' i.e. end products, but in raw material, parts, chemical intermediates, and subassemblies. Of the 87 highest rated ("AA") items, not one is an end product purchased directly by the Army or Navy."

30. "Integrated Concentration (Series) of Vital Chemical Intermediates," in "Exhibit 'K'" of Office of the Provost Marshal General, *Defense against Enemy Action*, 176, 177.

31. Ibid., 177, 178.

32. Alton C. Miller, memorandum for the director, Special Planning Division, WDSS, October 18, 1945, "Consideration of Vital Facilities, Exclusive of Manufacturing Facilities in Demobilization Planning," in "Exhibit 'L'" of Office of Provost Marshal General, *Defense against Enemy Action*, 205.

33. "Water Systems," in "Exhibit 'L'" of Office of Provost Marshal General, *Defense against Enemy Action*, 222, 245.

34. Ibid., 242.

35. "The Petroleum Industry," in "Exhibit 'L'" of Office of Provost Marshal General, *Defense against Enemy Action*, 250, 251.

36. "Electric Power," in "Exhibit 'L'" of Office of Provost Marshal General, *Defense against Enemy Action*, 222.

37. "Railroad Facilities," in "Exhibit 'L'" of Office of Provost Marshal General, *Defense against Enemy Action*, 212.

38. "Petroleum Industry," 255.

39. "Railroad Facilities," 209.

40. "Petroleum Industry," 255.

41. Some postwar systems analysts explicitly traced their approach back to the War Production Board. For example, David Novick, a WPB economist who pioneered systems analysis at the RAND Corporation—and served on the ODM Mobilization Program Advisory Committee (see chapter 5)—argued that the analyses conducted in the WPB could be regarded as the first program budgets, a key technique of systems analysis as it developed at RAND after World War II. See Novick, *Origins and History of Program Budgeting.*

42. "Railroad Facilities," 209; "Electric Power," 222.

43. "Petroleum Industry," 253; "Water Systems," 242. In some cases, the Resources Protection Board simply decided that ratings should be assigned to systems rather than individual facilities. "The determination of the sensitive points of water systems is so dependent upon the actual layout of the particular system," the report recounted, "that the Resources Protection Board was compelled in its exploration of this field to rate only systems, not specific installations within a system."

44. "Railroad Facilities," 210.

45. "Electric Power," 223, 225, 224.

46. Ibid., 223, 222.

47. "Scope and Application," 192.

48. Schmidt, *Targeting Organizations*, 31. Like the later Strategic Vulnerability Branch, JTG was organized into four sections. An Economic Vulnerability Section analyzed the Japanese war economy, working to identify profitable targets and to assess the economic effects of attacks by Allied air forces. The Physical Vulnerability Section studied specific targets, recommending the selection of weapons, fusing, and sorties required to destroy or disrupt a particular target. An Evaluation Section drew on the analyses of physical and economic vulnerability to recommended target priorities, and to prepare special studies and reports for the War and Navy Departments. Finally, a Production Section produced operational charts and target folders for Allied air forces. See Eden, *Whole World on Fire*; and Gentile, *How Effective Is Strategic Bombing?*.

49. Lowe, "Intelligence in the Selection of Strategic Target Systems," 12. Initially trained as a diplomatic historian, Lowe had been recruited by Haywood Hansell, one of the ACTS instructors who invented strategic target selection, to join the Army's Strategic Air Intelligence Section during the leadup to World War II. Like the other wartime air-targeting groups discussed in chapter 1, the Strategic Air Intelligence Section engaged in "the economic-industrial-social analysis of major foreign powers," as Hansell put it, "culminating in analysis and description of vital and vulnerable systems and, finally, target selection and preparation of target folders." Hansell, *Strategic Air War*, 22.

50. As noted in chapter 2, the military was in general late to adopt punched cards, tabulators, and other automated accounting techniques. Prior to World War II, military use of punched cards had been limited to medical and personnel records. It was not until the early stages of World War II that punched card technology was incorporated on a large scale into military production planning, when the military was prodded, at least in part, by civilian mobilization planners, who demanded detailed requirements information from military procurement agencies. The Army Air Forces were among the first to adopt such techniques, in a new Directorate of Management Control, whose Office of Statistical Planning was created, among other things, to "'provide machine tabulation and other statistical services for all subdivisions of the AAF' located at Headquarters, and to coordinate the activities of all machine-tabulation installations in the AAF." For a detailed account, see Lass, "Into a Wild New Yonder."

51. Along with the *United States Strategic Bombing Survey* itself, see Eden, *Whole World on Fire*; and Curatola, *Bigger Bombs for a Brighter Tomorrow*.

52. USSBS, *Summary Report*, 39.

53. Ibid., 117–18. See also Farquhar (*Need to Know, 37*), which describes the view of military analysts: "No time interval existed in modern warfare to gather information, select targets, and collect operational data needed for weapons delivery."

54. USSBS, *Summary Report*, 117.

55. This was likely at the urging of the Joint Intelligence Committee.

56. In 1947, the Strategic Vulnerability Branch (SVB) became part of the US Air Force, which was established by the National Security Act as an autonomous military branch within the Department of Defense. The SVB was renamed the Targets Directorate in 1950. For a detailed description of the organization of air intelligence and its changes over time, see Eden, *Whole World on Fire*, 101, 99. Eden points out that "the analysts who worked in target intelligence have been virtually invisible in the historiography." Her account focuses on analysts in the Physical Vulnerability Section.

57. Lowe, "Intelligence in the Selection of Strategic Target Systems," 2. The Air Force representative on the Joint Chiefs was General Henry "Hap" Arnold, and the head of air intelligence

was World War II air commander Elwood Quesada. See also Farquhar, *Need to Know*. On the continuity between JTG and SVB see Eden, *Whole World on Fire*, 103.

58. As Lynn Eden puts it, the Joint Target Group "prefigured the organization and problem-solving focus of air target intelligence in the postwar US Air Force." Eden, *Whole World on Fire*, 83.

59. Physical vulnerability specialist Richard Grassy recounted, "The principal members of the geographic sections were principally economists. . . . [T]hey studied the economics of target countries and would determine whether these particular systems were the ones that would do the most damage to the enemy's war effort." Quoted in Eden, *Whole World on Fire*, 99.

60. Lowe, "Theory of Strategic Vulnerability," 2–3. "We have concentrated our effort, not on a world-wide basis of target analysis, but primarily on the analysis of the USSR and that part of the economy of Europe and Asia which she could probably capture, plus the United States."

61. Lowe, "Intelligence in the Selection of Strategic Target Systems," 7.

62. Lowe, "Theory of Strategic Vulnerability," 9.

63. Lowe, "Intelligence in the Selection of Strategic Target Systems," 12, 3. Lowe argued that it was necessary to conduct "a pre-analysis" of the vulnerability of potential adversaries, and to "carry that analysis to the point where the right bombs could be put on the right targets concomitant with the decision to wage war without any intervening time period whatsoever." Lowe, "Theory of Strategic Vulnerability," 4.

64. Lowe, "Intelligence in the Selection of Strategic Target Systems," 11, 7.

65. Ibid., 12.

66. Ibid., 12, 7; Gregory, "Bombing Encyclopedia."

67. This is summarized in Lowe, "Intelligence in the Selection of Strategic Target Systems," 3; and Lowe, "Theory of Strategic Vulnerability," 7.

68. For a summary, see Eden, *Whole World on Fire*, 107.

69. See Clinard, "Developments in Air Targeting."

70. Indeed, much of the data in the Bombing Encyclopedia seems to have been gathered through joint projects of the Directorate of Intelligence—in which SVB was housed—and the Directorate of Management Analysis, which was responsible for Air Force mobilization planning.

71. This data collection process is summarized in Directorate of Management Analysis, DCR/Comptroller Headquarters, US Air Force, *Bomb Damage Problem*, 8.

72. Lowe, "Intelligence in the Selection of Target Systems," 13–14.

73. Lowe, "Theory of Strategic Vulnerability," 8.

74. Lowe, "Intelligence in the Selection of Target Systems," 13–14.

75. Lowe, "Theory of Strategic Vulnerability," 13.

76. Lowe, "Intelligence in the Selection of Target Systems," 3.

77. Coker, "Strategic Vulnerability," 5, 8.

78. Fowler was another graduate of the Air Corps Tactical School who had served as an air commander in the Pacific theater during World War II.

79. Fowler, "Vulnerability of the United States," Lecture delivered at Air War College, 10 January 1953, 6, 9, 11. Air Force Historical Research Agency, Air Force Historical Research Agency. Maxwel, Alabama.

80. Historian Gian Gentile refers to Hansell as an "architect" of precision bombing doctrine. Gentile, *How Effective Is Strategic Bombing?*, 84.

81. Supporters of this strategy included General Arnold, who had previously insisted that area attacks were "contrary to our national policy of attacking only military objectives." Fedman and Karacas, "Cartographic Fade to Black," 310.

82. Quoted in Sherry, *Rise of American Air Power*, 287.

83. USSBS, *Evaluation of Photographic Intelligence*, 4.01.

84. Gentile, *How Effective Is Strategic Bombing?*, reports that JTG analysts never investigated the civilian population, per se, as a target. When they used area analysis, it was to assess the effects of urban bombing on industrial production.

85. "The industrial activities were outlined on a controlled mosaic, and the ground area of each was determined by measurement with a planimeter or transparent grid." Three different targeting groups worked on criteria for selecting area targets. The most elaborated was developed by the JTG. USSBS, *Evaluation of Photographic Intelligence*, 4.03.

86. Ibid.

87. Ibid., 4.27.

88. Coker, "Economics of Strategic Target Selection," 136.

89. Ibid., 38.

90. Even though the military oversaw the project, information about atomic weapons' effects was not widely available to military intelligence specialists until the late 1940s.

91. Lapp's study examined atomic bomb effects on a hypothetical American city. The military would continue to conduct studies that modeled bomb and assessed targets in the U.S. E.g., in 1947 SVB was asked to conduct a "target study of the atomic bomb damage to certain industrial cities in the US." Eden, *Whole World on Fire*, 110.

92. Lapp, "Atomic Bomb Explosions," 8.

93. Eden, *Whole World on Fire*, 115, 119.

94. Lowe reported that the task of developing this technique had been contracted to a university, and his description refers to the method that emerged from this contract.

95. Lowe, "Theory of Strategic Vulnerability," 12–14.

96. Lowe, "Intelligence in the Selection of Target Systems," 7–8, 12.

97. Lowe, "Theory of Strategic Vulnerability," 3.

98. Coker, "Strategic Vulnerability," 11.

99. Fowler, "Strategic Vulnerability." 13, 14, 15, 17.

100. Galison refers to this understanding of the United States as a view through "the bombsite mirror" ("War against the Center," 12).

101. Oscar Sutermeister, memorandum to Mr. Presley Lancaster Jr., *Civil Defense Vulnerability Manual*, November 9, 1951, declassified NWDD 978112, entry 63A-260, Industrial Dispersion: 1951–2 folder, box 204, RG 396, Records of the Office of Emergency Preparedness, National Archives, College Park, MD. This request was made to George Marshall, the secretary of defense (1950–1951). Marshall handed the task to the Air Research Division of the Library of Congress.

102. Federal Civil Defense Administration, *Civil Defense Urban Analysis*, 1.

103. Ibid., 66–77.

104. Ibid., 8.

105. Ibid., 10.

106. Ibid., 8, 11.

107. Ibid., 19.

108. US Department of Defense Civil Liaison Office, *Fire Effects of Bombing Attacks*.

109. Radiation effects—including aiming point–type analyses—were exhaustively described in the study by the Los Alamos Scientific Laboratory, *Effects of Atomic Weapons*.

110. Federal Civil Defense Administration, *Civil Defense Urban Analysis*, 36.

111. Ibid., 12, 19, 50, 53.

112. As the mobilization specialist George Lincoln put it, military officials worked to develop techniques for analyzing "new weapons, capabilities, and attitudes of potential enemies, potential

speed of expansion," and other aspects of postwar defense. Cited in Sherry, *Preparing for the Next War*, 114. One vehicle for these efforts was the RAND Corporation, which was created in 1948 as an independent research body that could conduct studies of future air-atomic wars. Project SCOOP, established one year earlier, worked on new methods for rational planning that were precedents for the later development of practices such as systems analysis and program budgeting. See Kaplan, *Wizards of Armageddon*.

113. Novick, *Origin and History of Program Budgeting*. According to Dunaway, Wood and Dantzig's initial discussions took place in 1947, and SCOOP was operational by 1948. SCOOP became an important site for the application of digital computing technology to mobilization planning and, related to this, for assessing the potential effects of a future enemy attack on US military-industrial production capacity; Edward Dunaway, interview by Daniel R. Mortensen, April 17, 1980, transcript, 6–7, US Air Force Oral History, IRIS No. 01129703, Office of Air Force History. Air Force Historical Research Agency. Maxwell, Alabama. Historians have traced SCOOP's approach to wartime operations research, and have shown how it influenced work at the RAND Corporation (where Dantzig relocated in 1953) as well as postwar applications of systems analysis in contexts like the Department of Defense under Secretary of Defense Robert McNamara during the Kennedy administration (e.g., in Erickson et al., *How Reason Almost Lost Its Mind*). Our account points to an unexplored source of these techniques in wartime mobilization planning, as is indicated by the fact that two of the most important exponents of these techniques—Charles Hitch and David Novick—were production specialists in the high-level planning bodies of the War Production Board. See, for example, Novick, *Origins and History of Program Budgeting*.

114. Wood and Dantzig, "Programming of Interdependent Activities." A central concern for Air Force planners at this time was the formulation of requirements as a basis for credible budgetary requests to Congress. Wood and Dantzig hoped that Project SCOOP would ultimately yield techniques for optimizing the allocation of scarce resources by comparing alternate programs of industrial production or weapons development. Initially, however, optimization was not central to SCOOP's work, as SCOOP analysts did not have access to the computational resources that would be required to carry out optimization procedures. Wood reported in 1950 that "any wide-scale introduction of considerations of alternatives involves a very great increase in the computing requirement, which we cannot handle until we get the large-scale electronic computers that we hope to have working by midsummer." By this time SCOOP analysts had addressed some simple optimization problems, such as the choice between two bomber programs; Wood, "Use of Mathematical Techniques," 10–11.

115. Dunaway, interview transcript, 6–7.

116. Wood, "Use of Mathematical Techniques," 7.

117. Dunaway interview transcript, 5.

118. Wood, "Use of Mathematical Techniques," 7.

119. The Air Force's UNIVAC was one of the first three UNIVACs installed in US government agencies. See Johnson, "Coming to grips with UNIVAC."

120. Ibid., 42n11.

121. Dal Hitchcock, chairman of the Executive Office Committee on Interindustry Economics, minutes of the 3rd Meeting of the Executive Office Committee on Interindustry Economics, January 22, 1952, entry 31, box 109, National Security Resources Board, Security Classified, General Correspondence of the Board, July 1949–April 1953, RG 304, Records of the Office of Civil and Defense Mobilization.

122. Wood and Horton both served on the interagency Committee on Interindustry Economics. The Council of Economic Advisers was represented by Gerhard Colm.

123. Horton, "Some Uses of Interindustry Statistical Techniques," 582.

124. Wood, "PARM."

125. Ibid.; Directorate of Management Analysis, DCR/Comptroller Headquarters, US Air Force, *Bomb Damage Problem.*

126. Walter R. Ristow, "Introduction to the Sanborn Map Collection," Geography and Map Reading Room, Library of Congress, last updated July 15, 2014, https://www.loc.gov/rr/geogmap/sanborn/san4al.html. Sanborn maps were originally developed for the purposes of fire insurance.

127. In an analysis of government damage assessment systems, Colonel F. A. Ramsey later recognized this innovation: "Since the UTM grid system is a method of expressing locations on a map in terms of a standard arrangement of numerical digits, it also has the very obvious advantage of simplicity in programing computers or in otherwise being inserted into computer systems either on punched cards or on tapes." Ramsey, "Damage Assessment Systems," 207.

128. Directorate of Management Analysis, DCR/Comptroller Headquarters, US Air Force, *Bomb Damage Problem,* 9.

129. This analysis of "mobile targets" echoed the Strategic Bombing Survey's reference to "target occupancy"—the number of people located in a given facility at the time of an attack; USSBS, *Evaluation of Photographic Intelligence,* 4.30.

130. Directorate of Management Analysis, DCR/Comptroller Headquarters, US Air Force, *Bomb Damage Problem,* 12.

131. Ibid., 12.

132. Horton had written an article in 1948 on a method for generating random numbers, in which he explained that the "need for large quantities of random numbers to be used in sample design, subsampling, and other statistical problems is well known. . . . The following procedure may be of interest to those who wish to develop their own random series." Horton, "Method for Obtaining Random Numbers," 81–82.

133. Directorate of Management Analysis, DCR/Comptroller Headquarters, US Air Force, *Bomb Damage Problem,* 12.

134. Ibid., 16.

135. Ibid., 17. This assumption did not necessarily hold true for "blast, thermal, and radiation effects, taken separately," but the report posited that both blast damage and total damage should follow this rule "fairly closely."

136. Like the first "distance adjustment factor," this adjustment was introduced by means of a multiplier that either took the plant "farther away" from ground zero or, if the value of the adjustment factor was less than one, moved it closer in.

137. Directorate of Management Analysis, DCR/Comptroller Headquarters, US Air Force, *Bomb Damage Problem,* 13.

138. "Bomb Damage Assessment," 12–13, Comptroller of the Air Force, December 1954, call no. K131.04-27, IRIS No. 1016574, Air Force Historical Research Agency, Air University.

139. Different levels of sectoral detail could be investigated by two-digit, three-digit, or four-digit codes.

140. "Bomb Damage Assessment," 4.

141. Ibid., 14.

142. Directorate of Management Analysis, DCR/Comptroller Headquarters, US Air Force, *Bomb Damage Problem,* 14.

143. Ibid.

144. Under the leadership of Harold Huglin, the Air Force Management Analysis Directorate continued to conduct tests of the bomb damage assessment system, at least through the end of 1954.

145. Mobilization planning thus provided the tools through which the "issue" of vulnerability was turned into a well-specified problem—what Michel Callon has analyzed as a process of "problematization" (see Callon, "Civilizing Markets").

146. According to historian John Cloud, during World War II the German military developed overlay maps composed of multiple transparent sheets showing features like vegetation, soil, and road surfaces. After examples were captured by American forces near the end of the war, US military mapmakers adopted the technique, applying it first to physical maps and then to digital data. See the interview with Cloud in Charles, "Do Maps Have Morals?"

Chapter 4. Preparedness

1. Lapp, "Atomic Bomb Explosions."

2. Ibid., 50.

3. Ibid., 54.

4. On this transposition of knowledge forms from the military, see Galison, "War against the Center."

5. Light, *From Warfare to Welfare*, 12.

6. Masco, "Life Underground," 19, 16.

7. Farish, *Contours of America's Cold War*, 206.

8. Baldwin, *Price of Power*, 263. See also Bromage, "Public Administration in the Atomic Age," 947; and Rossiter, "Constitutional Dictatorship in the Atomic Age." Lasswell was Merriam's student at Chicago and was likely influenced by Merriam's concern with the fate of democratic institutions in light of economic crisis and total war.

9. James Forrestal, memorandum for the president, October 7, 1948, in National Security Council, Interagency Committee on Internal Security, *State, Army, Navy, Air Force Coordinating Committee on Internal Security: A Report to the National Security Council 17/3*, November 16, 1948, 24, U.S. Declassified Documents Online, Gale, accessed 3 May 2021.

10. USSBS, *Effects of the Atomic Bombings*, 38, 41.

11. Robert Patterson, memorandum, "Studies for Planning Activities of the Under Secretary of War," March 7, 1945, in "Exhibit 'K'" of Office of Provost Marshal General, *Defense against Enemy Action*, April 30, 1956, 195.

12. Robert Patterson to Mr. J. A. Krug, n.d., in "Exhibit 'K'" of Office of Provost Marshal General, *Defense against Enemy Action*, 197. Patterson wrote to Krug: "I understand that you have determined that the Board will have served your purposes and that it will be dissolved about 12 August 1945. If so, it would be advantageous for the War Department to continue the Board's activities on an appropriately reduced scale."

13. Archer L. Lerch, memorandum, "Consideration of Vital Industrial Facilities in Demobilization Planning," June 26, 1945, in "Exhibit 'K'" of Office of Provost Marshal General, *Defense against Enemy Action*, 138.

14. George K. Engelhart, enclosure to memorandum for Major General Archer L. Lerch, Tab D, "Basis of Resources Protection Board's Ratings," May 12, 1945, in "Exhibit 'K'" of Office of Provost Marshal General, *Defense against Enemy Action*, 166. Not all national security analysts shared this view. For example, Bernard Brodie and William Borden—even though they occupied opposite poles of a debate on nuclear strategy (Brodie thought that nuclear war was unwinnable and had to be avoided, Borden that the United States had to be prepared to win a nuclear war)— both thought that atomic bombs made the problems of total war obsolete, and they focused on the vulnerability of military powers of retaliation (see Herken, *Counsels of War*). By the late 1950s this latter view prevailed, and the questions of military-mobilization after an attack faded to the background (see chapter 5).

15. Miller memorandum, "Consideration of Vital Facilities," 206.

16. "The Application of Resources Protection Board Methods to Military Procurement Planning," in "Exhibit 'K'" of Office of Provost Marshal General, *Defense against Enemy Action*, 162; George K. Engelhart, memorandum for Major General Archer L. Lerch, "Critical War Production

Capacity Bottlenecks," May 12, 1945, in "Exhibit 'K'" of Office of Provost Marshal General, *Defense against Enemy Action*, 151.

17. "Application of Resources Protection Board Methods," 161.

18. Robert Patterson, memorandum to J. A. Krug, requesting transfer of the files and records of the Resources Protection Board, August 1, 1945, in "Exhibit 'L'" of Office of Provost Marshal General, *Defense against Enemy Action*, 202.

19. Archer L. Lerch, memorandum, "Policies and Assumptions Governing Industrial and Material Demobilization," June 26, 1945, in "Exhibit 'K'" of Office of Provost Marshal General, *Defense against Enemy Action*, 140. Meanwhile, the functions that had been central for the OPMG during World War II—guidance of "protection activities of the War and Navy Departments" and other agencies "having internal security missions"—would become a secondary concern. See George K. Engelhart, enclosure to memorandum for Major General Archer L. Lerch, "Summary of Tabs C to I, Inclusive," May 12, 1945, in "Exhibit 'K'" of Office of Provost Marshal General, *Defense against Enemy Action*, 156–57.

20. Engelhart, enclosure to memorandum, "Summary of Tabs C to I, Inclusive," 166. Resource evaluation would thus help to address what OPMG officers saw as the distressing number of critical vulnerabilities that were *produced by* carefully laid-out mobilization plans. "It can be assumed," wrote Engelhart, that had vulnerability been taken into account in mobilization planning decisions at the outset of WWII, "we would not now be completely dependent upon 87 'AA' rated plants." In this view, mobilization preparedness and economic vulnerability assessment were not to be understood as "distinct problems, but rather [as] two aspects of the same." See George K. Engelhart, enclosure to memorandum for Major General Archer L. Lerch, Tab C, "The Application of Resources Protection Board Methods to Military Procurement Planning," May 12, 1945, in "Exhibit 'K'" of Office of Provost Marshal General, *Defense against Enemy Action*, 161.

21. Miller memorandum, "Consideration of Vital Facilities," 205, 206, 207.

22. "Water Systems," 242.

23. Tyler, "Civil Defense," 12. Military officials debated whether nonmilitary defense should be housed within the Army or in a civilian office but agreed that if "the Army was to achieve either goal it would have to get the ball rolling," and supported War Department planning for nonmilitary defense.

24. Cooling, "US Army Support of Civil Defense," 10. Indeed, it seems that the military acted without any presidential directive (beyond Roosevelt's original order in 1942 relating to wartime internal security). The report that resulted from this study, *Defense against Enemy Action Directed at Civilians*, issued on April 30, 1946, has generally been analyzed as the first in a succession of studies of civil defense that culminated in NSRB's Blue Book, the basis for the 1950 Federal Civil Defense Act. But the scope of the report went far beyond the problems of local protection and relief that were the focus of the Blue Book and the Federal Civil Defense Administration. OPMG understood its charge to include "the entire civil sector including the economy and the government, not just the people." Yoshpe, *Our Missing Shield*, 78–79.

25. S. L. Scott, memorandum for the Provost Marshal General, "Civilian Defense against Enemy Action Directed at Civilians, Their Installations, and Communities," August 4, 1945, in "Exhibit 'A'" of *Defense against Enemy Action*, 18.

26. Tyler, "Civil Defense," 10. According to Tyler (32) postwar military planners associated civilian leadership with "LaGuardian chaos" (referring to Fiorello LaGuardia's management of the Office of Civilian Defense) and "Rooseveltian dance instructors" (referring to the activities Eleanor Roosevelt organized in the name of civil defense).

27. Ibid., 26.

28. Office of the Provost Marshal General, *Defense against Enemy Action*, 2 (underlined in original). "At Nagasaki, Colonel Beers had personally interviewed 'close to two hundred people who had been within one hundred yards of Ground Zero,' the point directly under the center of

the explosion. Nobody could ever thereafter convince him there was no defense against the atomic bomb." Tyler, "Civil Defense," 26.

29. Office of the Provost Marshal General, *Defense against Enemy Action*, 15 (underlined in original).

30. Ibid., 32–33. World War II "civilian" defense—a term that, Miller recalled, was "studiously avoided" in the OPMG's final report—focused on local measures to provide emergency relief to stricken civilians and to prepare civilians to evacuate cities or black out windows in anticipation of enemy attack. According to Yosphe, *Our Missing Shield*, 78–79, the OMPG study's reference to *civil* defense was meant to mark a difference with the wartime experience.

31. Office of the Provost Marshal General, *Defense against Enemy Action*, 6, 2–3.

32. Ibid., 7.

33. Dudley, "Sprawl as Strategy."

34. On concerns about the implications of urban dispersal for the American political and economic system, see Tobin, "Reduction of Urban Vulnerability." Our suggestion is that, partly in response to these concerns, planners shifted their focus from urban dispersal to the selective dispersal of vital nodes of production systems, essential services, and government facilities.

35. On the Atomic Scientists of Chicago, see *Guide to the Atomic Scientists of Chicago Records 1943–1955*, University of Chicago Library, 2007, https://www.lib.uchicago.edu/ead/rlg/ICU .SPCL.ASCHICAGO.pdf. In November 1945, a group of social scientists—the Office of Inquiry into the Social Aspects of Atomic Energy—was created as a "companion" organization. Its members included sociologist Edward Shils, anthropologists Robert Redfield and Fred Eggan, and economists T. W. Schultz, Jacob Viner, and Jacob Marschak.

36. The scientists who worked on the bomb had quite diverse reactions to its development and deployment (see Herken, *Counsels of War*). Our discussion refers to the collectively articulated views of the Atomic Scientists of Chicago (e.g., Atomic Scientists of Chicago, *Atomic Bomb*).

37. Advocates of airpower pushed for massive investments to maintain US superiority in weaponry, and, in some cases, preemptive bombing to prevent the Soviet Union from acquiring a nuclear stockpile.

38. Hill, Rabinowitch, and Simpson, "Atomic Scientists Speak Up," 25. The atomic scientists saw the problem as particularly acute for the United States. "Because of its tremendous concentrations of population, industry, government and transportation facilities in a relatively small number of metropolitan centers [the United States] is one of the most vulnerable of all to atomic weapons, which are most efficiently employed against such concentrations." Atomic Scientists of Chicago, *Atomic Bomb*, 24.

39. In their early writings, the atomic scientists doubted the value of "active" measures, but some civilian experts would later come to a different conclusion, pushing for massive investment in detection and interception of enemy bombers. See Edwards, *Closed World*.

40. In the 1950s, the center of gravity of this discussion shifted to Cambridge, Massachusetts, and to a group of scientific experts and science administrators who published influential reports advocating nonmilitary defense, ranging from Project Charles and Project East River in the early 1950s (discussed later in this chapter), to the 1957 Gaither Committee study, Security Resources Panel of the Science Advisory Committee, *Deterrence and Survival* (discussed in chapter 6).

41. Rabinowitch, "Only Real Defense," 243.

42. Atomic Scientists of Chicago, *Atomic Bomb*, 24, 48.

43. Hill, Rabinowitch, and Simpson, "Atomic Scientists Speak Up," 48.

44. Marschak, Teller, and Klein, "Dispersal of Cities and Industries," 20.

45. Meeting of Social Science Research Council, Committee on Social Aspects of Atomic Energy, February 10, 1946, 6, folder 37, box 6, Rockefeller Foundation Records, Projects, RG 1.2, series 216.S, Rockefeller Archive Center, Sleepy Hollow, NY. Members included sociologist William Ogburn, economist Jacob Marschak, demographer Frank Notestein, physicist Isidor Rabi,

psychologist Rensis Linkert, and Bernard Brodie, the military strategist who developed a theory of atomic war around the concept of deterrence. In a February 1946 meeting to discuss the Committee's agenda, Ogburn reflected that "statements issued by social scientists have been largely ethical" and urged that the social scientists focus on "the planning and support of research."

46. Riefler, preface, viii, x, xi, xii.

47. Ibid., xx. The rationale, he later explained, was that "most of the techniques required to study the reduction of vulnerability lie within the fields of competence of the social sciences."

48. Meeting of Social Science Research Council, Committee on Social Aspects of Atomic Energy, February 10, 1946, 11.

49. Dembitz et al., "Discussion," 109.

50. Hanson Baldwin cited Coale's book as the key source for his discussion of dispersal and civil defense in the influential book *Price of Power*. Tyler, "Civil Defense," 72n2, notes that the study was also circulated to NSRB staff working on civilian mobilization in spring 1950 and that Coale spent "two summers as consultant to the RAND Corporation in 1951 and 1953, and in 1953 was a consultant to the Weapons System Evaluation Group in the Pentagon," no doubt bringing his work to the attention of the elite civilian experts working on nuclear strategy. It was one of the only studies of nonmilitary defense cited by the Stanford Research Institute report on Post-Attack Industrial Rehabilitation and was cited by Carl Kaysen in 1954 as one of only two unclassified expositions of the economic theory of vulnerability. Kaysen, "Vulnerability of the United States."

51. This scenario was based on a limited number of atomic bombs delivered by long-range bombers. Coale also developed a prescient analysis of a future war with large numbers of more powerful, missile-delivered bombs. Following Brodie, he posited that in such a war the old problems of military-industrial production would be replaced by the military capacity to retaliate, the continued operation of the government, and the survival of the national population. These assumptions dominated American strategic thought by the late 1950s (see chapters 5 and 6).

52. Coale, *Problem of Reducing Vulnerability*, 17, 19.

53. Coale, "Problem of Reducing Vulnerability."

54. Coale, *Problem of Reducing Vulnerability*, 62–63, 114–15.

55. Ibid., 25–27.

56. Ibid., 36.

57. Ibid., 31, 32.

58. Ibid., 103, 107.

59. Ibid., 19–22; Coale, "Problem of Reducing Vulnerability," 90.

60. According to Jordan, *U.S. Civil Defense before 1950*, 65, the report was completed in February 1947 but not declassified and circulated (by Forrestal) until 1948.

61. Tyler, "Civil Defense," 59–60. In the so-called unification study that laid out the basic framework for the National Security Act (see chapter 2), nonmilitary defense was discussed only in a passing comment: "A substantial degree of industrial relocation may be required if the present concentration of industry proved unduly vulnerable to attack." Eberstadt, *Unification*, 21. See also Tyler, "Civil Defense," 119. A draft of the National Security Act debated in Congress in 1946 included no mention of such functions. The addition to the final law suggests that discussions during 1946 and 1947 played some role in shaping the language in the Bull report and the statutory language finally included in the National Security Act.

62. National Security Act of 1947, Pub. L. No. 253, 61 Stat. 496 (July 26, 1947).

63. Given NSRB's primary focus on industrial mobilization planning at this time, relocation remained relatively peripheral to its work. Thus, in a 1948 lecture to the Industrial College of the Armed Forces on the functions of the National Security Resources Board, Ralph Watkins, the director of NSRB's Office of Plans and Programs devoted only a single phrase to "measures to decrease vulnerability, such as relocation of industries and vital services, dispersion of plants,

duplication of key facilities, breaking of transportation bottlenecks, etc." The remainder of his lecture was devoted to traditional mobilization preparedness activities. See Yoshpe, *National Security Resources Board*.

64. National Security Resources Board, *National Security Factors*, 6.

65. Arthur M. Hill, report to the president, "A Recommendation to the President by the National Security Resources Board on Security for the Nation's Capital," October 27, 1948, 1, 3, 5, declassified NWDD 978112, entry 63A-260, Nation's Capital Program folder, box 204, RG 396, Records of the Office of Emergency Preparedness.

66. The Atomic Energy Commission, formed on January 1, 1947, was set up to house the Manhattan Project, which had previously been located in the Army Corps of Engineers. In the late 1940s the AEC had unique access to classified information about the effects of atomic weapons, which, as we have seen, was not widely shared across the government at this time. It seems likely that either the AEC study drew on Ralph Lapp's work or that Lapp's study was conducted for the AEC when it was part of the Army Corps of Engineers and Lapp was executive director of the Committee on Atomic Energy of the Research and Development Board.

67. US Atomic Energy Commission, *City of Washington*, 3–4.

68. On Augur's dispersal plan for Washington, DC, see Krugler, *This Is Only a Test*; and Tobin, "Reduction of Urban Vulnerability."

69. The history of congressional discussion of governmental dispersal is reviewed in "Outline Statement on Current Program and Historical Background of Federal Government Policy and Plans for Dispersal and Decentralization of Essential Agencies and Demolition of Temporary Building," Dispersion: General folder, box 204, declassified NWDD 978112, entry 63A-260, RG 296, General Records of the Economic Stabilization Agency, National Archives, College Park, MD.

70. National Security Resources Board, *National Security Factors*, iii.

71. E. T. Dickinson, "Analytical Review," 5.

72. On these pressures specifically as they concerned NSRB, see Funigiello, "Managing Armageddon"; and Tyler, "Civil Defense."

73. Lapp, "Industrial Dispersion," 258, thus referred to NSRB's dispersal program as "dormant" between 1948 and 1951.

74. Memorandum, White House to Charles W. Davis, clerk of the House Committee on Ways and Means, September 15, 1950, in response to a letter from September 12 requesting clarification on several issues relating to the DPA, reprinted in US House of Representatives Committee on Ways and Means, *Revenue Revision of 1951*, 2692. When the interest of national security made it "desirable to ask businesses to locate new plants in places other than those they would select on the basis of cost and market considerations," rapid amortization could compensate businesses that "participate in such industrial dispersal" (2692).

75. US House of Representatives Committee on Public Works, *Dispersal of Government Agencies*, 48.

76. US House of Representatives, *Certificates of Necessity*, 34–35.

77. US Congress Joint Committee on the Economic Report, *Need for Industrial Dispersal*, 56.

78. "News and Notes," 276–77.

79. Discussion of Rains amendment, 97 Cong. Rec. 97 H7985–86 (July 11, 1951).

80. Discussion on the Rains amendment, 97 Cong. Rec. H10231 (August 17, 1951).

81. Gorrie, "Federal Dispersion Policy," 270–71. NSRB piloted this model in Seattle, Washington, where the local government, according to acting chairman Jack Gorrie, identified dispersal areas located "well outside the territory that might be affected by any possible bombing" (270).

82. National Security Resources Board, "Discussion Paper on Strategic Dispersal of Industry," May 28, 1951, 1, Industrial Dispersal Program folder, box 11, ODM Central Files, 1950–1951, RG 304, Records of the Office of Civil and Defense Mobilization.

83. Lapp, "Industrial Dispersion." A few months earlier, Ramsay Potts, special assistant to the chair of NSRB and head of the Special Programs Office, had written to Lapp following a vote on the program for government dispersal: "Planning for the security of the nation's capital needs your support now more than ever since dispersal bill killed in Senate yesterday. . . . Anything you can do to point up the danger and the urgent need for this dispersal program will be appreciated and considered by all to be in the best interests of national security." Ramsay D. Potts to Dr. Ralph E. Lapp, telegram, April 24, 1951, box 44, stack area 650, row 38, compartment 25, shelf 6, NM 69/E, 14, RG 304, Records of the Office of Civil and Defense Mobilization.

84. Lapp, "Industrial Dispersion," 259.

85. Lapp, "Strategy of Civil Defense," 243.

86. Lapp, "Industrial Dispersion," 259.

87. Ibid.

88. Discussion on the Rains amendment, 97 Cong. Rec. H10231 (August 17, 1951).

89. Harry S. Truman, "Memorandum and Statement of Policy on the Need for Industrial Dispersion," August 10, 1951, American Presidency Project, https://www.presidency.ucsb.edu /documents/memorandum-and-statement-policy-the-need-for-industrial-dispersion.

90. National Security Resources Board, *Is Your Plant a Target?*, 6.

91. As summarized in National Security Resources Board, "Discussion Paper on Strategic Dispersal of Industry," 1–2.

92. Ramsay D. Potts to Mr. Dimmitt, May 24, 1951, Office Files of Special Assistant to Ramsay D. Potts, box 44, National Security Resources Board, stack area 650, row 38, compartment 25, shelf 6, NM 69/E, 14, RG 304, Records of the Office of Civil and Defense Mobilization.

93. Sutermeister, "Discussion of Papers," 129–31.

94. Oscar Sutermeister, "Initial Criteria to Be Used in Administration of the National Industrial Dispersion Policy," 1951, box 4, RG 304, Records of the Office of Civil and Defense Mobilization.

95. It was necessary, as Robinson Newcomb put it, to set up procedures that would make it possible to "accomplish the objectives [of the program] on paper and be workable in practice" so that an "operating staff can handle cases in a consistent, rapid, and readily explicable fashion." Robinson Newcomb, "Progress Report on Dispersal," August 17, 1951, box 4, RG 304, Records of the Office of Civil and Defense Mobilization.

96. The status of resource evaluation remained uncertain for many years after World War II. OPMG's demobilization studies had proposed that a postwar resource evaluation office could be located in the "upper echelons" of the military. In 1947, Secretary of Defense James Forrestal urged the National Security Council to take responsibility for internal security functions, including resource evaluation, which it did by creating an Interagency Committee on Internal Security. This Committee divided its work into two areas. An Interdepartmental Intelligence Conference took responsibility for the detection and interdiction of enemy agents and saboteurs, and the Industrial Security Committee was responsible for facility-level security measures. See National Security Council, Interagency Committee on Internal Security, *State, Army, Navy, Air Force Coordinating Committee on Internal Security*.

97. Ibid., 12.

98. According to Yoshpe (*National Security Resources Board*), the Facilities Protection Board was officially part of the Interagency Committee on Internal Security but, like the Industry Evaluation Board, was housed in the Department of Commerce.

99. See National Security Council, *Progress Report by NSC Representative on Internal Security, on the Implementation of Internal Security (NSC 17/4;176)*, March 26, 1951. The January 1951 order that established the Industry Evaluation Board in the Department of Commerce repeated, word for word, the War Production Board order establishing the Resources Protection Board a decade earlier. Department of Commerce, Order No. 129 (January 19, 1951). The War Production Board order was on May 8, 1942.

100. *Third Supplemental Appropriation Bill for 1951: Emergency Agencies Hearing, Part 1, Special Subcommittee on Emergency Defense Appropriations, Committee on Appropriations, House of Representatives*, 82nd Cong., 1st sess. (March 7–9 and 12–16, 1951).

101. *Supplemental Appropriation Bill for 1952: Emergency Agencies, Part 1, Committee on Appropriations, House of Representatives*, 82nd Cong., 1st sess. (July 30, August 3, 6–8, and 10, 1951).

102. National Security Resources Board, "Discussion Paper on Strategic Dispersal of Industry," 1–2. Nonmilitary defense planners continued to push to expand the purview of resource evaluation beyond plant-level security measures to include broader vulnerability reduction measures such as dispersal. Thus, a 1952 memorandum explained that President Truman's Executive Order 10421, which transferred jurisdiction over the Industry Evaluation Board to an agency in the executive office, was to "allow a broadening of its outlook and the development of a rating system that could be used for all defense mobilization purposes, rather than just for physical security." Memorandum for Dr. Flemming, "Organization for Non-Military Defense Activities," NWDD 978112, Wartime Organization folder, entry 63a-260, box 209, RG 396, Records of the Office of Emergency Preparedness.

103. Newcomb represented the Office of Defense Mobilization on NSRB's dispersal program and also served on the Council of Economic Advisers. See "Dr. Robinson Newcomb Oral History Interview," by James R. Fuchs, August 6, 1977, Harry S. Truman Library, https://www.trumanlibrary.gov/library/oral-histories/newcomb.

104. Memo to James F. King, "Notes on Dispersion Discussion," in "A Recommendation to the President by the National Security Resources Board on Security for the Nation's Capital," September 24, 1951, declassified NWDD 978112, Dispersion Policy up to June 1954 folder, entry 63A-260, box 204, RG 396, Records of the Office of Emergency Preparedness.

105. On SRI's project, see W. J. Platt, "Industrial Defense," 261–64. These standards were then adopted by the federal agencies involved in the mobilization control program: NSRB, ODM, the Department of Defense, and the Department of Commerce, where the Industrial Evaluation Board was housed.

106. Office of the Provost Marshal General, *Defense against Enemy Action*, 6.

107. National Military Establishment, Office of the Secretary of Defense, *Study of Civil Defense*, 3. This "distributed" model of organization corresponds to the concept of subsidiarity in contemporary discussions of federalism. See Collier and Lakoff, "Distributed Preparedness"; and Roberts, "Dispersed Federalism."

108. Quoted in Jordan, *U.S. Civil Defense before 1950*, 74.

109. Walter Binger, who served as a consultant for the OCDP, concluded that the Hopley report was "summarizing World War II rather than planning for the atomic age." Tyler, "Civil Defense," 113. See also Jordan, *U.S. Civil Defense before 1950*, 74. In the Committee's final report, industrial vulnerability was discussed only passingly; see Hopley, *Civil Defense for National Security*, 11.

110. Tyler, "Civil Defense," 114. For example, it dropped provisions such as mobile reserves of federal emergency response workers and coordination with the military.

111. Quoted in Jordan, *U.S. Civil Defense before 1950*, 174–75.

112. Tyler, "Civil Defense," 90–91.

113. Jordan argues that the Truman administration likely misread the military's intention, which was to provoke action by civilian authorities. Jordan, *U.S. Civil Defense before 1950*, 73–74.

114. V. L. Adams, *Eisenhower's Fine Group of Fellows*.

115. Yoshpe, *National Security Resources Board*, 122.

116. Jordan, *U.S. Civil Defense before 1950*, 91. Six months after Truman's order, this office had only eight staff members.

117. *Civil Defense Against Atomic Attack: Hearing, Part 1, Before the Joint Committee on Atomic Energy*, 81st Cong., 2nd sess. (March 17–December 4, 1950).

118. *Civil Defense: Hearing Before the Joint Committee on Atomic Energy*, 81st Cong., 2nd sess. (1950) (statement of Paul J. Larsen), 702.

119. Jordan, *U.S. Civil Defense before 1950*, 84.

120. National Resources Planning Board, *Civil Defense Planning Advisory Bulletin*, December 1, 1949, doc 121/1, reprinted in *Civil Defense against Atomic Attack*, 11.

121. Ibid.

122. Health resources was one of only two areas (housing was the other) run out of a separate division.

123. *Civil Defense Against Atomic Attack: Hearing, Part 2, Before the Joint Committee on Atomic Energy*, 81st Cong., 2nd sess. (March 17–December 4, 1950), 20.

124. National Resources Planning Board, *Civil Defense Planning Advisory Bulletin*, 17. The NSRB circulated the AEC's aiming point study of Washington, DC, and the AEC's *Effects of Atomic Weapons*, which became a best seller. The Board later asked the Air Force to prepare a technical manual for conducting aiming point studies (see chapter 3).

125. Federal Civil Defense Administration, *Health Services and Special Weapons Defense*, 18, 20.

126. These estimates were "developed by the Office of Medical Services of the Office of the Secretary of Defense and the three medical departments of the Department of Defense." Ibid., 109.

127. Ibid., 109–10, 104, 105.

128. Eric Biddle, a consultant to the Civilian Mobilization Office, proposed "a series of civil-defense test exercises in various cities" based on his observation of the British use of "war-gaming" for civil defense planning. Tyler, "Civil Defense," 245, 256.

129. Jordan, *U.S. Civil Defense before 1950*, 108.

130. Tyler, "Civil Defense," 245.

131. Chicago Civil Defense Committee, *Chicago Alerts*, 17, 134, 139.

132. See Tyler, "Civil Defense," 264; and Jordan, *U.S. Civil Defense before 1950*, 107.

133. Jordan, *U.S. Civil Defense before 1950*, 90–91.

134. Quoted in "Defense Lack Seen as 'Pearl Harbor,'" *New York Times*, October 9, 1949. Also writing in 1949, Clinton Rossiter lamented that, with Truman's order for further study, civil defense was being pushed "farther back on the shelf." "We may hope that at some point," he wrote, "*planning* will cease and *organizing* begin." Rossiter, "Constitutional Dictatorship in the Atomic Age," 417n22.

135. Jordan, *U.S. Civil Defense before 1950*, 97. Larsen had worked on high-profile technical projects during World War II and, after the war, served as associate director of Los Alamos Scientific Laboratory and as director of the Sandia Atomic Laboratory. His appointment was initially hailed as a "revival" of civil defense. See Tyler, "Civil Defense," 217.

136. Jordan, *U.S. Civil Defense before 1950*, 100, 129. The criticism of federal inaction "was severe and angry. The mayors and their representatives argued that they were more than ready to cooperate with Federal and State authorities in getting civil defense going, but that they were impotent unless the Federal Government assumed firm leadership and guidance." The Blue Book, the Federal Civil Defense Act, and the Federal Civil Defense Administration, according to Jordan, were "coerced upon these authorities against their will by public pressure" (129). This moment in the history of civil defense—which Jordan refers to as the rise of a "grass-roots" civil defense movement—stands in contrast to later periods in which federal authorities faced significant popular mobilization against civil defense measures. On the later episode, see Garrison, *Bracing for Armageddon*, 35–36.

137. Tyler, "Civil Defense," 284.

138. Rabinowitch, "Civil Defense," 226–29.

139. Lapp, "Strategy of Civil Defense," 241–43.

140. These were two of the "summer studies" conducted by Lloyd Berkner's American Universities, Incorporated. See Damms, "James Killian."

141. Project Charles, *Problems of Air Defense*, 205.

142. Jordan (*U.S. Civil Defense before 1950*, 129–30) refers to "behavioral evidence" for this view, including the loss of congressional and military interest in FCDA once it was created, and the proposal for another civil defense study—Project East River—immediately upon the passage of the Federal Civil Defense Act.

143. Large numbers of NSRB staff transferred to the newly formed Office of Defense Mobilization and Federal Civil Defense Administration between late 1950 and mid-1951, bringing NSRB staff from a peak of 540 to under 200. This remaining staff was much more narrowly focused on nonmilitary defense, which became a central concern of NSRB in 1951 and 1952. Yoshpe, *National Security Resources Board*, 169.

144. Yoshpe, *National Security Resources Board*, 125.

145. Ramsay D. Potts to Jack Gorrie, July 26, 1951, National Security Resources Board, Office Files of Special Assistant Ramsay D. Potts, box 44, stack Area 650, row 38, compartment 25, shelf 6, NM 69/E 14, RG 304, Records of the Office of Civil and Defense Mobilization. Potts reported to NSRB chair Jack Gorrie in July 1951 that the Board was "just starting to work really in this field." NSRB staff who worked on PAIR in some capacity include Dal Hitchcock, Tracy Augur, Oscar Sutermeister, Howard Bronson, Presley Lancaster, and Karl Tomfohrde.

146. See "Legends in the Law: Ramsay D. Potts," *Bar Report* (August/September 1999), http://old.dcbar.org/bar-resources/publications/washington-lawyer/articles/legend-potts.cfm.

147. William H. Stead, "Appendix to the Tasks of Nonmilitary Defense and the Present Status of Planning," reprinted in *Operations and Policies of the Civil Defense Program, Hearings, Part 2 and Appendix, Before the Subcommittee on Civil Defense, Committee on Armed Services, United States Senate*, 84th Cong., 1st sess. (March 9, 11, 22; April 5, 20, 21; May 18, 20; and June 20, 1955), 484.

148. George T. Hayes, William E. Hosken, and William J. Platt, *Report on Problems of Post-Attack Rehabilitation of Industry*, January 4, 1952, report prepared for the National Security Resources Board, Project Number 511, Stanford Research Institute. Platt was the director of research at SRI at this time.

149. A comparison of SRI's report—delivered on January 4, 1952—with NSRB's prior internal deliberations at this time suggests that much of the basic framework was provided by Board staff.

150. Executive Office of the President, National Security Resources Board, "Post-Attack Industrial Rehabilitation Conference," November 14, 1951, entry 8, box 13, RG 304, Records of the Office of Civil and Defense Mobilization. This distinction between the primary and secondary effects corresponded to the distinction between urban area analysis and industrial analysis in vulnerability assessment (described in chapter 2), and to the distinction economist John Maurice Clark drew between the "primary" and "secondary" effects of government interventions to address the Great Depression (described in chapter 1).

151. Hayes, Hosken, and Platt, *Report on Problems of Post-Attack Industrial Rehabilitation*, 3, 20.

152. Ibid., 4, 39, 40, 43.

153. Here, we suggest a genealogy of resilience that is quite different from the one that has been constructed by critical scholars (Walker and Cooper, "Genealogies of Resilience"; and Evans and Reid, *Resilient Life*) who have traced resilience to a conservative libertarian economics. Far from proposing a program of laissez-faire, NSRB asserted a capacious new government responsibility.

154. Hayes, Hosken, and Platt, *Report on Problems of Post-Attack Industrial Rehabilitation*, i, 1, 3, 38.

155. Ibid., i, 10.

156. Ibid., 11–12. On the embodied and affective dimensions of preparedness, see Adey and Anderson, "Anticipating Emergencies"; Davis, *Stages of Emergency*; Lakoff, "Preparing for the Next Emergency"; and Linnell, "Haptic Space of Disaster."

157. Hayes, Hosken, and Platt, *Report on the Problems of Post-Attack Industrial Rehabilitation*, 2.

158. Executive Office of the President, National Security Resources Board, "Post-Attack Industrial Rehabilitation Conference." This meeting arose from discussions with the ODM, which agreed that NSRB should take the lead in coordinating the project and convening an interagency group.

159. Among these were Tracy Augur (who had developed the NSRB plan for the dispersal of Washington), Oscar Sutermeister (who had formulated the NSRB position on dispersal), Dal Hitchcock, Oscar Endler, Presley Lancaster Jr., and William Stead.

160. Among these were the Defense Electric Power Administration (Interior), the Defense Manpower Administration (Labor), the Defense Solid Fuels Administration (Interior), as well as the Defense Transport Administration and the Industry Evaluation Board (Commerce).

161. Executive Office of the President, National Security Resources Board, "Post-Attack Industrial Rehabilitation Conference," 1.

162. Ibid., 3–6.

163. Ibid., 6.

164. Division of Post-Attack Rehabilitation, "Planning for Post-Attack Industrial Rehabilitation: Summary of Agency Reports," 5–7, National Security Resources Board, Office of the Vice Chairman, Security: Classified Office File of Edward T. Dickinson, 1951–1953, Projects: Post-Attack Rehabilitation, entry 8, box 13, RG 304, Records of the Office of Civil and Defense Mobilization.

165. It is likely no coincidence that the SRI report was released at almost the same time as NSRB's internal report on the survey's results.

166. US House of Representatives Committee on Public Works, *Dispersal of Government Agencies*, 46.

167. "Outline Statement on Current Program and Historical Background of Federal Government Policy and Plans for Dispersal and Decentralization of Essential Agencies and Demolition of Temporary Building," Dispersion: General folder, box 204, declassified NWDD 978112, entry 63A-260, RG 296, General Records of the Economic Stabilization Agency, National Archives, College Park, MD.

168. "Legends in the Law."

169. Potts to Lamphier, memorandum, March 22, 1951, National Security Resources Board, Office Files of Special Assistant Ramsay D. Potts, 1950–1951, box 44, stack area 650, row 38, compartment 35, shelf 6, NM 69/E 14, RG 304, Records of the Office of Civil and Defense Mobilization.

170. NSRB staff used the template developed for essential functions analysis in its survey of federal agency preparedness for the PAIR project.

171. "Continuity of Government Functions: Outline of Planning Factors," appended to memorandum from Lawton to heads of executive departments and establishments, "Planning for Continuity of Operations under Emergency Conditions," November 9, 1950, declassified NWDD 978112, entry 63A-260, Nation's Capital Program folder, RG 396, Records of the Office of Emergency Preparedness.

172. Ibid.

173. Funigiello, "Managing Armageddon," 413.

174. Exec. Order No. 10,346, 3 C.F.R. (April 17, 1952), https://www.presidency.ucsb.edu/documents/executive-order-10346-preparation-federal-agencies-civil-defense-emergency-plans.

175. Central Task Force on Post-Attack Rehabilitation, meeting minutes, June 26, 1952. Initially Dickinson assumed the role on an interim basis, but subsequent meetings suggest that he remained in this role and co-led the Central Task Force with Shaw Livermore. Stead departed to take over the directorship of NSRB's Office of Natural Resources, in which capacity he worked on the Paley Commission, which produced a seminal report on security resources.

176. Division of Post-Attack Rehabilitation, "Planning for Post-Attack Industrial Rehabilitation," 9–10.

177. National Security Resources Board, special meeting of senior staff, Post-Attack Industrial Rehabilitation Program, October 22, 1952, National Security Resources Board, Office of the Vice Chairman, Security: Classified Office File of Edward D. Dickinson, 1951–1953, Projects: Post-Attack Rehabilitation, entry 8, box 13, RG 304, Records of the Office of Civil and Defense Mobilization. A sixth group, headed by NSRB, worked on finance and legal questions.

178. This committee was "to develop procedures required to meet this vital problem of damage assessment." Central Task Force on Post-Attack Industrial Rehabilitation, meeting minutes, October 3, 1952, National Security Resources Board, Office of the Vice Chairman, Security: Classified Office File of Edward T. Dickinson, 1951–1953, Projects: Post-Attack Rehabilitation, entry 8, box 13, RG 304, Records of the Office of Civil and Defense Mobilization.

179. Central Task Force on Industrial Rehabilitation, meeting minutes, June 26, 1952, National Security Resources Board, Office of the Vice Chairman, Security: Classified Office File of Edward T. Dickinson, 1951–1953, Projects: Post-Attack Rehabilitation, entry 8, box 13, RG 304, Records of the Office of Civil and Defense Mobilization.

180. The Federal Civil Defense Administration had taken over the program of local test exercises that was initially organized under NSRB's civilian mobilization program. Along with the Office of Defense Mobilization—with which NSRB was merged in 1953—it would take responsibility for the well-known program of Operation Alert exercises in the 1950s (see chapter 5). Thus, the much-studied national test exercise program had its origins in the National Security Resources Board's work on alert planning for localities and the federal government.

181. *National Security Resources Board Bulletin* 53/2, December 18, 1952, National Security Resources Board, Office of the Vice Chairman, Security: Classified Office File of Edward T. Dickinson, 1951–1953, Projects: Post-Attack Rehabilitation, entry 8, box 13, RG 304, Records of the Office of Civil and Defense Mobilization.

182. Shaw Livermore, oral history interview by James R. Fuchs, March 4, 1974, Harry S. Truman Library, Independence, MO.

183. *National Security Resources Board Bulletin* 53/2.

184. W. J. Platt, "Industrial Defense," 264. As Platt noted, the change in administration and the transfer of NSRB authorities to a reconstituted Office of Defense Mobilization in 1953 "delayed federal action on the Committee's recommendations" (264). But staff who had worked on PAIR continued to push for action in this area after the reconstitution of ODM. Presley Lancaster Jr., who took over the Special Security Programs Office, thus included postattack industrial rehabilitation in a memorandum on "Major Non-Military Defense Projects Requiring Immediate Action and Administrative Requirements Therefore," National Security Resources Board, Security: Classified Correspondence of the Board, July 1949–1953, Transfer of NSRB Functions to ODM folder, entry 31, box 64, RG 304, Records of the Office of Civil and Defense Mobilization. Although ODM director Arthur Flemming did not act on Lancaster's recommendation to reconvene the Central Task Force of the PAIR project, the basic template of preparedness outlined in the project shaped ODM's entire approach to preparedness, which was guided by specialists who had worked on the PAIR project, such as Shaw Livermore.

185. Project East River, *General Report*, 1. On the influence of NSRB on the Project East River report, see C. McKim Norton, who served on the Project East River study committee. Norton reports that after a three-day briefing with the FCDA, the "operational problems involved in coping with disaster situations of the magnitude of an atomic explosion seemed almost overwhelming," and a briefing with the Department of Defense on the atomic threat "was equally overwhelming and depressing." It was not until the group was briefed by NSRB that there was

"a gleam of hope that some long range programs for dispersion and structural protection could effectively reduce today's appalling vulnerability of most American cities." Norton, "Report on Project East River," 88.

186. Bowie and Immerman, *Waging Peace*, 37–38. See also Needell, *Science, Cold War, and the American State*, 237. Shaw Livermore, who worked on the PAIR project, served on the NSC Planning Board at this time.

187. V. L. Adams, *Eisenhower's Fine Group of Fellows*, 88.

188. Paul H. Nitze and Carlton Savage of the policy planning staff, memorandum, "Continental Defense," May 6, 1953, *Foreign Relations of the United States, 1952–1954, National Security Affairs*, vol. 2, pt. 1, PPS files, lot 64 D 563, Office of the Historian, State Department, https://history .state.gov/historicaldocuments/frus1952-54v02p1/d61.

Chapter 5. Enacting Catastrophe

1. Rossiter, "Constitutional Dictatorship," 396.

2. Ibid., 395.

3. Ibid., 409.

4. Ibid., 410.

5. Yoshpe, *National Security Resources Board*, 67. Eisenhower's message accompanying the reorganization plan stated that although in principle, NSRB was "charged with planning for the future" and ODM was "charged with programs of the present," the "progress of the current mobilization effort has made plain how artificial is the present separation of these functions." Message of the President, Reorganization Plan No. 3 of 1953, 18 F.R., 3375, 67 Stat. 634 (June 12, 1953), https://uscode.house .gov/view.xhtml?req=granuleid:USC-prelim-title5a-node84-leafl36&num=0&edition=prelim.

6. Flemming memorandum, February 1953, cited in Yoshpe, *National Security Resources Board*, 69–70.

7. Yoshpe, *National Security Resources Board*, 70–71.

8. Nitze and Savage, memorandum, "Continental Defense."

9. Elliott was also the author of 1950 report to Congress, *Mobilization Planning and the National Security*.

10. *NSC 159: A Report to the National Security Council by the Continental Defense Committee on Continental Defense*, July 22, 1953, 58.

11. Ibid, 16–17.

12. The functions it specified for ODM were "the reduction of urban vulnerability, the continuity of government, the continuity of industry, and physical security of facilities." Ibid., 15.

13. Ibid., 62. This shift in assumptions corresponded to a shift in organization. Up to this point, nonmilitary defense had been separate from mobilization planning. In NSRB, nonmilitary defense was run out of a Special Security Programs Office, and NSRB staff who transferred to ODM following the merger of the two agencies were initially housed in a separate Non-Military Defense Division. As we will see, ODM soon worked to integrate nonmilitary defense considerations across its mobilization programs.

14. "A Report to the National Security Council by the Executive Secretary (Lay)," September 25, 1953, NSC files, lot 63 D 351, NSC 159, Office of the Historian, Department of State, Washington, DC, https://history.state.gov/historicaldocuments/frus1952-54v02p1/d92.

15. "A Report to the National Security Council by the Executive Secretary (Lay) on Interim Defense Mobilization Planning Assumptions," November 20, 1953, NSC files, lot 63 D 351, NSC 172, Office of the Historian, Department of State, Washington, DC, https://history.state.gov /historicaldocuments/frus1952-54v02p1/d105.

16. The minutes from a November 1953 meeting of the NSC describe Flemming's position: "Since some kind of guidance was now essential for the ODM, he expressed the hope that the Council would adopt the present assumptions as guidance to the ODM in the formulation of its plans and programs." The assumptions were formally adopted in early 1954. "Memorandum of Discussion at the 171st Meeting of the National Security Council, Thursday, November 19, 1953 [Extracts]," Office of the Historian, US Department of State, https://history.state.gov /historicaldocuments/frus1952-54v02p1/d104.

17. "Report to the National Security Council by the National Security Council Planning Board, Note by the [Acting] Executive Secretary to the National Security Council on Continental Defense," February 11, 1954, NSC files, lot 63 D 351, NSC 5408, Office of the Historian, Department of State, Washington, DC, https://history.state.gov/historicaldocuments/frus1952-54v02p1/d109.

18. Arthur S. Flemming, memorandum to all assistant directors, "Development of Mobilization Plans for Global War Involving Attack on the Continental United States," July 14, 1954, Records of Mobilization Plans for Total War, NWDD 978112, entry 63a-260, box 209, RG 396, Records of the Office of Emergency Preparedness, 2.

19. M. J. Collett, memorandum for General Willard S. Paul, "Thoughts on Executive Office Organization," October 8, 1954, NWDD 978112, entry 63a-260, box 209, Records of Wartime Organization, RG 396, Records of the Office of Emergency Preparedness.

20. Livermore, "Resources Management," 276.

21. The Committee included the economists William Truppner, Melvin Anshen, David Novick, Ernest Tupper, Walter Skuce, Lincoln Gordon, and Bertrand Fox. All had worked—often in high-level capacities—on the War Production Board. Both Anshen and Novick subsequently worked on program budgeting at RAND and were one of the key links between wartime mobilization planning and systems analysis of the postwar period. Another can be traced through the work of Project SCOOP (see chapter 2).

22. E. George, "Management of a Wartime Economy," 28. The "Bedsheet" (and its official name) are discussed in Livermore, "Resources Management."

23. Mobilization Program Advisory Committee, Records of Mobilization Program Advisory Committee 1954–1958, Meeting 19, October 14, 1955, 4, folder 3, box 17, Public Policy Papers, J. Douglas Brown Papers, Department of Rare Books and Special Collections, Princeton University Library.

24. E. George, "Management of a Wartime Economy," 16.

25. Livermore, "Resources Management," 13. Other problem areas addressed by the Bedsheet included "material resources and industrial production," "economic stabilization," "civil government and public morale," and "relations with allies."

26. E. George, "Management of a Wartime Economy," 6.

27. See discussion in chapter 3 as well as Directorate of Management Analysis, DCR/Comptroller Headquarters, US Air Force, *Bomb Damage Problem*, 2.

28. Office of Emergency Planning, Executive Office of the President, *National Plan for Emergency Preparedness*; Livermore, "Resources Management." Speaking at a 1967 symposium on Post-Attack Recovery, Livermore reflected: "If we glance at the sixteen chapters of the National Plan, we can see the imprint of the Big Seven categories which we as a committee finalized in 1957."

29. Livermore, "Resources Management," 276.

30. The key operational assumption for Plan D-Minus was that there would be a sudden enemy attack with no time for mobilization. As Willard S. Paul put it in a March 28, 1955, memo, "Emergency actions for Plan D-minus must be specific and have immediate operational effect. They must be those actions which you would take tomorrow if the President declared a state of emergency." Lt. Gen. Willard S. Paul, memorandum to all assistant directors, "Subject: Mobilization Plan

D-Minus," March 28, 1955, Innis Harris Papers 1952–1958, entry 63A-260, File: Plans and Program Division, 1956–1958, box 199, RG 396, Records of the Office of Emergency Preparedness.

31. Director, Office of Defense Mobilization, untitled memorandum to all departments and agencies, March 13, 1957, Innis Harris Papers 1952–1958, entry 63A-260, File: Plans and Program Division, 1956–1958, box 199, RG 396, Records of the Office of Emergency Preparedness. The exact reference of the term "D-Minus" is not clear from the archival documents we have reviewed. It may refer to the assumption that the United States would be operating with a *lower* level of total resources after an attack than on D-Day—i.e., the day of the attack. At this time, ODM continued to develop more traditional assumptions in a mobilization Plan C, based on a future war without an attack on the United States. "The primary purpose of Plan C," according to Flemming, was to "enable the Government to attain industrial mobilization" on the model of World War II or the Korean War.

32. Memorandum for Mr. Flemming, "Organization for Non-Military Defense Activities," NWDD 978112, entry 63a-260, box 209, RG 396, Records of the Office of Emergency Preparedness.

33. Ibid. A 1955 ODM report to the National Security Council on the status of the Mobilization Program defined administrative readiness as a set of "activities to assure the continuity of government, the emerging development of a standby wartime organizational structure, the difficult and important new program for damage assessment and prediction, and the testing of all of our mobilization plans and preparations." National Security Council, *Status on 6/30/55 of the Mobilization Program (NSC 5525, Part 4)*, June 30, 1955, 2, U.S. Declassified Documents Online, Gale.

34. As Arthur Flemming reported in early 1954, ODM was "devoting a considerable amount of time to the kind of organizational structure we should have in the Government in the event of general war." Flemming, "Impact of Atomic War," 15.

35. Yoshpe, *National Security Resources Board*, 99–100.

36. Lt. Gen. Willard S. Paul, memorandum to Dr. Arthur S. Flemming, June 14, 1954, NWDD 978112, entry 63a-260, box 209, RG 396, Records of the Office of Emergency Preparedness.

37. Arthur S. Flemming, memorandum to all assistant directors, "Development of Mobilization Plans for Global War Involving Attack on the Continental United States," July 14, 1954, Records of Mobilization Plans for Total War, NWDD 978112, entry 63a-260, box 209, RG 396, Records of the Office of Emergency Preparedness.

38. Ibid.

39. The DMO seems to have been soon replaced by detailed executive orders that assigned emergency responsibilities, which are still used today. See, for example, Exec. Order No. 11490 (Assigning Emergency Preparedness Functions to Federal Departments and Agencies), 34 Fed. Reg. 17567 (October 30, 1969), issued by President Richard Nixon on October 28, 1969; and President Barack Obama's Executive Order on the Assignment of National Security and Emergency Preparedness Communications Functions, July 6, 2012, Exec. Order No. 13,618, 77 Fed. Reg. 40779 (July 11, 2012).

40. Todd, Paul, and Peterson, "National Defense against Atomic Attack," 241.

41. Vincent P. Rock, Innis D. Harris, and William S. Royce, "Draft DMO on Assignment of Mobilization Responsibilities to Federal Civil Defense Administration," March 8, 1954, NWDD 978112, entry 63A-260, Dispersion folder, box 204, RG 396, Records of the Office of Emergency Preparedness.

42. Arthur S. Flemming to Secretary of Agriculture Ezra Taft Benson, April 29, 1955, Binder (Black) Untitled folder, OCDM Organization in the Event of War, accession 396-65C0204, box 15, RG 396, Records of the Office of Emergency Preparedness.

43. "Report to the National Security Council by the National Security Council Planning Board." The determination of each agency's essential functions would serve two purposes. First, it would "provide the knowledge required before detailed plans can be fully developed" for the

ongoing process of establishing a "realistic wartime organization structure." Second, it would serve "as a planning base for final decisions on emergency relocation problems." Memorandum for the National Security Council, "Progress Reports on Continental Defense," June 14, 1954, 12, NWDD 978112, entry 63a-260, Mobilization Plans for Total War folder, box 209, RG 396, Records of the Office of Emergency Preparedness. Work on essential functions had been initiated by NSRB in 1950 as part of its program on governmental continuity but lapsed with the dissolution of the PAIR project in late 1952 (see chapter 4).

44. Memorandum for the National Security Council, "Progress Reports on Continental Defense." One of the major relocation sites—where ODM's damage assessment center would later be sited—was in a secret location in Mount Weather in Virginia, where a deep bunker was built to protect essential government personnel. See Krugler, *This Is Only a Test*.

45. Memorandum for the National Security Council, "Progress Reports on Continental Defense," 12–14.

46. Ibid. In July 1954, Flemming reported: "As monitor and coordinator, ODM is providing guidance to departments and agencies in the determination of their essential functions. . . . A uniform approach to the agency-by-agency appraisal of functions has been established as a result of pilot studies in all of the constituent bureaus of the Department of Commerce. This approach has now been applied to all but a few of the key mobilization agencies. . . . A task force of personnel, with predominant interests in key mobilization areas, has been assembled to help: (1) In the over-all review and analysis of the submissions from the agencies; (2) In the assemblage of all significantly related segments of organizations and functions." Flemming, memorandum to all assistant directors, "Development of Mobilization Plans for Global War."

47. Memorandum, "Discussion at the 238th Meeting of the National Security Council, Thursday, February 24th, 1955," February 25, 1955, 20, U.S. Declassified Documents Online, Gale.

48. Ibid., 19–20.

49. A cover letter to the secretary of labor from Arthur Flemming stated: "These determinations are the result of analysis by your staff, subsequent review by the Office of Defense Mobilization and the Bureau of the Budget, and whatever negotiation has been necessary between our respective agencies." Flemming to the secretary of labor, May 5, 1955, NWDD 978112, entry 63a-260, Folder: Essential Wartime Functions, box 209, RG 396, Records of the Office of Emergency Preparedness.

50. Department of Health, Education, and Welfare, "List of Essential Wartime Functions," 2–3, Binder (Black) Untitled folder, OCDM Organization in the Event of War, accession 396-65C0204, box 15, RG 396, Records of the Office of Emergency Preparedness.

51. During World War II, this single function had been the concern of massive bureaucracies, employing thousands of federal employees.

52. Department of Commerce, "List of Essential Functions," 2–3, NWDD 978112, entry 63a-260, Essential Wartime Functions folder, box 209, RG 396, Records of the Office of Emergency Preparedness.

53. NSRB's planning for emergency action steps was initiated in the late 1940s, when the office undertook planning for setting up a World War II–style emergency resource management agency. As noted in chapter 2, these plans were enacted with the establishment of the Office of Defense Mobilization to manage the mobilization during the Korean War.

54. Jack Gorrie to Millard Caldwell, November 6, 1952, Priority Status of Civil Defense folder, OCDM Project Files, accession 304-62A574, 650-39-34, RG 396, Records of the Office of Emergency Preparedness. As Yoshpe summarizes it, the project sought to "assemble in a single document summary check lists of the essential emergency actions to be taken and programs to be implemented; provide a comprehensive reference table indicating necessary measures and points at which coordination would be required; serve as a basis for assembling and organizing

materials in support of the mobilization measures listed in the document; and assist in uncovering weaknesses in our security position in terms of the possibility of war." Yosphe, *National Security Resources Board*, 101. In late 1952, NSRB vice chairman Edward Dickinson charged an interagency committee with assembling a "composite check list of executive actions" to be taken under three different scenarios: full mobilization without a declaration of war; a declaration of war but with the United States not under attack; and the United States at war and under attack. The committee included Shaw Livermore of ODM as well as representatives from multiple executive branch agencies. Executive Office of the President, National Security Resources Board, "First Interagency Meeting on the Compilation of Emergency Action Steps," November 13, 1952, U.S. Declassified Documents Online, Gale.

55. Although not much progress was made under NSRB, a basic plan of work was established that persisted later. A second meeting of the emergency action steps group took place in July 1953; participants included veteran mobilizers George Steiner and Vincent Rock. Yosphe explains this process: "As an initial step in the planning, the NSRB staff prepared a check list of actions that should be taken in the event of a sudden transition to war, the enactments and orders then in effect that would permit such actions, and the preparations made and still needed in support of the actions." Yoshpe, *National Security Resources Board*, 101.

56. Gorrie to Caldwell, November 6, 1952.

57. Ibid. Harold Yoshpe writes that the emergency action steps project served as "a basis for assembling and organizing materials" in support of mobilization measures; Yoshpe, *National Security Resources Board*, 101.

58. Executive Office of the President, Office of Defense Mobilization, "Briefing for Operation Alert 1956," July 11, 1956, 8, 11, U.S. Declassified Documents Online, Gale.

59. Ibid., 51, 24, 56, 30.

60. Davis, *Stages of Emergency*; Garrison, *Bracing for Armageddon*; Oakes, *Imaginary War*; Krugler, *This Is Only a Test*.

61. Indeed, while the mass exercises waned in the late 1950s in the face of increasing public scorn, exercises oriented to elite decision makers and officials continued as an ongoing governmental practice. As Dee Garrison writes, "Each year, the discussion of Operation Alert by the presidential staff focused more and more on the elite shelter problems and the continuity of government until this issue all but dominated the White House's connection to civil defense." Garrison, *Bracing for Armageddon*, 76–77.

62. Vincent Rock, memorandum to Dr. Arthur S. Flemming, "Schedule for Maintaining Mobilization Plans," June 29, 1954, Records of Mobilization Plans for Total War, NWDD 978112, entry 63-260, box 209, RG 396, Records of the Office of Emergency Preparedness.

63. Executive Office of the President, Office of Defense Mobilization, "Mobilization Program Advisory Committee, Recommended War Problems Exercise," July 9, 1954, MPAC-14 (revised), folder 3, box 17, Public Policy Papers, J. Douglas Brown Papers, Department of Rare Books and Special Collections, Princeton University Library, Princeton, NJ.

64. Livermore, "Resources Management," 277.

65. Harris, "Lessons Learned," 3. Such exercises were not entirely new: as we discuss in chapter 4, NSRB had developed civil defense test exercises in the late 1940s. But the NSRB exercises focused on urban problems of disaster response rather than centralized mobilization tasks.

66. Flemming memorandum, "Development of Mobilization Plans."

67. Harris, "Lessons Learned," 4.

68. Flemming memorandum, "Development of Mobilization Plans."

69. ODM staff meeting minutes, July 14, 1954, Records of Office of Defense Mobilization Director's Weekly Staff Meetings, 1951–1956, entry 64A-126, box 17, RG 304, Records of the Office of Civil and Defense Mobilization.

70. ODM staff meeting minutes, November 9, 1954, Records of Office of Defense Mobilization Director's Weekly Staff Meetings, 1951–1956, entry 64A-126, box 17, RG 304, Records of the Office of Civil and Defense Mobilization.

71. ODM staff meeting minutes, "Global War Plan—Materials," staff paper, November 9, 1954, Records of Office of Defense Mobilization Director's Weekly Staff Meetings, 1951–56, entry 64A-126, box 17, RG 304, Records of the Office of Civil and Defense Mobilization.

72. Oakes, "Cold War Conception," 349.

73. Flemming memorandum, "Development of Mobilization Plans."

74. Merrill J. Collett, memorandum to Lt. Gen. Willard S. Paul, "Federal Organization to Deal with the Bombing of Urban Target Centers," January 22, 1955, 3, Innis D. Harris Papers, 1952–1958, entry 63A-260, Emergency Action Task Force file, box 199, RG 396, Records of the Office of Emergency Preparedness.

75. Ibid., 3, 6, 8, 9.

76. Ibid., 17, 5, 1.

77. The "regions" in question were defined as areas of resource management and in this sense should be understood to lie in the tradition of regional planning, discussed in chapter 1, and in the tradition of Progressive political reform, discussed in chapter 2. In fact, key participants in the earlier episode, such as Luther Gulick, continued to be involved in these postwar discussions of emergency organization. US emergency management has continued to be organized into this kind of regional structure. Today's Federal Emergency Management Agency is organized into ten territorial regions, which "serve as the primary organizational unit for liaison to states and local governments within each region, and non-governmental and private sector entities within each Regional Office's geographical area." Zaffar, *Understanding Homeland Security Policy*, 349.

78. Lt. Gen. Willard S. Paul, memorandum to Arthur S. Flemming, "Governmental Chain of Command to Deal with Urban Problems in an Attack Emergency," January 27, 1955, Innis D. Harris Papers, 1952–1958, entry 63A-260, Emergency Action Task Force file, box 199, RG 396, Records of the Office of Emergency Preparedness.

79. "Emergency Action Task Force Update," August 24, 1955, Innis Harris Papers, 1952–1958, entry 63A-260, Emergency Action Task Force file, box 199, RG 396, Records of the Office of Emergency Preparedness.

80. Office of Defense Mobilization, memorandum to the president, April 7, 1955, Innis Harris Papers, 1952–1958, entry 63A-260, Emergency Action Task Force file, box 199, RG 396, Records of the Office of Emergency Preparedness.

81. President Eisenhower to Arthur S. Flemming, dictated "Memorandum for the Director of Defense Mobilization," April 7, 1955, Innis Harris Papers, 1952–1958, entry 63A-260, Emergency Action Task Force file, box 199, RG 396, Records of the Office of Emergency Preparedness.

82. Attorney General Herbert Brownell, memorandum to Arthur S. Flemming, "Assumptions in the Event of a Bombing Attack," April 20, 1955, Innis D. Harris Papers, 1952–1958, entry 63A-260, Emergency Action Task Force file, box 199, RG 396, Records of the Office of Emergency Preparedness.

83. Assistant Attorney General Lee Rankin to Charles Kendall, general counsel, Office of Defense Mobilization, March 27, 1956, Innis D. Harris Papers, 1952–1958, entry 63A-260, Emergency Action Task Force file, box 199, RG 396, Records of the Office of Emergency Preparedness.

84. ODM general counsel Charles Kendall to Assistant Attorney General Lee Rankin, May 4, 1956, Innis D. Harris Papers, 1952–1958, entry 63A-260, Emergency Action Task Force file, box 199, RG 396, Records of the Office of Emergency Preparedness.

85. Clinton Rossiter had raised precisely such questions, asking, "What sort of government will step in to restore order and bring people to safety and their senses" in the wake of an atomic

attack? Rossiter anticipated that government in this situation would be a strictly "executive-military" affair. The president would "rule under the terms of a nation-wide declaration of martial law, and the agents who execute their commands will be men in uniform." Rossiter, "Constitutional Dictatorship in the Atomic Age," 398.

86. The vague formulation of this clause in the Federal Civil Defense Act stands in contrast to the detailed explication of emergency mobilization powers in the Defense Production Act. "Title III [of the Federal Civil Defense Act] was the defense production act of civil defense, but without the previous history and comprehensiveness that characterized its analogue and gave rise to its more detailed policies, priorities, and procedural instruments." Gessert, Jordan, and Tashjean, *Federal Civil Defense Organization*, 27.

87. This renewed criticism of civil defense was part of a broader attention to continental defense in elite circles of Eisenhower's national security establishment at this time. Much of the conversation revolved around active defense measures—such as bomber interception and radar detection—that had long been championed by civilian experts. Many of these civilian experts were gathered in ODM's Science Advisory Committee. In mid-1954, Eisenhower had instructed this Committee's chair, Lee DuBridge, to conduct a study of American vulnerability with a focus on technological capabilities. The resulting study, *Meeting the Threat of Surprise Attack*, which was presented to the NSC on March 17, 1955, was apparently influential in Eisenhower's thinking. See V. L. Adams, *Eisenhower's Fine Group of Fellows*; and Damms, "James Killian."

88. Thomas R. Phillips, "Decisions to Establish a Joint Continental Air Command Is a Long Step in Defense of U.S.," *St. Louis Post-Dispatch*, August 8, 1954.

89. *Hearings held before the Committee on Armed Services, Interim Report on Civil Defense by the Subcommittee on Civil Defense*, H.R. Rep. No. 3885 (1955), 10.

90. "A Program for the Nonmilitary Defense of the United States, prepared by the National Planning Association," exhibit 11 in the appendix to Hearings Before the Subcommittee on Civil Defense, Committee on the Armed Services, 84 Cong. Rec. (1955), 814.

91. The NPA's study group included Shaw Livermore and Melvin Anshen of ODM's Mobilization Program Advisory Committee. The study's lead author, William Stead, also consulted with officials such as General Willard Paul and Vincent Rock, who led ODM's work on administrative readiness.

92. "Program for the Nonmilitary Defense of the United States," 815.

93. *Operations and Policies of the Civil Defense Program* (June 20, 1955) (statement of Hon. Val Peterson, Federal Civil Defense Administration), 725.

94. Gessert, Jordan, and Tashjean, *Federal Civil Defense Organization*, 29.

95. Harris, "Lessons Learned," 5. On the public context of Operation Alert 1955, Guy Oakes writes, "Participants included most agencies of the federal government, with the President and members of the Cabinet playing leading and highly visible roles, scores of cities that had been marked for 'destruction,' businesses that had developed their own civil defense preparedness plans, organized labor, and thousands of small towns across the country that did not intend to be left out of an event that appealed to the passion of patriotism as well as the interests of civic pride and the competitiveness of community spirit." Oakes, "Cold War Conception," 349–50.

96. Garrison, *Bracing for Armageddon*, 78; Oakes, "Cold War Conception," 350.

97. Fairman, "Government under Law."

98. "The President's News Conference," July 6, 1955, American Presidency Project, UC Santa Barbara, https://www.presidency.ucsb.edu/documents/the-presidents-news-conference-340.

99. FCDA general counsel Archambault, memorandum, "Declaration of Martial Law in Operation Alert 1955," January 13, 1956, Innis Harris Papers, 1952–1958, entry 63A-260, Emergency Action Task Force file, box 199, RG 396, Records of the Office of Emergency Preparedness.

100. As we discuss below, the meaning of "limited martial law" was ambiguous. An FCDA official later explained that it meant that the declaration did not give unlimited power to the military. Rather, the authority of the military varied from the need to provide troops to support state police in "control of panic-stricken evacuees fighting for egress from a stricken city" to "absolute military control of a critical target complex where local government had totally collapsed under the weight of the panic of its citizenry, and political boundaries had been erased by the ensuing chaos." Archambault memorandum, "Declaration of Martial Law."

101. "President's News Conference," July 6, 1955.

102. The text of Eisenhower's martial law declaration is reprinted in the June 1955 hearings held by the Kefauver Subcommittee on Civil Defense, *Operations and Policies of the Civil Defense Program* (June 20, 1955), 748.

103. "President's News Conference," July 6, 1955.

104. *Operations and Policies of the Civil Defense Program* (June 20, 1955) (Peterson testimony), 750.

105. Harris, "Lessons Learned," 5.

106. Rossiter, *American Presidency*, 36.

107. US House of Representatives Committee on Government Operations, *Civil Defense for National Survival*, 68. Flemming later insisted in congressional hearings that the declaration was never meant to displace civilian authority, a claim that was viewed with extreme skepticism by members of Congress.

108. Harris, "Lessons Learned," 5.

109. Cecil Holland, "Operation Alert Leaves Big Question Marks," *Sunday Star*, June 19, 1955, quoted in June 21, 1955, edition of FCDA Daily News Digest (no. 1063), cited in Fairman, "Government under Law."

110. From June 21, 1955, edition of FCDA Daily News Digest (no. 1063), cited in Fairman, "Government under Law."

111. A 1955 ODM report on the mobilization program to the National Security Council referred to the results of Operation Alert 1955, stating that "of particular value from the standpoint of mobilization planning" was the "decision in the absence of Congress to proclaim limited martial law," which "led to the assignment to an Emergency Action Task Force of responsibility for developing necessary implementing organization and procedure." Office of Defense Mobilization, NSC 5525: Status of United States National Security Programs on June 30, 1955, Part 4: The Mobilization Program, 11, U.S. Declassified Documents Online, Gale.

112. Notes on task force meeting, September 13, 1955, Innis Harris Papers, 1952–1958, entry 63A-260, Emergency Action Task Force file, box 199, RG 396, Records of the Office of Emergency Preparedness.

113. Notes on task force meeting, September 27, 1955, Innis Harris Papers, 1952–1958, entry 63A-260, Emergency Action Task Force file, box 199, RG 396, Records of the Office of Emergency Preparedness.

114. Harris, "Lessons Learned," 5.

115. Notes from task force meeting, June 23, 1955, Innis Harris Papers, 1952–1958, entry 63A-260, Emergency Action Task Force file, box 199, RG 396, Records of the Office of Emergency Preparedness.

116. Cited in Charles Fairman to Chet Holifield, February 29, 1956, 13, Military Assistance to Civilian Authority folder, Box 15, RG 369, Records of the Office of Emergency Preparedness.

117. Arthur Krock, "Case against Nation-Wide Martial Law," *New York Times*, June 28, 1955.

118. Douglas B. Cornell, "Emergency Military Dictatorship Facing United States if Real Air Raid Comes," *Washington Post and Times Herald*, June 19, 1955.

119. Editorial: "If Bombs Fall," *Washington Post and Times Herald*, July 12, 1955.

120. Archambault's argument recalled Schmitt's claim that the inability to envision the future under an emergency situation makes the device of dictatorship necessary. "If the concrete means of achieving a goal can, under normal circumstances, be predicted with regularity," wrote Schmitt, "in cases of emergency we can only say this much: the dictator is entitled to do everything that is appropriate in the actual circumstances." Schmitt, *Dictatorship*, 8.

121. US House of Representatives Committee on Government Operations, *Civil Defense for National Survival*, part 4, 1423–24.

122. Fairman, "Law of Martial Rule," 1301.

123. US House of Representatives Committee on Government Operations, *Civil Defense for National Survival*, 68.

124. Fairman, "Government under Law," 3.

125. Ibid., 5. Fairman also argued that the military was not organized or well adapted to direct national recovery after a nuclear attack, nor did it wish to be assigned the task. Fairman cited Beers on this point: "The military attitude toward this is not any desire to apply martial rule, but on the contrary to avoid it as much as necessary and their doctrine is to the effect that virtually any war that we would be fighting would be in defense of our form of government and, therefore, we do not consider an automatic application of martial rule when we think so highly of the civil rule." Cited in Fairman to Holifield, February 29, 1956. Folder: Military Assistance to Civilian Authority, RG 396, Records of the Office of Emergency Preparedness.

126. Fairman, "Government under Law," 3.

127. Ibid., 39. "Our great need is to assure effective civil government, under the direction of the President, in event of a nuclear attack. 'Martial law' should be a last resort. Operation Alert took a wrong turn, and has generated an erroneous mode of thought."

128. US House of Representatives Committee on Government Operations, *Civil Defense for National Survival*, 67–68.

129. Ibid., 1426.

130. Notes on task force meeting, October 6, 1955, Innis Harris Papers, 1952–1958, entry 63A-260, Emergency Action Task Force file, box 199, RG 396, Records of the Office of Emergency Preparedness.

131. Ibid.

132. Barnett Beers, memorandum to the chairman of the Emergency Action Task Force, "Revision of Statement of Conclusions, Contained in Memorandum for the President," April 24, 1956, Innis Harris Papers, 1952–1958, entry 63A-260, Emergency Action Task Force file, box 199, RG 396, Records of the Office of Emergency Preparedness. If federal assistance in discharging responsibilities was refused by local governments, the Federal Civil Defense administrator was empowered to "assume and discharge on behalf of the Government of the United States of America, the functions of civil government in such areas where civil control has broken down."

133. "Emergency Action Document: Executive Order, Assigning Major Areas of Responsibilities for Defense," OCDM Organization in the Event of War, accession 396-65C0204, Military Assistance to Civilian Authority folder, box 15, RG 396, Records of the Office of Emergency Preparedness.

134. This provision was an exercise of the Title III authorities of the Civil Defense Act, which were largely unelaborated in the Act itself. "Emergency Action Document: Proclaiming Assistance to Civil Authority by the Armed Forces by the President of the United States of America, a Proclamation," OCDM Organization in the Event of War, accession 396-65C0204, Military Assistance to Civilian Authority folder, box 15, RG 396, Records of the Office of Emergency Preparedness.

135. Lt. Gen. Willard S. Paul to USAF Brig. Gen. Harold Q. Huglin, "Military Assistance to Civil Authority in a D-Minus Situation," June 1, 1956, OCDM Organization in the Event of War, accession 396-65C0204, Military Assistance to Civilian Authority folder, box 15, RG 396, Records of the Office of Emergency Preparedness.

136. "Procedure for Agencies to Follow in Submitting Emergency Action Documents to the President," April 27, 1956, Presidential Emergency Action Documents, Brennan Center for Justice, accessed May 10, 2021, https://www.brennancenter.org/sites/default/files/2020-05/1%29%20 CIA-RDP80B01676R001100090094-0%20%281%29.pdf.

137. Executive Office of the President, Office of Defense Mobilization, "Mobilization Plans Group: Wartime Organization for Resource Mobilization," October 1, 1956, NWDD 978112, entry 63a-260, Wartime Organization folder, box 209, RG 396, Records of the Office of Emergency Preparedness.

138. Executive Office of the President, Office of Defense Mobilization, "(Proposed Emergency Document) Executive Order: Providing for the Mobilization of the Nation's Resources," NWDD 978112, entry 63a-260, Wartime Organization folder, box 209, RG 396, Records of the Office of Emergency Preparedness.

139. Executive Office of the President, Office of Defense Mobilization, "Mobilization Plan D-Minus," May 1, 1957, vii, 72, U.S. Declassified Documents Online, Gale.

140. Office of Defense Mobilization, *Mobilization Plan D-Minus*, 30, 32.

141. Ibid., 30, 72.

142. Ibid., 146, 107, 118.

Chapter 6. Survival Resources

1. DeCoursey, "Foreword," vii.

2. Horton, "National Aspects."

3. Ibid., 27–28.

4. Ibid., 29–30.

5. Ibid., 31, 29.

6. Miller, "Activities of the Health Resources Advisory Committee." NSRB used the term "health resources" to refer to health manpower, health facilities, and medical supplies and equipment.

7. Kiefer, "Role of Health Services," 1489.

8. Aufranc, "Mobilization of Health Resources."

9. *Hearing on Civil Defense against Atomic Attack, Before the Joint Committee on Atomic Energy, Part II*, 81st Cong., 2nd sess. (March 30, 1950) (statement of Norvin C. Kiefer, director, Health Resources Division, Office of Civilian Mobilization, National Security Resources Board), 24.

10. Other proposals to address this vulnerability included the use of rapid amortization certificates and procurement contracts to encourage manufacturers to maintain a production base for health supplies. For instance, one study suggested that antibiotics production would be a problem in the event of an emergency. To address this gap, NSRB instituted a program to help the industry obtain accelerated amortization for expanding its production facilities. See Yoshpe, *National Security Resources Board*, 136–37.

11. Federal Civil Defense Administration, *Health Services*, 96–97.

12. At the same time, since it would take several hours for these federal supplies to arrive in a bombed city after an attack, "each critical target area would need to have on hand, well scattered throughout the community, health supplies sufficient to last through the early hours of civil defense operations and casualty care." Ibid., 99, 97.

13. Ibid., 109.

14. *Civil Defense for National Survival, Hearings before a Subcommittee of the Committee on Government Operations, Part IV, House of Representatives*, 84th Cong., 2nd sess. (1956) (testimony of John Whitney). The 1951 FCDA *Annual Report* noted that the agency had prepared specifications for two hundred individual medical items, which were used for both state and federal stockpiling

efforts. By June 1952, $90 million in federal funds had been procured for stockpiling medical supplies and equipment. Federal Civil Defense Administration, *Annual Report*, 55.

15. *Civil Defense for National Survival* (testimony of John Whitney), 1434.

16. FCDA hoped to eventually maintain roughly one hundred warehouses around the country stocked with medical supplies.

17. Whitney, "Federal Civil Defense Administration," 261.

18. *Civil Defense Program, Hearings before the Subcommittee of the Senate Armed Services Committee, House of Representatives*, 84th Cong., 1st sess. (April 1955) (Hoover Commission report to Congress), 575.

19. Ibid., 574. Here the commission suggested that perhaps the military, as "the only existing force which has the experience and resources to cope with" (575) postattack problems, should assume authority for directing emergency medical care.

20. Fallout from the Castle Bravo blast spread over seven thousand square miles in the Pacific Ocean and fell on Pacific Islanders as well as members of the crew of a Japanese fishing vessel, causing acute radiation sickness among those exposed and inciting an international incident.

21. *A Report by the United States Atomic Energy Commission on the Effects of High-Yield Nuclear Weapons, Hearings before the Subcommittee on Civil Defense of the Committee on Armed Services, United States Senate*, Part I, Appendix I, 84th Cong., 1st sess. (February 15, 1955) (statement of Lewis L. Strauss), 234, 233.

22. "Exhibit 5: Report on Operation Alert, 1955," in *Civil Defense for National Survival*, 1490.

23. Ibid., 1493, 1492.

24. Ibid., 1490, 1430, 1431–32.

25. Ibid., 1435.

26. Ibid., 1435–36.

27. *Civil Defense for National Survival: Twenty-Fourth Intermediate Report of the Committee on Government Operations, House of Representatives* (Washington, DC: Government Printing Office, July 27, 1956), 45.

28. Technical Subcommittee of the Interagency Committee on Dispersal Policy, "Interim Recommendations for Revisions of the National Dispersion Policy," March 31, 1955, 2, declassified NWDD 978112, entry 63A-260, Black Binder: Mr. H. F. Hurley—Dispersion, box 204, RG 396, Records of the Office of Emergency Preparedness.

29. *Civil Defense for National Survival, Hearings before a Subcommittee of the Committee on Government Operations, Part III, House of Representatives*, 84th Cong., 2nd sess. (1956), 705.

30. Project East River Review Committee, *1955 Review*, 3, 5.

31. Ibid., 25.

32. Ibid., 25, 30.

33. Mobilization Program Advisory Committee Meeting No. 19, Discussion on the Scope of War Planning, October 14, 1955, Records of Mobilization Program Advisory Committee 1954–1958, folder 3, box 17, Public Policy Papers, J. Douglas Brown Papers, Department of Rare Books and Special Collections, Princeton University Library, Princeton, NJ.

34. Office of Defense Mobilization, staff comments on Mobilization Program Advisory Committee's "Tentative Outline of the Scope of Planning for Full War," Mobilization Program Advisory Committee Report No. 22, November 2, 1955, 4–5, folder 3, box 17, Public Policy Papers, J. Douglas Brown Papers, Department of Rare Books and Special Collections, Princeton University Library, Princeton, NJ.

35. Livermore, "Economic Mobilization Planning," 10–11.

36. Ibid., 21–22.

37. Mobilization Program Advisory Committee, Meeting No. 19, Discussion on the Scope of War Planning, October 14, 1955, Records of Mobilization Program Advisory Committee

1954–1958, folder 3, box 17, Public Policy Papers, J. Douglas Brown Papers, Department of Rare Books and Special Collections, Princeton University Library, Princeton, NJ.

38. Office of Defense Mobilization, staff comments on Mobilization Program Advisory Committee's "Tentative Outline."

39. Mobilization Program Advisory Committee Meeting No. 19, October 14, 1955.

40. *Civil Defense for National Survival* (testimony of Arthur Flemming), 1041.

41. Harris "Lessons Learned," 7.

42. Meeting with representatives of the agencies, on the summary evaluation of Operation Alert, August 2, 1956, NWDD 978112, entry 63A-260, Activation of Relocation Sites Under Plan "C" Situation folder, box 209, RG 396, Records of the Office of Emergency Preparedness.

43. US Congress Joint Committee on Defense Production, *Seventh Annual Report*, 7.

44. Cited in Joint Committee on Defense Production, S. Rep. No. 85-1172 (1958), 7–8. In early 1958, ODM reported to the Joint Committee on Defense Production that the essential survival item project "is being carried out through the Interagency Committee on Essential Survival Items, which was established by ODM early in 1956. This Committee, operating through a series of task forces, has completed a determination of the items which would be essential to survival in a postattack situation." Ibid., 132.

45. Office of Civil and Defense Mobilization, "Essential Survival Items."

46. Security Resources Panel of the Science Advisory Committee, *Deterrence and Survival*, 54.

47. *Civil Defense for National Survival* (testimony of Arthur Flemming), 1041.

48. Horton, "Computing Hazards of Nuclear Attack," 34.

49. Flemming, "Impact of Atomic War," 3, 4.

50. Lawrence, "Role of the Office of Defense Mobilization," 9.

51. Executive Office of the President, Office of Defense Mobilization, "MPAC Paper no. 26: Bomb Damage Assessment Program. Background Material for MPAC Meeting of November 10, 1955," November 1, 1955, folder 3, box 17, Public Policy Papers, J. Douglas Brown Papers, Department of Rare Books and Special Collections, Princeton University Library, Princeton, NJ. When they began work on a bomb damage assessment procedure in the early 1950s, mobilization planners imagined that it would be used in the immediate aftermath of an attack. At the time, according to Shaw Livermore, the planners had "only vague notions of what resources management would mean after a wide and devastating attack." Given the anticipated difficulty of assessing damage through field surveys, having a system in place that could model damage based on known detonation points would provide "burden-laden managers, in some never-never time in the future, a partial set of tools for their unwelcome task" of allocating remaining resources. Livermore, "Resources Management," 276.

52. Office of Defense Mobilization, Executive Office of the President, "Position Description: Chief, Damage Assessment Division," 1956, ODM-OCDM Records, RG 304, Records of the Office of Civil and Defense Mobilization.

53. Flemming, "Impact of Atomic War," 21.

54. Ramsey, "Damage Assessment Systems," 216. As part of its damage assessment work, FCDA contracted with the Stanford Research Institute to develop automated methods for simulating attacks. In 1956, SRI delivered an operational system for modeling nuclear attack damage to the population and other civilian resources. SRI's modeling was later incorporated into ODM's modeling program. Stanford Research Institute, *Damage Assessment System*.

55. Specifically, they borrowed time on UNIVAC computers located in Air Force Headquarters; the David Taylor Model Basin (naval research); and the Army Map Service. These were among the handful of digital computers then in existence. See Air Force Comptroller, "Bomb Damage Assessment," December 28, 1954, call no. K131.04–27, reel 32063, Air Force Historical Research Agency, Air University, Maxwell Air Force Base, AL.

56. Horton, "National Damage Assessment Program," 491; Krugler, *This Is Only a Test*.

57. Coker, "Role of NREC," 6.

58. Air Force Comptroller, "Bomb Damage Assessment," December 28, 1954, call no. K131.04–27, reel 32063, Air Force Historical Research Agency, Air University, Maxwell Air Force Base, AL.

59. Horton, "National Aspects," 28.

60. Ibid., 28; Ramsey, "Damage Assessment Systems," 216.

61. Horton, "National Damage Assessment Program," 791.

62. Horton, "Computing Hazards of Nuclear Attack," 34.

63. Air Force Comptroller, "Bomb Damage Assessment," December 28, 1954, call no. K131.04–27, reel 32063, Air Force Historical Research Agency, Air University, Maxwell Air Force Base, AL; *Civil Defense for National Survival* (testimony of Arthur Flemming), 1040.

64. US Congress Joint Committee on Defense Production, *Sixth Annual Report*, 69.

65. E. George, "Management of a Wartime Economy," 13.

66. Truppner, "Industrial Readiness Planning," 3.

67. US Congress Joint Committee on Defense Production, *Defense Production Act Progress Report No. 40*, 11.

68. US Congress Joint Committee on Defense Production, *Defense Production Act Progress Report No. 43*, 40–41.

69. Horton, "National Aspects," 28–29.

70. *Civil Defense for National Survival* (testimony of Arthur Flemming), 1040.

71. Horton, "National Damage Assessment Program," 790.

72. Coker, "Nuclear Attack Hazards," 8.

73. Horton, National Aspects," 28.

74. Ibid.

75. Horton, "National Damage Assessment Program," 791.

76. Horton, "National Aspects," 27; Directorate of Management Analysis, DCR/Comptroller Headquarters, US Air Force, *Bomb Damage Problem*, 12.

77. Horton, "National Damage Assessment Program," 791–92.

78. A report on the damage assessment system identified this limitation as "a serious contributing factor to the weaknesses that currently exist in the damage assessment system." Ramsey, "Damage Assessment Systems," 213–14.

79. Mobilization Program Advisory Committee, "Bomb Damage Assessment Program." The paper added that, despite this limitation, the new method did yield "very valuable information concerning the over-all magnitude of the fallout problem."

80. Horton, "Computing Hazards of Nuclear Attack," 34.

81. Coker, "Nuclear Attack Hazards," 14.

82. Ibid., 9.

83. Coker, "Role of NREC," 13; Pettee, "RISK II Model," 43.

84. Horton, "National Damage Assessment Program," 792.

85. Horton, "National Aspects," 28; Coker, "Role of NREC," 12–13.

86. Coker, "Role of NREC," 13.

87. Pettee, "RISK II Model," 43.

88. Coker, "Nuclear Attack Hazards," 14.

89. Pettee, "RISK II Model," 43.

90. Coker, "Role of NREC," 13 (emphasis in the original).

91. Horton, "National Aspects," 28.

92. Coker, "Nuclear Attack Hazards," 7. As Coker explained, mobilization officials had gathered a committee of "outstanding radiologists" to provide them with a definition of "effective biological dose" (5) that could be used for nuclear preparedness planning.

93. Horton, "Computing Hazards of Nuclear Attack," 34.

94. Coker, "Role of NREC," 13.

95. Ibid., 10–11.

96. Horton, "Computing Hazards of Nuclear Attack," 34.

97. Office of Defense Mobilization, *Mobilization Plan D-Minus*, 42–43.

98. Office of Defense Mobilization, *Mobilization Plan D-Minus*, viii.

99. US Congress Joint Committee on Defense Production, *Defense Production Act Progress Report No. 40*, 8.

100. Harris, "Lessons Learned," 10.

101. US Congress Joint Committee on Defense Production, *Seventh Annual Report*, 8, 7.

102. See, for instance, Kaplan, *Wizards of Armageddon*; V. L. Adams, *Eisenhower's Fine Group of Fellows*; and Snead, *Gaither Committee*.

103. Ramsey, "Damage Assessment Systems."

104. See V. L. Adams, *Eisenhower's Fine Group of Fellows*, 159. Adams notes that the NSC planning board recommended what became the Gaither study. She also notes that Nitze came on late and—from the perspective of the NSC's Robert Cutler—hijacked the panel, pushing it to look at broader national security strategy (rather than just civil defense).

105. The legislation also proposed a new Department of Civil Defense. See ibid., 156.

106. Security Resources Panel of the Science Advisory Committee, *Deterrence and Survival*, 12.

107. The passive defense group's findings were issued as a lengthy compilation of staff papers in a separate volume.

108. The year after the Gaither report was released, Kahn submitted a proposal on behalf of RAND for a major research project on the costs and benefits of nonmilitary defense measures. Kahn's proposal acknowledged the contributions of several ODM officials, including Burke Horton, Joseph Coker, Harold Huglin and Vincent Rock. RAND Corporation, *Report on a Study of Non-Military Defense*.

109. Strope, "Autobiography of a Nerd," 145. Strope's discussion of casualty estimates drew on the work that FCDA and the Stanford Research Institute were doing as part of the coordinated bomb damage assessment program.

110. Security Resources Panel of the Science Advisory Committee, *Deterrence and Survival*, 18.

111. Ibid., 19, 8–9, 10.

112. Ibid., 51, 59.

113. Ibid., 60, 61.

114. Ibid., 21, 6, 64, 23, 64–65.

115. The study estimated that such a program would cost from $5 to $10 billion.

116. Funigiello, "Managing Armageddon."

117. US House of Representatives, Subcommittee on Government Operations, *Civil Defense for National Survival*, 2.

118. Ibid., v.

119. Reorganization Plan No. 1, 23 Fed. Reg. 4991 (July 1, 1958).

120. *Civil Defense, Hearings Before a Subcommittee of the Committee on Government Operations, House of Representatives*, 85th Cong., 2nd sess. (July 10, 1958) (testimony of William Finan), 326.

121. US Congress Joint Committee on Defense Production, *Defense Production Act Progress Report No. 42* (testimony of Gordon Gray, director of Office of Defense Mobilization), 70.

122. McKinsey and Company, "Report on Nonmilitary Defense Organization," 420, 450–51.

123. US Congress Joint Committee on Defense Production. *Seventh Annual Report*, 161.

124. Harris, "Lessons Learned," 12.

125. US Congress Joint Committee on Defense Production, *Defense Production Act Progress Report No. 42* (testimony of Gordon Gray), 67.

126. US Congress Joint Committee on Defense Production, *Seventh Annual Report*, 8.

127. US Congress Joint Committee on Defense Production, *Defense Production Act Progress Report No. 42*, 38.

128. Ibid., 39.

129. US Congress Joint Committee on Defense Production, *Seventh Annual Report*, 176.

130. Truppner, "Industrial Readiness Planning," 17.

131. US Congress Joint Committee on Defense Production, *Defense Production Act Progress Report No. 42*, 68. In 1958, the office reported that it was currently engaged in vertical analyses for "approximately 20 individual items, almost all of which are survival items."

132. US Congress Joint Committee on Defense Production, *Seventh Annual Report*, 176.

133. US Congress Joint Committee on Defense Production, *Defense Production Act Progress Report No. 42*, 38.

134. Ibid., 67.

135. Ibid., 38.

136. Truppner, "Industrial Readiness Planning," 15.

137. Lawrence, "Balancing National Requirements," 14.

138. As Gaither's passive defense group had put it, to assess proposals to ensure "adequate facilities for national survival," it was first necessary to establish "requirements for survival items and for rehabilitation." Security Resources Panel of the Science Advisory Committee, *Deterrence and Survival*, 69.

139. Lawrence, "Balancing National Requirements," 14.

140. Ibid.

141. Coker, "Economics of Strategic Target Selection." In 1953, Carl Kaysen ("Vulnerability of the United States") cited Coker's dissertation as one of two unclassified accounts of the economics of target selection that had been developed during the war. The other was Ansley's Coale's *The Problem of Reducing Vulnerability to Atomic Bombs*.

142. Özgöde, "Logistics of Survival."

143. Oliver, "Survival Model," 13.

144. Ibid., 15.

145. Coker, "Application of Statistics," 5.

146. Oliver, "Survival Model," 19.

Epilogue: From Nuclear War to Climate Change

1. US Congress Joint Committee on Defense Production, *Defense Production Act Progress Report No. 42*, 38.

2. Lawrence, "Balancing National Requirements," 16.

3. Whitney, "Federal Civil Defense Administration," 261.

4. Price, "Use of the Packaged Disaster Hospital," 664.

5. Harold M. Schmieck Jr., "U.S. to Dispose of Huge Medical Supplies," *New York Times*, February 19, 1973, 21.

6. Foucault, *Security, Territory, Population*, 31.

7. Hacking, *Taming of Chance*.

8. The mobilization planner Shaw Livermore ("Resources Management," 279) referred to OCDM as a "dim and unregretted memory."

9. Exec. Order No. 11,725, "Transfer of Certain Functions of the Office of Emergency Preparedness," June 27, 1973.

10. S. Knowles, *Disaster Experts*, 271.

11. Haddow, Bullock, and Coppola, *Introduction to Emergency Management*.

12. Collier and Lakoff, "Vulnerability of Vital Systems"; S. Knowles, *Disaster Experts*.

13. On the evolution from mobilization planning and civil defense to contemporary practices of emergency management in these domains, see Collier, "Enacting Catastrophe" and Gray, "Hazardous Simulations" (on catastrophe modeling); Lakoff, "Preparing for the Next Emergency" (on

exercises); Folkers, "Business Continuity Management" (on continuity of operations planning); and Collier and Lakoff, "Vulnerability of Vital Systems" (on critical infrastructure protection).

14. A traditional approach sought to minimize the overall number of infections, rather than reducing them to a level that could be managed by the health care system.

15. For example, as Covid-19 cases surged in California in early April, Governor Gavin Newsom detailed the precise number of available intensive care unit beds around the state. "Why do I start with the number 774?" he asked. "Because that's the number that I wake up to that I'm most focused on in the state of California." Kerry Crowley, "Governor Newsom: COVID-19 Cases Have Quadrupled, Hospitalizations Have Tripled in Last Six Days," *San Jose Mercury News*, April 1, 2020.

16. As it turned out, the need for ventilators was much lower than initially anticipated.

17. Cybersecurity and Infrastructure Security Agency, "Guidance on the Essential Critical Infrastructure Workforce: Ensuring Community and National Resilience in COVID-19 response," March 19, 2020, https://www.cisa.gov/sites/default/files/publications/CISA-Guidance-on-Essential-Critical-Infrastructure-Workers-1-20-508c.pdf.

18. Bill Gates, "COVID-19 Is Awful: Climate Change Could Be Worse," *GatesNotes*, August 4, 2020, https://www.gatesnotes.com/Energy/Climate-and-COVID-19; See also Ilana Cohen, "Covid-19 and Climate Change Will Remain Inextricably Linked, Thanks to the Parallels (and the Denial)," *Inside Climate News*, January 1, 2021; Adam Tooze, "We Are Living through the First Economic Crisis of the Anthropocene," *Guardian* (London), May 7, 2020.

19. Dickon Pinner, Matt Rogers, and Hamid Samandari, "Addressing Climate Change in a Post-Pandemic World," McKinsey and Company, April 7, 2020, https://www.mckinsey.com/business-functions/sustainability/our-insights/addressing-climate-change-in-a-post-pandemic-world.

20. Adam Frank, "Coronavirus and Climate Change: The Pandemic Is a Fire Drill for Our Planet's Future," NBC News, March 27, 2020, https://www.nbcnews.com/think/opinion/coronavirus-climate-change-pandemic-fire-drill-our-planet-s-future-ncna1169991.

21. Scheppele, "Small Emergencies," 861–62. Scheppele points to how emergency provisions have suffused throughout normal politics, a development she finds troubling. Our analysis views these institutions from a somewhat different perspective—the middle of the twentieth century, when many observers doubted that democratic governments could manage recurrent crises of economic depression and war.

22. Quoted in "It's Official: Democratic Party Calls for WWII-Scale National Mobilization to Combat 'Global Climate Emergency,'" Climate Mobilization, July 27, 2016, https://www.theclimatemobilization.org/blog/2016/07/27/2016-7-27-dnc-party-platform-mobilization/.

23. "Climate Mobilization: A Whole-Society Transformation," Climate Mobilization, accessed April 19, 2021, https://www.theclimatemobilization.org/climate-mobilization/.

24. Ripple, "World Scientists' Warning"; "Climate Emergency Declaration in 1,926 Jurisdictions and Local Governments Cover 826 Million Citizens," Climate Emergency Declaration, May 14, 2021, https://climateemergencydeclaration.org/climate-emergency-declarations-cover-15-million-citizens/.

25. Margaret Klein Salamon, "Leading the Public into Emergency Mode: Introducing the Climate Emergency Movement," *Noteworthy: The Journal Blog*, May 24, 2019, https://blog.usejournal.com/leading-the-public-into-emergency-mode-b96740475b8f.

26. Nils Gilman, "The Coming Avocado Politics," *Breakthrough Journal*, February 7, 2020, https://thebreakthrough.org/journal/no-12-winter-2020/avocado-politics.

27. Beck, "Terrorist Threat."

28. Rossiter, *Constitutional Dictatorship*, 1, 3.

BIBLIOGRAPHY

Archival Sources

Air Force Historical Research Agency. Air University, Maxwell Air Force Base, AL.

American Presidency Project. University of California, Santa Barbara.

Atomic Scientists of Chicago Records 1943–1955. Special Collections Research Center. University of Chicago Library.

Economic Stabilization Agency. General Records. National Archives, College Park, MD.

Harry S. Truman Library, Independence, MO.

Historical Manuscripts Collection. Office of Chief of Military History Manuscripts. US Army Center of Military History, Washington, DC.

Innis Harris Papers. National Archives Building, Washington, DC.

J. Douglas Brown Papers. Department of Rare Books and Special Collections. Princeton University Library, Princeton, NJ.

Luther Halsey Gulick III Papers. Baruch College Archives, NY.

National Defense University Archives. National Defense University Library, Washington, DC.

National Security Council. Records. National Archives, College Park, MD.

National Security Resources Board. Records. National Archives, College Park, MD.

Office of Civil and Defense Mobilization. Records. National Archives, College Park, MD.

Office of Emergency Preparedness. Records. National Archives, College Park, MD.

Robert H. Kupperman Papers. Hoover Institution Library and Archives, Stanford, CA.

Rockefeller Foundation Records, Projects. Rockefeller Archive Center, Sleepy Hollow, NY.

Security Studies Program. Massachusetts Institute of Technology, Cambridge, MA.

U.S. Classified Documents Online. Gale Primary Sources.

Published Sources

Adams, Thomas. Preface to *Major Economic Factors in Metropolitan Growth and Arrangement: Regional Survey of New York and its Environs*. Vol. 1. New York: Regional Plan of New York and Its Environs, 1927.

Adams, Valerie L. *Eisenhower's Fine Group of Fellows: Crafting a National Security Policy to Uphold the Great Equation*. Landham, MD: Lexington Books, 2006.

Adey, Peter, Ben Anderson, and Stephen Graham. "Governing Emergencies: Beyond Exceptionality." *Theory, Culture and Society* 32, no. 2 (2015): 3–17.

Adey, Peter, and Ben Anderson. "Anticipating Emergencies: Technologies of Preparedness and the Matter of Security." *Security Dialogue* 43, no. 2 (2012): 99–117.

Agamben, Giorgio. *State of Exception*. Chicago: University of Chicago Press, 2005.

Alford, Ryan. *Permanent State of Emergency: Unchecked Executive Power and the Demise of the Rule of Law*. Kingston, ON: McGill-Queen's University Press, 2017.

Amadae, S. M. *Rationalizing Capitalist Democracy: The Cold War Origins of Rational Choice Liberalism*. Chicago: University of Chicago Press, 2003.

Amoore, Louise. *The Politics of Possibility: Risk and Security beyond Probability*. Durham, NC: Duke University Press, 2013.

Anderson, Ben. "What Kind of Thing Is Resilience?" *Politics* 35, no. 1 (2015): 60–66.

Anderson, Ben, and Peter Adey. "Affect and Security: Exercising Emergency in 'UK Civil Contingencies.'" *Environment and Planning D: Society and Space* 29, no. 6 (2011): 1092–1109.

Anderson, Perry. *Lineages of the Absolutist State*. London: Verso, 1979.

Aradau, Claudia, and Rens van Munster. "Exceptionalism and the 'War on Terror': Criminology Meets International Relations." *British Journal of Criminology* 49, no. 5 (2009): 686–701.

Aradau, Claudia, and Rens van Munster. *Politics of Catastrophe: Genealogies of the Unknown*. New York: Routledge, 2011.

Atomic Scientists of Chicago. *The Atomic Bomb: Facts and Implications*. Chicago: Atomic Scientists of Chicago, 1946.

Aufranc, W. H. "Mobilization of Health Resources for Defense." *Radiology* 56, no. 5 (1951): 641–44.

Baker, Tom, and Jonathan Simon. *Embracing Risk: The Changing Culture of Insurance and Responsibility*. Chicago: University of Chicago Press, 2010.

Baldwin, Hanson W. "Oil Strategy in World War II." *American Petroleum Institute Quarterly*, Centennial Issue, reprinted (1959; August 2007): 10–11.

Baldwin, Hanson W. *The Price of Power*. New York: Harper and Brothers, 1948.

Barnett, Clive. "On the Milieu of Security: Situating the Emergence of New Spaces of Public Action." *Dialogues in Human Geography* 5, no. 3 (2015): 257–70.

Barry, Andrew. *Political Machines: Governing a Technological Society*. New York: Athlone Press, 2001.

Barry, John M. *Rising Tide: The Great Mississippi Flood of 1927 and How It Changed America*. New York: Simon and Schuster, 1997.

Beck, Ulrich. *Risk Society: Towards A New Modernity*. London: Sage, 1992.

Beck, Ulrich. "The Terrorist Threat: World Risk Society Revisited." *Theory, Culture and Society* 19, no. 4 (2002): 39–55.

Beck, Ulrich. *World at Risk*. Malden, MA: Polity Press, 2013.

Bernstein, Michael A. *A Perilous Progress: Economists and Public Purpose in Twentieth-Century America*. Princeton, NJ: Princeton University Press, 2001.

Biddle, Tami D. *Rhetoric and Reality in Air Warfare: The Evolution of British and American Ideas about Strategic Bombing, 1914–1945*. Princeton, NJ: Princeton University Press, 2002.

Black, Russell V. N. *Criteria and Planning for Public Works*. Washington, DC: National Planning Board, 1934.

Bowie, Robert Richardson, and Richard H. Immerman. *Waging Peace: How Eisenhower Shaped an Enduring Cold War Strategy*. Oxford: Oxford University Press, 1998.

Boyle, Philip J., and Shannon T. Speed. "From Protection to Coordinated Preparedness: A Genealogy of Critical Infrastructure in Canada." *Security Dialogue* 49, no. 3 (2018): 1–15.

Breslau, Daniel. "Economics Invents the Economy: Mathematics, Statistics, and Models in the Work of Irving Fisher and Wesley Mitchell." *Theory and Society* 32, no. 3 (2003): 379–411.

Brigante, John E. *The Feasibility Dispute: Determination of War Production Objective for 1942 and 1943*. Washington, DC: Committee on Public Administration, 1950.

Brinkley, Alan. *The End of Reform: New Deal Liberalism in Recession and War*. New York: Knopf, 1995.

Brinkley, Alan. "The National Resources Planning Board and the Reconstruction of Planning." In *The American Planning Tradition: Culture and Policy*, edited by Robert Fishman, 173–92. Washington, DC: Woodrow Wilson Center Press, 2000.

Brinkley, Alan. "The New Deal and the Idea of the State." In *The Rise and Fall of the New Deal Order, 1930–1980*, edited by Steve Fraser and Gary Gerstle, 85–121. Princeton, NJ: Princeton University Press, 1989.

Bromage, Arthur W. "Public Administration in the Atomic Age." *American Political Science Review* 41, no. 5 (1947): 947–54.

Cabantous, Laure, and Théodora Dupont-Courtade. "What Is a Catastrophe Model Worth?" In *Making Things Valuable*, edited by Martin Kornberger, Jan Mouritsen, Lise Justesen, and Anders Koed Madsen, 167–86. Oxford: Oxford University Press, 2015.

Calhoun, Craig. "A World of Emergencies: Fear, Intervention, and the Limits of Cosmopolitan Order." *Canadian Review of Sociology* 41, no. 4 (November 2004): 373–95.

Callon, Michel. "Civilizing Markets: Carbon Trading between in Vitro and in Vivo Experiments." *Accounting, Organizations and Society* 34, nos. 3–4 (2009): 535–48.

Callon, Michel, Pierre Lascoumes, and Yannick Barthe. *Acting in an Uncertain World: An Essay on Technical Democracy*. Cambridge, MA: MIT Press, 2011.

Carse, Ashley. "Keyword: Infrastructure—How a Humble French Engineering Term Shaped the Modern World." In *Infrastructures and Social Complexity: A Companion*, edited by Penelope Harvey, Casper Bruun Jensen, and Atsuro Morita. New York: Routledge, 2017.

Carson, Carol S. "The History of the United States National Income and Product Accounts: The Development of an Analytical Tool." *Review of Income and Wealth* 21 (1975): 153–81.

Charles, Daniel. "Do Maps Have Morals?" *MIT Technology Review*, June 1, 2005.

Chicago Civil Defense Committee. *Chicago Alerts: A City Plans Its Civil Defense against Atomic Attack*. Chicago: Chicago Civil Defense Corps, 1950.

Civilian Production Administration. *Industrial Mobilization for War: History of the War Production Board and Predecessor Agencies, 1940–1945*. Washington, DC: Government Printing Office, 1947.

Clark, Jane Perry. "Emergencies and the Law." Political Science Quarterly 49, no. 2 (1934): 268–83.

Clark, John Maurice. *Economics of Planning Public Works*. Washington, DC: Government Printing Office, 1935.

Clark, John Maurice. *Strategic Factors in Business Cycles, with an introduction by the Committee on Recent Economic Changes*. New York: National Bureau of Economic Research, 1935.

Clarke, Lee. *Worst Cases: Terror and Catastrophe in Popular Imagination*. Chicago: University of Chicago Press, 2006.

Clinard, Outten J. "Developments in Air Targeting: Data Handling Techniques." *Studies in Intelligence* 3, no. 2 (1959): 95–104.

Coale, Ansley J. "The Problem of Reducing Vulnerability to Atomic Bombs." *American Economic Review* 37, no. 2 (1947): 87–97.

Coale, Ansley J. *The Problem of Reducing Vulnerability to Atomic Bombs*. Princeton, NJ: Princeton University Press, 1947.

Coker, Joseph D. "The Application of Statistics to the Resource Management Program." Proceedings of the Social Statistics Section, American Statistical Association Annual Meeting, December 27–30, 1964, Chicago, IL. http://www.asasrms.org/Proceedings/y1964/y1964.html.

Coker, Joseph D. "The Economics of Strategic Target Selection." PhD diss., George Washington University, 1949.

Coker, Joseph D. "Nuclear Attack Hazards and Damage Assessment." Paper presented for the Scientific Working Party of the Civil Defense Committee, North Atlantic Treaty Organization, May 31–June 1, 1960.

Coker, Joseph D. "The Role of NREC in Emergency Preparedness." Paper Presented to the Bi-Regional Meeting of Manpower Mobilization Coordinators, September 9, 1965, New Orleans, LA.

Coker, Joseph D. "Strategic Vulnerability of the USSR." Lecture to the Air War College, Maxwell Air Force Base, Montgomery, AL, March 8, 1951.

Collier, Stephen J. "Enacting Catastrophe: Preparedness, Insurance, Budgetary Rationalization." *Economy and Society* 37, no. 2 (2008): 224–50.

Collier, Stephen J. *Post-Soviet Social: Neoliberalism, Social Modernity, Biopolitics.* Princeton, NJ: Princeton University Press, 2011.

Collier, Stephen J. "Topologies of Power: Foucault's Analysis of Political Government beyond 'Governmentality.'" *Theory, Culture and Society* 26, no. 6 (2009): 78–108.

Collier, Stephen J., and Andrew Lakoff. "Distributed Preparedness: Notes on the Genealogy of Homeland Security." *Environment and Planning D: Society and Space* 26, no. 1 (2008): 7–28.

Collier, Stephen J., and Andrew Lakoff. "The Problem of Securing Health." In *Biosecurity Interventions: Global Health and Security in Question,* edited by Andrew Lakoff and Stephen J. Collier, 7–32. New York: Columbia University Press, 2008.

Collier, Stephen J., and Andrew Lakoff. "Vital Systems Security: Reflexive Biopolitics and the Government of Emergency." *Theory, Culture and Society* 32, no. 2 (2015): 19–51.

Collier, Stephen J., and Andrew Lakoff. "The Vulnerability of Vital Systems: How 'Critical Infrastructure' Became a Security Problem." In *Securing the Homeland: Critical Infrastructure, Risk, and (In)security,* edited by Myriam Dunn Cavelty and Søby Kristensen. New York: Routledge, 2008.

Collier, Stephen J., Andrew Lakoff, and Paul Rabinow. "Biosecurity: Towards an Anthropology of the Contemporary." *Anthropology Today* 20, no. 5 (2004): 3–7.

Comfort, Louise, Arjen Boin, and Chris C. Demchak, eds., *Designing Resilience: Preparing for Extreme Events.* Pittsburgh, PA: University of Pittsburgh Press, 2010.

Committee on Elimination of Waste in Industry. *Waste in Industry.* Washington, DC: Federated American Engineering Societies, 1921.

Committee on Regional Plan of New York and Its Environs. *Regional Plan of New York and Its Environs: The Graphic Regional Plan.* New York: Regional Plan of New York and Its Environs, 1929.

Conn, Stetson, Rose C. Engelman, and Byron Fairchild. *Guarding the United States and Its Outposts.* Vol. 12. Washington, DC: Office of the Chief of Military History, Department of the Army, 1964.

Cooling, B. Franklin. "US Army Support of Civil Defense: The Formative Years." *Military Affairs* 35 (1971): 7–11.

Cooper, Christopher. *Disaster: Hurricane Katrina and the Failure of Homeland Security.* New York: Holt Paperbacks, 2007.

Cuff, Robert. "Ferdinand Eberstadt, the National Security Resources Board, and the Search for Integrated Mobilization Planning, 1947–1948." *Public Historian* 7, no. 4 (1985): 37–52.

Cuff, Robert. "From the Controlled Materials Plan to the Defense Materials System, 1942–1953." *Journal of Military History* 51, no. 1 (1987): 1–6.

Cuff, Robert. "Organizational Capabilities and U.S. War Production: The Controlled Materials Plan of World War II." *Business and Economic History* 19 (1990): 103–12.

Culbertson, Charles N. *Air Intelligence and the Search for the Center of Gravity.* Air War College Research Report. Maxwell Air Force Base, AL: Air War College, Air University, 1988.

Curatola, John M. *Bigger Bombs for a Brighter Tomorrow: The Strategic Air Command and American War Plans at the Dawn of the Atomic Age, 1945–1950.* Jefferson, NC: McFarland, 2016.

Curley, Tyler M. "Emergency-War Machine: National Crisis, Democratic Governance, and the Historical Construction of the American State." PhD diss., University of Southern California, 2016.

Curley, Tyler M. "Models of Emergency Statebuilding in the United States." *Perspectives on Politics* 13, no. 3 (2015): 697–713.

Damms, Richard V. "James Killian, the Technological Capabilities Panel, and the Emergence of President Eisenhower's 'Scientific-Technological Elite.'" *Diplomatic History* 24, no. 1 (2000): 57–78.

Davis, Tracy C. *Stages of Emergency: Cold War Nuclear Civil Defense*. Durham, NC: Duke University Press, 2007.

DeCoursey, Elbert. "Foreword to the First Edition." In *Symposium on the Management of Mass Casualties*. Fort Sam Houston, TX: Army Medical Service, 1958.

De Goede Marieke. "Beyond Risk: Premediation and the Post-9/11 Security Imagination." *Security Dialogue* 39, nos. 2–3 (2008): 155–76.

Dembitz, Lewis N., Philip Sporn, Sam H. Schurr, and Jacob Marschak. "Discussion." *The American Economic Review* 37, no. 2 (May 1947): 109–17.

Dickinson, Matthew J. *Bitter Harvest: FDR, Presidential Power, and the Growth of the Presidential Branch*. Cambridge: Cambridge University Press, 1997.

Didier, Emmanuel. "Cunning Observation: US Agricultural Statistics in the Time of Laissez-Faire." *History of Political Economy* 44, suppl. 1 (December 2012): 27–45.

Directorate of Management Analysis, DCR/Comptroller Headquarters, US Air Force. *The Bomb Damage Problem*. Washington, DC: DCS/Comptroller Headquarters, US Air Force, 1954.

Donzelot, Jacques. *The Policing of Families*. New York: Random House, 1979.

Dorwart, Jeffery M. *Eberstadt and Forrestal: A National Security Partnership, 1909–1949*. College Station: Texas A&M University Press, 1991.

Douhet, Giulio. *Command of the Air*. 1921. Dehradu, India: Natraj, 2003.

Dreyfus, Hubert L., and Paul Rabinow. *Michel Foucault: Beyond Structuralism and Hermeneutics*. Chicago: University of Chicago Press, 1982.

Dudley, Michael Quinn. "Sprawl as Strategy: City Planners Face the Bomb." *Journal of Planning Education and Research* 21, no. 1 (2001): 52–67.

Duncan, Joseph W., and William C. Shelton. *Revolution in United States Government Statistics, 1926–1976*. Washington DC: Department of Commerce, 1978.

Dunn Cavelty, Myriam, and Jennifer A. Giroux. "The Good, the Bad, and the Sometimes Ugly: Complexity as Both Threat and Opportunity in the Vital Systems Security Discourse." In *World Politics at the Edge of Chaos: Reflections on Complexity and Global Life*, edited by Emilian Kavalski, 209–27. Albany: State University of New York Press, 2015.

Eberstadt, Ferdinand. *Unification of the War and Navy Departments and Postwar Organization for National Security*. Report to Hon. James Forrestal, Secretary of the Navy. Washington, DC: Government Printing Office, 1945.

Eden, Lynn. *Whole World on Fire: Organizations, Knowledge, and Nuclear Weapons Devastation*. Ithaca, NY: Cornell University Press, 2004.

Edwards, Paul N. *The Closed World: Computers and the Politics of Discourse in Cold War America*. Cambridge, MA: MIT Press, 1997.

Edwards, Paul N. "Infrastructure and Modernity: Force, Time, and Social Organization in the History of Sociotechnical Systems." In *Modernity and Technology*, edited by Thomas J. Misa, Philip Brey, and Andrew Feenberg. Cambridge: MIT Press, 2003.

Edwards, Paul N. "Knowledge Infrastructures for the Anthropocene." *Anthropocene Review* 4, no. 1 (2017): 34–43.

Edwards, Paul. *A Vast Machine: Computer Models, Climate Data, and the Politics of Global Warming*. Cambridge, MA: MIT Press, 2010.

Elliott, William Y. *Mobilization Planning and the National Security*. Washington, DC: Government Printing Office, 1950.

Emmerich, Herbert. "The Administrative Legacy of Franklin D. Roosevelt." *International Review of Administrative Sciences* 36, no. 2 (1970): 122–32.

Emmerich, Herbert. *Federal Organization and Administrative Management*. Tuscaloosa: University of Alabama Press, 1971.

Engelhart, George K. "Coordinating Wartime Internal-Security Programs." *American Petroleum Institute: Proceedings* 24 (1943): 54–59.

Ericson, Richard V., and Aaron Doyle. "Catastrophe Risk, Insurance and Terrorism." *Economy and Society* 33 (2004): 135–73.

Erickson, Paul, Judy L. Klein, Lorraine Daston, Rebecca Lemov, Thomas Sturm, and Michael D. Gordin. "The Bounded Rationality of Cold War Operations Research." In *How Reason Almost Lost Its Mind: The Strange Career of Cold War Rationality*, 51–80. Chicago: University of Chicago Press, 2013.

Erickson, Paul, Judy L. Klein, Lorraine Daston, Rebecca Lemov, Thomas Sturm, and Michael D. Gordin. *How Reason Almost Lost Its Mind: The Strange Career of Cold War Rationality*. Chicago: University of Chicago Press, 2013.

Escobar, Arturo. *Encountering Development: The Making and Unmaking of the Third World*. Princeton, NJ: Princeton University Press, 1995.

Evans, Brad, and Julian Reid. *Resilient Life: The Art of Living Dangerously*. Cambridge: Wiley, 2014.

Ewald, François. "Insurance and Risk." In *The Foucault Effect: Studies in Governmentality,* edited by Graham Burchell, Colin Gordon, and Peter Miller, 197–210. Chicago: University of Chicago Press.

Ewald, François. "The Return of Descartes's Malicious Demon: An Outline of a Philosophy of Precaution." In *Embracing Risk: The Changing Culture of Insurance and Responsibility*, edited by Tom Baker and Jonathan Simon, 273–301. Chicago: University of Chicago Press, 2010.

Executive Office of the President. *The National Plan for Emergency Preparedness*. Washington, DC: Office of Emergency Planning, 1964.

Eyal, Gil. *The Crisis of Expertise*. Cambridge: Polity Press, 2019.

Eyal, Gil, and Moran Levy. "Economic Indicators as Public Interventions." *History of Political Economy* 45, suppl. (2013): 220–53.

Fairman, Charles. "Government under Law in the Shadow of a Nuclear War." Lecture presented at the Industrial College of the Armed Forces, Washington, DC, April 13, 1956.

Fairman, Charles. "The Law of Martial Rule and National Security." *Harvard Law Review* 55, no. 8 (1942): 1253–1302.

Farish, Matthew. *The Contours of America's Cold War*. Minneapolis: University of Minnesota Press, 2010.

Farquhar, John T. *A Need to Know: The Role of Air Force Reconnaissance in War Planning, 1945–1953*. Maxwell Air Force Base, Montgomery, AL: Air University Press, 2004.

Fearnley, Lyle. "Epidemic Intelligence: Langmuir and the Birth of Disease Surveillance." *Behemoth—A Journal on Civilization* 3 (2010): 36–56.

Federal Civil Defense Administration. *Annual Report*. Washington, DC: Government Printing Office, 1951.

Federal Civil Defense Administration. *Civil Defense Urban Analysis: Technical Manual*. Washington, DC: Government Printing Office, 1953.

Federal Civil Defense Administration. *Health Services and Special Weapons Defense*. Washington, DC: Government Printing Office, 1950.

Federal Emergency Management Administration. *National Preparedness Goal*. Washington, DC: Department of Homeland Security, 2015.

Fedman, David, and Cary Karacas. "A Cartographic Fade to Black: Mapping the Destruction of Urban Japan during World War II." *Journal of Historical Geography* 38, no. 3 (2012): 306–28.

Fesler, James W. "The Brownlow Committee Fifty Years Later." *Public Administration Review* 47, no. 4 (July–August 1987): 291–96.

Fesler, James W. *Principles of Administration*, vol. 4 of *Emergency Management of the National Economy*. Washington, DC: Industrial College of the Armed Forces, 1954.

Fishman, Robert. "The Regional Plan and the Transformation of the Industrial Metropolis." In *The Landscape of Modernity: New York City, 1900–1940*, edited by David Ward and Oliver Zunz. Baltimore, MD: Johns Hopkins University Press, 1992.

Flemming, Arthur. Foreword to *Emergency Management of the National Economy*, by Industrial College of the Armed Forces. Washington, DC: Industrial College of the Armed Forces, 1954.

Flemming, Arthur. "Impact of Atomic War on Methods and Processes of Economic Mobilization." Lecture to the Industrial College of the Armed Forces, Washington, DC, May 1954.

Folkers, Andreas. "Business Continuity Management and the Security of Financial Operations." *Economy and Society* 46, no. 1 (2017): 103–27.

Folkers, Andreas. "Existential Provisions: The Technopolitics of Public Infrastructure." *Environment and Planning D: Society and Space* 35, no. 5 (2017): 855–74.

Foucault, Michel. *The Birth of Biopolitics: Lectures at the College de France, 1978–79*. Translated by Graham Burchell. New York: Palgrave Macmillan, 2008.

Foucault, Michel. "The Concern for Truth." In *Michel Foucault: Politics, Philosophy, Culture. Interviews and Other Writings*. Edited by L. D. Kritzman. New York: Routledge, 1988.

Foucault, Michel. "Nietzsche, Genealogy, History." In *The Foucault Reader*, edited by Paul Rabinow, 76–100. New York: Pantheon, 1984.

Foucault, Michel. "Polemics, Politics, and Problematizations." In *Ethics: Subjectivity and Truth*, vol. 1 of *Essential Works of Michel Foucault*, edited by Paul Rabinow, 111–19. New York: New Press, 1997.

Foucault, Michel. *Security, Territory, Population: Lectures at the Collège de France, 1977–78*. Edited by Michel Senellart. Translated by Graham Burchell. London: Palgrave MacMillan, 2007.

Foucault, Michel. *Society Must Be Defended: Lectures at the Collège de France, 1975–76*. New York: Picador, 2003.

Freeman, Douglas Southall. *George Washington: A Biography*. Vol. 4, *Leader of the Revolution*. New York: Scribner, 1951.

Friedmann, John, and Clyde Weaver. *Territory and Function: The Evolution of Regional Planning*. Berkeley: University of California Press, 1979.

Funigiello, Philip J. "Managing Armageddon: The Truman Administration, Atomic War, and the National Security Resources Board." *Journal of Policy History* 2, no. 4 (1990): 403–24.

Galbraith, J. K. *The Economic Effects of the Public Works Expenditures, 1933–1938*. Prepared for the National Resources Planning Board. Washington, DC: Government Printing Office, 1940.

Galison, Peter. "War against the Center." *Grey Room* 4 (2001): 5–33.

Garrison, Dee. *Bracing for Armageddon: Why Civil Defense Never Worked*. New York: Oxford University Press, 2006.

Gayer, Arthur D. *Public Works in Prosperity and Depression*. New York: National Bureau of Economic Research, 1935.

Gentile, Gian P. *How Effective Is Strategic Bombing? Lessons Learned from World War II and Kosovo*. New York: New York University Press, 2001.

George, Edwin. "Management of a Wartime Economy." Publication No. L56-130. Washington, DC: Industrial College of the Armed Forces, 1956.

George, Hal. "An Inquiry in to the Subject 'War.'" Lecture to the Air Force Course, 1935, in *Lectures of the Air Corps Tactical School and American Strategic Bombing in World War II*, edited by Phil Haun, 42–43. Lexington: University Press of Kentucky, 2019.

Gessert, Robert, Nehemiah Jordan, and John D. Tashjean. *Federal Civil Defense Organization: The Rationale of Its Development*. Alexandria, VA: Institute for Defense Analyses, 1965.

Ghamari-Tabrizi, Sharon. *The Worlds of Herman Kahn: The Intuitive Science of Thermonuclear War*. Cambridge, MA: Harvard University Press, 2005.

Gladwell, Malcolm. *The Bomber Mafia: A Dream, a Temptation, and the Longest Night of the Second World War*. New York: Little, Brown, 2021.

Gorrie, Jack. "Federal Dispersion Policy Stresses Local Initiative." *Bulletin of the Atomic Scientists* 7, no. 9 (1951): 270–71.

Graham, Otis L., Jr. *An Encore for Reform: The Old Progressive and the New Deal*. Oxford: Oxford University Press, 1967.

Graham, Stephen, ed. *Disrupted Cities: When Infrastructure Fails*. London: Routledge, 2010.

Graham, Stephen, and Simon Marvin. *Splintering Urbanism: Networked Infrastructures, Technological Mobilities, and the Urban Condition*. London: Routledge, 2001.

Gray, Ian. "Hazardous Simulations: Pricing Climate Risk in US Coastal Insurance." *Economy and Society* 50, no. 2 (2021): 196–223.

Gregory, Derek. "Bombing Encyclopedia of the World." *Geographic Imaginations: War Space and Security* (blog), August 3, 2012.

Grossi, Patricia, and Howard Kunreuther, ed. *Catastrophe Modeling: A New Approach to Managing Risk*. New York: Springer Science and Business Media, 2005.

Guglielmo, Mark. "The Contribution of Economists to Military Intelligence during World War II." *Journal of Economic History*, 68, no. 1 (March 2008): 113.

Gulick, Luther. "Conclusion." In "The Executive Office of the President: A Symposium," *Public Administration Review* 1, no. 2 (Winter 1941): 138–40.

Hacking, Ian. *Historical Ontology*. Cambridge, MA: Harvard University Press, 2004.

Hacking, Ian. "'Style' for Historians and Philosophers." *Studies in History and Philosophy of Science Part A* 23, no. 1 (1992): 1–20.

Hacking, Ian. *The Taming of Chance*. Cambridge: Cambridge University Press, 1990.

Haddow, George D., Jane A. Bullock, and Damon P. Coppola. *Introduction to Emergency Management*. 5th ed. New York: Butterworth-Heinemann, 2014.

Haig, Robert Murray, and Roswell C. McCrea. *Major Economic Factors in Metropolitan Growth and Arrangement*. Vol. 1 of *Regional Survey of New York and Its Environs*. New York: Regional Plan of New York and Its Environs, 1927.

Hansell, Haywood S. *The Air Plan that Defeated Hitler*. Atlanta: Higgins-McArthur/Longino & Porter, Inc., 1972.

Hansell, Haywood S. *The Strategic Air War against Germany and Japan: A Memoir*. Washington DC: Office of Air Force History, US Air Force, 1986.

Harris, Innis D. "Lessons Learned from Operations Alert 1955–1957." Lecture to the Industrial College of the Armed Forces, Washington, DC, April 30, 1958.

Harrison, Robert L. "Introduction to Monte Carlo Simulation." *American Institute of Physics Conference Proceedings* 1204, no. 1 (2010): 17–21.

Haun, Phil. *Lectures of the Air Corps Tactical School and American Strategic Bombing in World War II*. Lexington: University Press of Kentucky, 2019.

Hecht, Gabrielle. *The Radiance of France: Nuclear Power and National Identity after World War II*. Cambridge, MA: MIT Press, 2009.

Hecht, Gabrielle, and Paul N. Edwards. *The Technopolitics of Cold War: Toward a Transregional Perspective*. Washington, DC: American Historical Association, 2007.

Herken, Gregg. *Counsels of War*. New York: Knopf, 1985.

Herring, E. Pendleton. *The Impact of War: Our Democracy under Arms*. New York: Farrar and Rinehart, 1941.

Hill, David L., Eugene Rabinowitch, and John A. Simpson Jr. "The Atomic Scientists Speak Up." Reprinted in *Bulletin of the Atomic Scientists* 37, no. 1 (1981): 23–25.

Hobbs, Edward Henry. *Behind the President: A Study of Executive Office Agencies*. New York: Public Affairs Press, 1954.

Hogan, Michael J. *A Cross of Iron: Harry S. Truman and the Origins of the National Security State, 1945–1954*. Cambridge: Cambridge University Press, 2000.

Hoover, Herbert. 1921. *Report of the President's Commission on Unemployment*. Washington, DC: Government. Printing Office, 1921.

Hopley, Russell J. *Civil Defense for National Security*. Washington, DC: Government Printing Office, 1948.

Horton, H. Burke. "Computing Hazards of Nuclear Attack." *Underwater Engineering* 2 (1960–1961): 32–34.

Horton, H. Burke. "A Method for Obtaining Random Numbers." *Annals of Mathematical Statistics* 19, no. 1 (1948): 81–85.

Horton, H. Burke. "National Aspects of Defense Mobilization." In *Symposium on the Management of Mass Casualties*, 27–32. Fort Sam Houston, TX: Army Medical Service, 1958.

Horton, H. Burke. "National Damage Assessment Program and Critique of Operation Alert 1957." *Journal of the American Medical Association* 166, no. 7 (1958): 790.

Horton, H. Burke. "Some Uses of Interindustry Statistical Techniques." *Journal of the American Society for Naval Engineers* 65, no. 3 (1953): 581–90.

Hughes, Agatha C., and Thomas Parke Hughes. *Systems, Experts, and Computers: The Systems Approach in Management and Engineering, World War II and After*. Cambridge, MA: MIT Press, 2000.

Hughes, Thomas P. *Networks of Power: Electrification in Western Society, 1880–1930*. Baltimore, MD: Johns Hopkins University Press, 1983.

Huthmacher, J. Joseph. "Urban Liberalism and the Age of Reform." In *Progressivism: The Critical Issues*, edited by David Kennedy, 78–86. Boston, MA: Little, Brown, 1971.

Ickes, Harold. *Back to Work: The Story of PWA*. New York: Macmillan, 1935.

Industrial College of the Armed Forces. *Emergency Management of the National Economy*. Washington, DC: Industrial College of the Armed Forces, 1954.

Jasanoff, Sheila. *Science and Public Reason*. New York: Routledge, 2012.

Jasanoff, Sheila. "Technological Risk and Cultures of Rationality." In *Incorporating Science, Economics, and Sociology in Developing Sanitary and Phytosanitary Standards in International Trade*, 65–84. Washington, DC: National Academies Press, 2000.

Jasanoff, Sheila. "Technologies of Humility: Citizen Participation in Governing Science." In *Wozu Experten?*, edited by Alexander Bogner, 370–89. Wiesbaden: VS Verlag für Sozialwissenschaften, 2005.

Johnson, Lyle R. "Coming to Grips with Univac." *IEEE Annals of the History of Computing* 28, no. 2 (2006): 32–42.

Jones, Byrd L. "The Role of Keynesians in Wartime Policy and Postwar Planning, 1940–1946." *American Economic Review* 62, nos. 1/2 (1972): 125–33.

Jordan, Nehemiah. *U.S. Civil Defense before 1950: The Roots of Public Law 920*. Alexandria, VA: Institute for Defense Analyses, 1966.

Kahn, Herman. *On Thermonuclear War*. Princeton, NJ: Princeton University Press, 1960.

Kahn, Herman. *Thinking about the Unthinkable*. New York: Horizon Press, 1962.

Kantor, Harvey. "Charles Dyer Norton and the Origins of the Regional Plan of New York." In *The American Planner: Biographies and Recollections*, edited by Donald Krueckeberg, 163–82. London: Methuen, 1983.

Kaplan, Fred. *The Wizards of Armageddon*, Stanford, CA: Stanford University Press, 1983.

Karl, Barry Dean. "Executive Reorganization and Presidential Power." *Supreme Court Review* (1977): 1–37.

Karl, Barry Dean. *Executive Reorganization and Reform in the New Deal: The Genesis of Administrative Management, 1900–1939*. Cambridge, MA: Harvard University Press, 1963.

Katz, Barry. *Foreign Intelligence: Research and Analysis in the Office of Strategic Services, 1942–1945*. Cambridge, MA: Harvard University Press, 1989.

Katznelson, Ira. *Fear Itself: The New Deal and the Origins of Our Time*. New York: Norton, 2013.

Kaysen, Carl. "The Vulnerability of the United States to Enemy Attack." *World Politics* 6, no. 2 (1954): 190–208.

Kennedy, David M., ed. *Progressivism: The Critical Issues*. Boston: Little, Brown, 1971.

Kiefer, Norvin C. "Role of Health Services in Civil Defense." *American Journal of Public Health and the Nation's Health* 40, no. 12 (1950): 1486–90.

Kindleberger, Charles P. *The Life of an Economist: An Autobiography*. Cambridge, MA: Basil Blackwell, 1991.

Kindleberger, Charles P. "Some Economic Lessons from World War II." In *Essays in History: Financial, Economic, Personal*, 179–203. Ann Arbor: University of Michigan Press, 1999.

Knowles, Morris. "Engineering Problems of Regional Planning." In *Planning Problems of Town, City, and Region: Proceedings of the Eleventh National Convention on City Planning, Niagara Falls and Buffalo, N.Y. May 26–28, 1919*, 115–38. Cambridge, MA: University Press, 1920.

Knowles, Scott Gabriel. *The Disaster Experts: Mastering Risk in Modern America*. Philadelphia: University of Pennsylvania Press, 2011.

Koistinen, Paul A. C. *Arsenal of World War II: The Political Economy of American Warfare, 1940–1945*. Lawrence: University Press of Kansas, 2004.

Koopman, Colin. *Genealogy as Critique: Foucault and the Problems of Modernity*. Bloomington: Indiana University Press, 2013.

Krugler, David. *This Is Only a Test: How Washington, DC, Prepared for Nuclear War*. New York: Palgrave Macmillan, 2006.

Kuhn, Thomas S. *The Structure of Scientific Revolutions*. Chicago: University of Chicago Press, 2012.

Kupperman, Robert H. *Technological Advances and Consequent Dangers: Growing Threats to Civilization*. Washington, DC: Center for Strategic and International Studies, 1984.

Kuznets, Simon. *Commodity Flow and Capital Formation*. New York: National Bureau of Economic Research, 1938.

Lacey, Jim. *Keep from All Thoughtful Men: How US Economists Won World War II*. Annapolis, MD: Naval Institute Press, 2011.

Lakoff, Andrew. "The Generic Biothreat, or, How We Became Unprepared." *Cultural Anthropology* 23, no. 3 (2008): 399–428.

Lakoff, Andrew. "Preparing for the Next Emergency." *Public Culture* 19, no. 2 (2007): 247–71.

Lakoff, Andrew. *Unprepared: Global Health in a Time of Emergency*. Oakland: University of California Press, 2017.

Lakoff, Andrew, and Eric Klinenberg. "Of Risk and Pork: Urban Security and the Politics of Objectivity." *Theory and Society* 39, no. 5 (2010): 503–25.

Landis Dauber, Michele. *The Sympathetic State: Disaster Relief and the Origins of the American Welfare State*. Chicago: University of Chicago Press, 2013.

Langley, Paul. *Liquidity Lost: The Governance of the Global Financial Crisis*. Oxford: Oxford University Press, 2015.

Lapp, Ralph E. "Atomic Bomb Explosions—Effects on an American City." *Bulletin of the Atomic Scientists* 4, no. 2 (1948): 49–54.

Lapp, Ralph E. "Industrial Dispersion in the United States." *Bulletin of the Atomic Scientists* 7, no. 9 (1951): 256–60.

Lapp, Ralph E. "The Strategy of Civil Defense." *Bulletin of the Atomic Scientists* 6, nos. 8–9 (1950): 241–43.

Lass, Robert H. "Into a Wild New Yonder: The United States Air Force and the Origins of Its Information Age." PhD diss., University of South Carolina–Columbia, 2013.

Lawrence, William. "Balancing National Requirements against National Resources." Lecture, Industrial College of the Armed Forces, Washington, DC, November 5, 1959.

Lawrence, William. "The Role of the Office of Defense Mobilization in the Determination and Use of Requirements Data." Lecture to the Industrial College of Armed Forces, Washington, DC, December 7, 1954.

Leuchtenberg, William E. "The New Deal and the Analogue of War." In *Change and Continuity in Twentieth Century America*, edited by John Braeman, Robert H. Bremner, and Everett Walters, 81–143. Columbus: Ohio State University Press, 1964.

Levinson, Sanford. "Constitutional Norms in a State of Permanent Emergency." *Georgia Law Review* 40, no. 3 (Spring 2006): 699–751.

Lewis, Harold M. *The Transit and Transportation Problem*. New York: Regional Plan of New York and Its Environs, 1926.

Light, Jennifer S. *From Warfare to Welfare: Defense Intellectuals and Urban Problems in Cold War America*. Baltimore, MD: Johns Hopkins University Press, 2003.

Link, Arthur S. "What Happened to the Progressive Movement in the 1920's?" *American Historical Review* 64, no. 4 (1959): 833–51.

Linnell, Mikael. "The Haptic Space of Disaster." *Space and Culture*, April 9, 2019. https://doi.org /10.1177/1206331219840292.

Livermore, Shaw. "Economic Mobilization Planning: A Critical Appraisal." Lecture to the Industrial College of the Armed Forces, Washington, DC, May 1957.

Livermore, Shaw. "Resources Management." In *Proceedings of the Symposium on Post-Attack Recovery*, Fort Monroe, Virginia, November 6–9, 1967, 275–90. Washington, DC: Office of Civil Defense, 1968.

Los Alamos Scientific Laboratory. *The Effects of Atomic Weapons*. Washington, DC: Government Printing Office, 1950.

Lovins, Amory B., and L. Hunter Lovins. *Brittle Power: Energy Strategy for National Security*. Andover, MA: Brick House, 1982.

Lowe, James T. "Intelligence in the Selection of Strategic Target Systems." Lecture, Air War College, Maxwell Field, AL, December 13, 1946. File K239.716246-22(S), Historical Research Agency, Maxwell Air Force Base, AL.

Lowe, James T. "The Theory of Strategic Vulnerability." Lecture to the Air War College, Maxwell Air Force Base, Montgomery, AL, 1948. File K239.716250-43. Historical Research Agency, Maxwell Air Force Base, Montgomery, AL.

Luhmann, Niklas. *Observations on Modernity*. Stanford, CA: Stanford University Press, 1998.

Mann, Michael. "The Autonomous Power of the State: Its Origins, Mechanisms, and Results." *European Journal of Sociology* 25, no. 2 (1984): 185–213.

Marschak, Jacob, Edward Teller, and Lawrence R. Klein. "Dispersal of Cities and Industries." *Bulletin of the Atomic Scientists* 1, no. 9 (1946): 13–15, 20.

Marshall, T. H. "Citizenship and Social Class." In *Class, Citizenship, and Social Development: Essays by T. H. Marshall*, 65–122. Westport, CT: Greenwood Press, 1950.

Masco, Joseph. "Life Underground: Building the Bunker Society." *Anthropology Now* 1, no. 2 (2009): 13–29.

McCormick, John P. *Carl Schmitt's Critique of Liberalism: Against Politics as Technology*. Cambridge: Cambridge University Press, 1999.

McDermott, Edward. *The Office of Emergency Planning in National Security Planning*. Publication no. L63-35. Washington, DC: Industrial College of the Armed Forces, 1962.

McFarland, Stephen L. *America's Pursuit of Precision Bombing, 1910–1945.* Washington, DC: Smithsonian Institution Press, 1995.

McKinsey and Company. "Report on Nonmilitary Defense Organization." In *Civil Defense: Hearings Before a Subcommittee of the Committee on Government Operations, House of Representatives.* Washington, DC: Government Printing Office, July 10, 1958.

McReynolds, William H. "The Office for Emergency Management." *Public Administration Review* 1, no. 2 (1941): 131–38.

Merriam, Charles Edward. "Government and Society." In *Recent Social Trends in the United States: Report of the President's Research Committee on Social Trends,* 1488–1541. New York: McGraw-Hill, 1933.

Merriam, Charles Edward. "The National Resources Planning Board: A Chapter in American Planning Experience." *American Political Science Review* 38, no. 6 (1944): 1075–88.

Merriam, Charles Edward. *New Aspects of Politics.* Chicago: University of Chicago Press, 1925.

Merriam, Charles Edward. *The New Democracy and the New Despotism.* New York: McGraw-Hill, 1939.

Metcalf, Evan B. "Secretary Hoover and the Emergence of Macroeconomic Management." *Business History Review* 49, no. 1 (1975): 60–80.

Miller, Seward E. "Activities of the Health Resources Advisory Committee, Office of Defense Mobilization—Rusk Committee." *American Journal of Public Health and the Nation's Health* 43, no. 3 (1953): 322–25.

Mitchell, Timothy. *Carbon Democracy: Political Power in the Age of Oil.* London: Verso, 2011.

Mitchell, Timothy. *Rule of Experts: Egypt, Techno-Politics, Modernity.* Berkeley: University of California Press, 2002.

Mitchell, William. "Airplanes in National Defense." *Annals of the American Academy of Political and Social Science* 131, no. 1 (1927): 38–42.

Morgan, Ted. *FDR: A Biography.* New York: Simon and Schuster, 1985.

Nathan, Robert R. "GNP and Military Mobilization." *Journal of Evolutionary Economics* 4, no. 1 (1994): 1–16.

Nathan, Robert R. "The Keynesian Revolution and Its Pioneers: Discussion." *American Economic Review* 62, no. 2 (1972): 138–39.

National Military Establishment, Office of the Secretary of Defense. *A Study of Civil Defense.* Washington, DC: Office of the Secretary of Defense, 1948.

National Planning Board. *National Planning Board: Final Report 1933–1934.* Washington, DC: Government Printing Office, 1934.

National Resources Board. *A Report on National Planning and Public Works in Relation to Natural Resources and Including Land Use and Water Resources, with Findings and Recommendations.* Washington, DC: Government Printing Office, 1934.

National Resources Committee. *The Structure of the American Economy, Part 1: Basic Characteristics.* A report prepared by the industrial section under the direction of Gardiner C. Means. Washington, DC: Government Printing Office, 1939.

National Resources Planning Board. *Federal Emergency Administration of Public Works: Final Report–1933–34.* Washington, DC: Government Printing Office, 1934.

National Resources Planning Board. *Security, Work, and Relief Policies.* Washington, DC: Government Printing Office, 1942.

National Security Resources Board. *Is Your Plant a Target?* Washington, DC: National Security Resources Board, 1951.

National Security Resources Board. *National Security Factors in Industrial Location.* Washington, DC: Government Printing Office, 1948.

National Security Resources Board. *United States Civil Defense.* Washington, DC: Government Printing Office, 1950.

Needell, Allan A. *Science, Cold War, and the American State: Lloyd V. Berkner and the Balance of Professional Ideals*. New York: Routledge, 2012.

Nelson, Donald M. *Arsenal of Democracy*. New York: Harcourt, Brace, 1946.

Neocleous, Mark. *Critique of Security*. Edinburgh: Edinburgh University Press, 2008.

"News and Notes: Dispersal of Washington." *Bulletin of the Atomic Scientists* 7, no. 9 (September 1951): 276–77.

Norton, C. McKim. "Report on Project East River." *Journal of the American Institute of Planners* 19, no. 2 (1953): 87–94.

Novick, David. *Origin and History of Program Budgeting*. Talk filmed on August 11, 1966, for course sponsored by US Bureau of the Budget and the US Civil Service Commission. Santa Monica, CA: Rand Corporation, 1966.

Novick, David. "Statistical Materials Collected by the War Production Board." *Review of Economics and Statistics* 28, no. 3 (1946): 131–34.

Novick, David, Melvin Anshen, and William Truppner. *Wartime Production Controls: The Story of Industrial Production Control in World War II, with Mistakes Pointed Out and Recommendations Made for the Future*. New York: Columbia University Press, 1949.

Novick, David, and George A. Steiner. "The War Production Board's Statistical Reporting Experience." *Journal of the American Statistical Association* 43, no. 242 (1948): 201–30.

Oakes, Guy. "The Cold War Conception of Nuclear Reality: Mobilizing the American Imagination for Nuclear War in the 1950s." *International Journal of Politics, Culture, and Society* 6, no. 3 (1993): 339–63.

Oakes, Guy. *The Imaginary War: Civil Defense and American Cold War Culture*. New York: Oxford University Press, 1994.

Office of the Assistant Chief of Air Staff, Intelligence Headquarters and the Photographic Intelligence Center, Division of Naval Intelligence. *Evaluation of Photographic Intelligence in the Japanese Homeland, Part Four: Urban Area Analysis*. Washington, DC: Government Printing Office, 1954.

Office of Civil and Defense Mobilization. "Appendix: Essential Survival Items." In *National Plan for Civil Defense and Defense Mobilization*. Washington, DC: Government Printing Office, 1960.

Office of Defense Mobilization. *Mobilization Plan D-Minus*. Washington, DC: Government Printing Office, 1957.

Office of Emergency Planning, Executive Office of the President. *The National Plan for Emergency Preparedness*. Washington, DC: Office of Emergency Planning, 1964.

Oliver, Wallace B. "The Survival Model: Gaming the Logistics of Survival." In *NREC Programs for Gaming the Logistics of National Survival*, 12–21. Washington, DC: National Resource Evaluation Center, 1965.

Olson, Mancur. "The Economics of Target Selection for the Combined Bomber Offensive." *Royal United Services Institution Journal* 107, no. 628 (1962): 308–14.

Opitz, Sven, and Ute Tellmann. "Future Emergencies: Temporal Politics in Law and Economy." *Theory, Culture and Society* 32, no. 2 (2015): 107–29.

Özgöde, Onur. "Governing the Economy at the Limits of Neoliberalism: The Genealogy of Systemic Risk Regulation in the United States, 1922–2012." PhD diss., Columbia University, 2015.

Özgöde, Onur. "Institutionalism in Action: Balancing the Substantive Imbalances of 'the Economy' through the Veil of Money." *History of Political Economy* 52, no. 2 (2020): 307–39.

Özgöde, Onur. "Logistics of Survival: Genealogical Origins of Systemic Risk Regulation in Nuclear War Preparedness." Lecture at the Society for the Advancement of Socio-Economics, the New School, New York City, June 27, 2019.

Patterson, James T. *Congressional Conservatism and the New Deal: the Growth of the Conservative Coalition in Congress, 1933–1939*. Lexington: University of Kentucky Press, 1967.

Perlman, Mark. "Political Purpose and the National Accounts." In *The Politics of Numbers*, edited by William Alonso and Paul Starr, 133–51. New York: Russell Sage Foundation, 1986.

Perrow, Charles. *Normal Accidents: Living with High-Risk Technologies*. Princeton, NJ: Princeton University Press, 1984.

Pettee, James C. "The RISK II Model: The Spectrum of Solutions in Logistics Gaming." In *NREC Programs for Gaming the Logistics of National Survival*, 32–51. Washington, DC: National Resource Evaluation Center, 1965.

Pew Research Center. *Beyond Distrust: How Americans View Their Government*. Washington, DC: Pew Research Center, 2015.

Pierpaoli, Paul G. "Truman's Other War: The Battle for the American Homefront, 1950–1953." *OAH Magazine of History* 14, no. 3 (2000): 15–19.

Platt, Harold L. "World War I and the Birth of American Regionalism." In *Urban Public Policy: Historical Modes and Methods,* edited by Martin V. Melosi, 128–52. University Park: Pennsylvania State University Press, 1993.

Platt, William J. "Industrial Defense: A Community Approach." *Bulletin of the Atomic Scientists* 9, no. 7 (1953): 261–64.

Polanyi, Karl. *The Great Transformation*. Boston: Beacon Press, 1944.

Polenberg, Richard. *Reorganizing Roosevelt's Government: The Controversy over Executive Reorganization, 1936–1939*. Cambridge, MA: Harvard University Press, 1966.

Poovey, Mary. *A History of the Modern Fact: Problems of Knowledge in the Sciences of Wealth and Society*. Chicago: University of Chicago Press, 1998.

Porter, Ted. *Trust in Numbers: The Pursuit of Objectivity in Science and Public Life*. Princeton, NJ: Princeton University Press, 1996.

Pozen, David E., and Kim Lane Scheppele. "Executive Underreach, in Pandemics and Otherwise." *American Journal of International Law* 114, no. 4 (2020): 608–17.

President's Committee on Administrative Management. *Report of the Committee*. Washington, DC: Government Printing Office, 1937.

Price, Robert L. "Use of the Packaged Disaster Hospital in Nigeria." *Public Health Reports* 85, no. 8 (1970): 659–65.

Project Charles. *Problems of Air Defense: Final Report of Project Charles*. Cambridge, MA: MIT Press, 1951.

Project East River. *General Report: Part of 1 of the Report of Project East River*. New York: Associated Universities, 1952.

Project East River Review Committee. *1955 Review of the Report of Project East River*. Washington, DC: Government Printing Office, 1955.

Quarantelli, E. L., and Russel R. Dynes. "Response to Social Crisis and Disaster." *Annual Review of Sociology* 3 (1977): 23–49.

Rabinow, Paul. *Anthropos Today: Reflections on Modern Equipment*. Princeton, NJ: Princeton University Press, 2003.

Rabinow, Paul. *French Modern: Norms and Forms of the Social Environment*. Chicago: University of Chicago Press, 1995.

Rabinowitch, Eugene. "Civil Defense: The Long-Range View." *Bulletin of the Atomic Scientists* 6, nos. 8–9 (1950): 226–30.

Rabinowitch, Eugene. "The Only Real Defense." *Bulletin of the Atomic Scientists* 7, no. 9 (1951): 242–43.

Ramsey, F. A., Jr. "Damage Assessment Systems and Their Relationship to Post-Nuclear Attack Damage and Recovery." *Naval Logistics Quarterly* 5, no. 3 (1958): 199–219.

RAND Corporation. *Report on a Study of Non-Military Defense*. Santa Monica, CA: RAND Corporation, 1958.

Reagan, Patrick D. *Designing a New America: The Origins of New Deal Planning, 1890–1943*. Amherst: University of Massachusetts Press, 1999.

Relyea, Harold C. *The Executive Office of the President: A Historical Overview*. Report for Congress. Washington, DC: Congressional Research Service, 2008.

Riefler, Winfield W. "Government and the Statistician." *Journal of the American Statistical Association* 37, no. 217 (1942): 1–11.

Riefler, Winfield W. "Our Economic Contribution to Victory." *Journal of Foreign Affairs* 26, no. 1 (1947): 90–103.

Riefler, Winfield W. Preface to *The Problem of Reducing Vulnerability to Atomic Bombs*, by Ansley J. Coale. Princeton, NJ: Princeton University Press, 1947.

Ripple, William J., Christopher Wolf, Thomas M. Newsome, Phoebe Barnard, William Moomaw, Matjaž Mikoš, et al. "World Scientists' Warning of a Climate Emergency." *BioScience* 70, no. 1 (2020): 8–12.

Roberts, Patrick S. "Dispersed Federalism as a New Regional Governance for Homeland Security." *Publius: The Journal of Federalism* 38, no. 3 (2008): 416–43.

Rogers, Daniel T. *Atlantic Crossings: Social Politics in a Progressive Age*. Cambridge, MA: Harvard University Press, 2000.

Roitman, Janet. *Anti-Crisis*. Durham, NC: Duke University Press, 2014.

Roosevelt, Franklin Delano. *The Court Disapproves, 1935*. Vol. 4 of *The Public Papers and Addresses of Franklin D. Roosevelt, with a Special Introduction and Explanatory Notes by President Roosevelt*. New York: Random House, 1938.

Roosevelt, Franklin Delano. *The Public Papers and Addresses of Franklin D. Roosevelt, with a Special Introduction and Explanatory Notes by President Roosevelt*. New York: Random House, 1938.

Roosevelt, Franklin Delano. *The Year of Crisis, 1933*. Vol. 2 of *The Public Papers and Addresses of Franklin D. Roosevelt, with a Special Introduction and Explanatory Notes by President Roosevelt*. New York: Random House, 1938.

Rose, Nikolas. "Governing 'Advanced' Liberal Democracies." In *Foucault and Political Reason: Liberalism, Neo-Liberalism, and Rationalities of Government,* edited by Andrew Barry, Thomas Osborne, and Nikolas Rose, 37–64. Chicago: University of Chicago Press, 1996.

Rose, Nikolas. *Powers of Freedom: Reframing Political Thought*. Cambridge: Cambridge University Press, 1999.

Rose, Nikolas, and Peter Miller. *Governing the Present: Administering Economic, Social, and Personal Life*. Cambridge: Polity Press, 2008.

Rossiter, Clinton L. *The American Presidency*. New York: Harcourt, Brace, 1956.

Rossiter, Clinton L. "Constitutional Dictatorship in the Atomic Age." *Review of Politics* 11, no. 4 (1950): 395–418.

Rossiter, Clinton L. *Constitutional Dictatorship: Crisis Government in the Modern Democracies*. Princeton, NJ: Princeton University Press, 1948.

Rossiter, Clinton L. "The Constitutional Significance of the Executive Office of the President." *American Political Science Review* 43, no. 6 (1949): 1206–17.

Rossiter, Clinton. "War, Depression, and the Presidency, 1933–1950." *Social Research* 17, no. 4 (1950): 417–40.

Rostow, Walt Whitman. *Concept and Controversy: Sixty Years of Taking Ideas to Market*. Austin: University of Texas Press, 2003.

Rostow, Walt Whitman. *The Division of Europe after World War II, 1946*. Austin: University of Texas Press, 1981.

Rostow, Walt Whitman. *Pre-Invasion Bombing Strategy: General Eisenhower's Decision of March 25, 1944*. Austin: University of Texas Press, 1981.

Samimian-Darash, Limor. "Practicing Uncertainty: Scenario-Based Preparedness Exercises in Israel." *Cultural Anthropology* 31, no. 3 (2016): 359–86.

Sandilands, Roger J. *The Life and Political Economy of Lauchlin Currie: New Dealer, Presidential Advisor, and Development Economist*. Durham, NC: Duke University Press, 1990.

Sawicki, Jana. "Heidegger and Foucault: Escaping Technological Nihilism." *Philosophy and Social Criticism* 13, no. 2 (April 1987): 155–73.

Schaffel, Kenneth. "Genesis of the Air Defense Mission." In *The Emerging Shield: The Air Force and the Evolution of Continental Air Defense, 1945–1960*. Washington, DC: Office of Air Force History, United States Air Force, 1991.

Scheppele, Kim Lane. "Small Emergencies." *Georgia Law Review* 40, no. 3 (2006): 835–62.

Scheuerman, William E. "The Economic State of Emergency." *Cardozo Law Review* 21 (1999): 1869–94.

Scheuerman, William E. *Liberal Democracy and the Social Acceleration of Time*. Baltimore, MD: Johns Hopkins University Press, 2004.

Scheuerman, William. "Time to Look Abroad? The Legal Regulation of Emergency Powers." *Georgia Law Review* 40, no. 3 (2006): 863–76.

Schmidt, Edward B. *Targeting Organizations: Centralized or Decentralized?* Maxwell Air Force Base, Montgomery, AL: Air University Press, 1993.

Schmitt, Carl. *Dictatorship*. 1921. Malden, MA: Polity Press, 2014.

Scott, Mel. *American City Planning Since 1890: A History Commemorating the Fiftieth Anniversary of the American Institute of Planners*. Berkeley: University of California Press, 1969.

Sebald, W. G. *On the Natural History of Destruction*. New York: Random House, 2003.

Security Resources Panel of the Science Advisory Committee, Office of Defense Mobilization. *Deterrence and Survival in the Nuclear Age*, vol. 2, *Passive Defense*. Washington, DC: Government Printing Office, 1957.

Shaw, Carroll K. *Field Organization and Administration of the War Production Board and Predecessor Agencies, May 1940 to November 1945*. Washington, DC: Civilian Production Administration, 1946.

Sherman, William C. *Air Warfare*. New York: Ronald Press, 1926.

Sherry, Michael S. *Preparing for the Next War: American Plans for Postwar Defense, 1941–1945*. New Haven, CT: Yale University Press, 1977.

Sherry, Michael S. *The Rise of American Air Power: The Creation of Armageddon*. New Haven, CT: Yale University Press, 1987.

Silvast, Antti. *Making Electricity Resilient: Risk and Security in a Liberalized Infrastructure*. New York: Routledge, 2017.

Smith, Jason Scott. *Building New Deal Liberalism: The Political Economy of Public Works, 1933–1956*. Cambridge: Cambridge University Press, 2006.

Smith, R. Elberton. *The Army and Economic Mobilization*. Washington, DC: Government Printing Office, 1959.

Snead, David L. *The Gaither Committee: Eisenhower and the Cold War*. Columbus: Ohio State University Press, 1999.

Stanford Research Institute. *The Damage Assessment System*. Washington, DC: Federal Civil Defense Administration, 1957.

Steinberg, Ted. *Acts of God: The Unnatural History of Disasters in America*. London: Oxford University Press, 2006.

Strope, Walmer E. "Autobiography of a Nerd." Unpublished memoir, n.d. Accessed July, 7, 2021 at https://stanford.app.box.com/s/d7u2zudqnk30hke1fk0r9odak7wr40rz/file/573377499254.

Stuart, Douglas T. "Present at the Legislation: The 1947 National Security Act." In *Organizing for National Security*, edited by Douglas T. Stuart, 5–24. Carlisle, PA: US Army War College, Strategic Studies Institute, 2000.

Sutermeister, Oscar. "Discussion of Papers." Presented at the Symposium on Human Problems in the Utilization of Fallout Shelters, Disaster Research Group, National Academy of Sciences, Washington, DC, February 1960.

Sweezy, Alan R. "The Government's Responsibility for Full Employment." In "Papers and Proceedings of the Fifty-fifth Annual Meeting of the American Economic Association," supplement, *American Economic Review* 33, no. 1, part 2 (March 1943): 9–26.

Taureck, Rita. "Securitization Theory and Securitization Studies." *Journal of International Relations and Development* 9, no. 1 (2006): 53–61.

Tilly, Charles. *Coercion, Capital, and European States, AD 990–1990*. Oxford: Blackwell, 1990.

Tobin, Kathleen. "The Reduction of Urban Vulnerability: Revisiting 1950s American Suburbanization as Civil Defence." *Cold War History* 2, no. 2 (2002): 1–32.

Todd, Walter E., Willard S. Paul, and Val Peterson. "National Defense against Atomic Attack." *Scientific Monthly* 80 no. 4 (1955): 240–49.

Truppner, William C. "Industrial Readiness Planning." Lecture to the Industrial College of the Armed Forces, Washington, DC, January 13, 1960.

Tyler, Lyon G., Jr. "Civil Defense: The Impact of the Planning Years, 1945–1950." PhD diss., Duke University, 1967.

Udley, Michael Q. "Sprawl as Strategy: City Planners Face the Bomb." *Journal of Planning Education and Research* 21, no. 1 (2001): 52–63.

Unger, David C. *The Emergency State: America's Pursuit of Absolute Security at All Costs*. New York: Penguin, 2012.

US Atomic Energy Commission. *The City of Washington and an Atomic Bomb Attack*. Washington, DC: US Atomic Energy Commission, 1949.

US Congress Joint Committee on Defense Production. *Defense Production Act, Progress Report No. 30*. Washington, DC: United States Government Printing Office, 1955.

US Congress Joint Committee on Defense Production. *Defense Production Act Progress Report No. 40: Report on Reducing Our Vulnerability to Attack*. Washington, DC: Government Printing Office, 1957.

US Congress Joint Committee on Defense Production. *Defense Production Act Progress Report No. 42, Hearing to Discuss the Adequacy of Preparedness Programs to Meet Nuclear Attack and Limited-Scale War, Before the Joint Committee on Defense Production*. Washington, DC: Government Printing Office, 1958.

US Congress Joint Committee on Defense Production. *Defense Production Act Progress Report No. 43*. Washington, DC: Government Printing Office, June 24, 1959.

US Congress Joint Committee on Defense Production. *Sixth Annual Report of the Activities of the Joint Committee on Defense Production*. Washington, DC: Government Printing Office, January 22, 1957.

US Congress Joint Committee on Defense Production. *Seventh Annual Report of the Activities of the Joint Committee on Defense Production together with Materials on National Defense Production*. Report No. 1172. Washington, DC: Government Printing Office, 1958.

US Congress Joint Committee on the Economic Report. *The Need for Industrial Dispersal*. Washington, DC: Government Printing Office, 1951.

US Department of Commerce. *Industrial Dispersion Guidebook for Communities*. Domestic Commerce Series No. 31. Washington DC: Government Printing Office, 1956.

US Department of Defense Civil Liaison Office. *Fire Effects of Bombing Attacks*. Washington, DC: Government Printing Office, 1950.

US House of Representatives. *Certificates of Necessity and Government Plant Expansion Loans: Fifth Intermediate Report of the Committee on Expenditures in the Executive Departments*. Washington, DC: Government Printing Office, 1951.

US House of Representatives Committee on Government Operations. *Civil Defense for National Survival*. Washington, DC: Government Printing Office, 1956.

US House of Representatives Committee on Public Works. *Dispersal of Government Agencies, Vicinity of the District of Columbia: Hearings on H.R. 9864, 2nd sess., Before the House of Representatives Committee on Public Works*. Washington, DC: Government Printing Office, 1950.

US House of Representatives Committee on Ways and Means. *Revenue Revision of 1951: Hearings before the Committee on Ways and Means, House of Representatives, Eighty Second Congress,*

First Session on Revenue Revision of 1951, Part 3, March 5, 6, 7, 8, 9, 12, 13, 14, 15, 16, 19, 21, and April 2, 1951. Washington, DC: Government Printing Office, 1951.

US Planning Committee of the War Production Board. *Meeting Minutes of the Planning Committee of the War Production Board.* Washington, DC: Civilian Production Administration, 1946.

US Senate National Resources Committee. *Progress Report: Statement of the Advisory Committee.* Washington D.C.: Government Printing Office, 1939.

US Strategic Bombing Survey. *The Effects of the Atomic Bombs on Hiroshima and Nagasaki.* Washington, DC: US Government Printing Office, 1946.

US Strategic Bombing Survey. *Evaluation of Photographic Intelligence in the Japanese Homeland, Part Four: Urban Area Analysis.* Washington, DC: Photographic Intelligence Section, June 1946.

US Strategic Bombing Survey. *The United States Strategic Bombing Survey Summary Report: European War, Pacific War.* Washington D.C.: Government Printing Office, 1945; reprint, Maxwell Air Force Base, Montgomery, AL: Air University Press, 1987.

US War Production Board. *Catalog of War Production Board Reporting and Application Forms, as of November 2, 1945.* Washington, DC: Government Printing Office, 1947.

Valero, Larry A. "The Joint Intelligence Committee and Estimates of the Soviet Union, 1945–1947." In *Studies in Intelligence,* edited by Henry Appelbaum, 65–80. Washington, DC: Center for the Study of Intelligence, Summer 2000.

Waldo, Dwight. *The Administrative State: A Study of the Political Theory of American Public Administration.* New York: Ronald Press, 1948.

Walker, Jeremy, and Melinda Cooper. "Genealogies of Resilience: From Systems Ecology to the Political Economy of Crisis Adaptation." *Security Dialogue* 42, no. 2 (2011): 143–60.

Walters, William. *Governmentality: Critical Encounters.* Vol. 3. New York: Routledge, 2012.

Warken, Philip W. *A History of the National Resources Planning Board, 1933–1943.* New York: Garland, 1979.

Weber, Max. *The Protestant Ethic and the Spirit of Capitalism.* Translated by Peter Baehr and Gordon C. Wells. New York: Penguin Classics, 2002.

Whitington, Jerome. "Modernist Infrastructure and the Vital Systems Security of Water: Singapore's Pluripotent Climate Futures." *Public Culture* 28, no. 2 (79) (2016): 415–41.

Whitney, John M. "Civil Defense—Medical Aspects from the Federal Standpoint." *California Medicine* 83, no. 4 (1955): 309–13.

Whitney, John M. "The Federal Civil Defense Administration Medical Stockpile." *Military Medicine* 118, no. 4 (1956): 260–61.

Wilson, Donald. "Origin of a Theory for Air Strategy." *Aerospace Historian* 18, no. 1 (1971): 19–25.

Wittfogel, Karl. *Oriental Despotism: A Comparative Study of Total Power.* New Haven, CT: Yale University Press, 1957.

Wood, Marshall K. "PARM—An Economic Programming Model." *Management Science* 11, no. 7 (1965): 619–80.

Wood, Marshall K. "Use of Mathematical Techniques: Electronic Devices in the Determination of Requirements." Lecture to the Industrial College of the Armed Forces, 1950, Washington, DC.

Wood, Marshall K., and George B. Dantzig. "Programming of Interdependent Activities: I General Discussion." *Econometrica, Journal of the Econometric Society* 17, no. 3–4 (1949): 193–99.

Yergin, Daniel. *Shattered Peace: The Origins of the Cold War and the National Security State.* Boston: Houghton Mifflin, 1977.

Yoshpe, Harry B. *The National Security Resources Board, 1947–1953: A Case Study in Peacetime Mobilization.* Washington, DC: Government Printing Office, 1953.

Yoshpe, Harry B. *Our Missing Shield: The US Civil Defense Program in Historical Perspective.* Washington, DC: Federal Emergency Management Agency, 1981.

Zaffar, Ehsan. *Understanding Homeland Security Policy.* London: Routledge, 2019.

A NOTE ON THE TYPE

This book has been composed in Adobe Text and Gotham.
Adobe Text, designed by Robert Slimbach for Adobe,
bridges the gap between fifteenth- and sixteenth-century
calligraphic and eighteenth-century Modern styles.
Gotham, inspired by New York street signs, was designed
by Tobias Frere-Jones for Hoefler & Co.

GPSR Authorized Representative: Easy Access System Europe - Mustamäe tee
50, 10621 Tallinn, Estonia, gpsr.requests@easproject.com